New and Future Development in Biopesticide Research: Biotechnological Exploration

Surajit De Mandal • G. Ramkumar • S. Karthi •
Fengliang Jin
Editors

New and Future Development in Biopesticide Research: Biotechnological Exploration

Springer

Editors
Surajit De Mandal
Key Laboratory of Bio-Pesticide
Innovation and Application of Guangdong
Province, College of Plant Protection
South China Agricultural University
Guangzhou, China

G. Ramkumar
Division of Biotechnology
ICAR- Indian Institute of Horticultural Research
(IIHR)
Bengaluru, India

S. Karthi
Sri Paramakalyani Centre for Excellence in
Environmental Sciences
Manonmaniam Sundaranar University
Alwarkurichi, Tamil Nadu, India

Fengliang Jin
Key Laboratory of Bio-Pesticide
Innovation and Application of Guangdong
Province, College of Plant Protection
South China Agricultural University
Guangzhou, China

ISBN 978-981-16-3988-3 ISBN 978-981-16-3989-0 (eBook)
https://doi.org/10.1007/978-981-16-3989-0

This Springer imprint is published by the registered company Springer Nature Singapore Pte Ltd.
The registered company address is: 152 Beach Road, #21-01/04 Gateway East, Singapore 189721, Singapore

Contents

About the Editors

Surajit De Mandal is a postdoctoral fellow at the College of Plant Protection, South China Agricultural University, Guangzhou, P. R. China. He has several years of research experience and published many research articles in international journals. He also acts as an editorial board member/reviewer for various international journals. His field of interest includes microbial diversity and metagenomics, molecular phylogeny, bioinformatics and microbial control of insect pests. He is presently working on microbial community analysis using next-generation sequencing methods.

G. Ramkumar is a Research Associate at the Indian Institute of Horticultural Research, Bangalore, India. He has completed his Ph.D. from Periyar University, India. His main areas of research are vector control, insecticides resistance monitoring, RNAi, gene editing, etc. He has published many scientific articles and edited several books on entomology. He is the editorial board member/reviewer of several journals. Presently he is actively involved in the research on Insect Molecular Toxicology.

S. Karthi is currently working as a postdoctoral fellow (UGC-DSKPDF, New Delhi) at the Centre for Environmental Sciences, Sri Paramakalyani Centre for Environmental Sciences, Manonmaniam Sundaranar University, India. He also worked as a Post-Doctoral Fellow (DST-SERB, N-PDF) at Manonmaniam Sundaranar University, India, and Assistant Professor in Rangasamy College of Arts and Science, India. He has published several high-quality research articles and served as a reviewer for many journals. Presently he is working on the Natural Products for Mosquito and Disease Control.

Fengliang Jin is a Professor in the Department of Entomology, College of Plant Protection, South China Agricultural University, Guangzhou, 510642, P. R. China. His major research interest is on the role of non-coding RNAs in the regulation of insect immune signal transduction, exploring the interaction mechanism of insects to entomopathogens using next-generation sequencing and bioinformatic analysis and RNAi-based functional analysis of immune-related genes and their role in the regulation of antimicrobial peptides of insects. Prof. Jin served as a Principal Investigator for several research projects related to the development of biopesticide from various agencies. He has several patents, scientific articles and presently serves as a reviewer and editor of many international journals.

Chapter 1
Biopesticides in Sustainable Agriculture: Current Status and Future Prospects

Emmanuel O. Fenibo, Grace N. Ijoma, and Tonderayi Matambo

Abstract Intensive application of synthetic pesticides was the routine practice of commercial agriculture during the Green Revolution to boost agricultural productivity to meet global food demand. Alongside this, the application of chemical pesticides caused adverse effects on the environment and its ecoreceptors including human health. Negative externalities arising from conventional farming instigated the call for sustainable development during the sixties to promote and balance the nexus between socially acceptable economic growth and environmental protection. Consequently, a blueprint of 17 Sustainable Development Goals (SDGs) and 169 targets including ecological stewardship and food security was drafted. Eight out of the 17 SDGs are directly linked to sustainable agriculture based on the direct impact of agriculture, judicious use of critical resources and conservation and the Principles of green chemistry. As a green chemical agent, biopesticides have been shown to have the potentials to substitute chemical pesticides with equal agricultural productivity. The adoption of bio-based pesticides via integrated pest management (IPM) has proven to be the most effective option to influence most dimensions of sustainable agriculture. Therefore, biopesticide-driven IPM if utilized with requisite education, skills and research would boost sustainable agriculture. This chapter reviews the prospects, importance, and limitations of biopesticides to sustainable agriculture and how sustainable agriculture is connected to sustainable development, Green Chemistry, and integrated pest management.

Keywords Biopesticides · Green chemistry · IPM · SDGs · Sustainable agriculture · Sustainable development

---

E. O. Fenibo (✉)
World Bank Africa Centre of Excellence, Centre for Oilfield Chemical Research, University of Port Harcourt, Port Harcourt, Nigeria
e-mail: feniboe1478@gmail.com

G. N. Ijoma · T. Matambo
Institute for the Development of Energy for African Sustainability, University of South Africa, Roodepoort, South Africa

S. De Mandal et al. (eds.), *New and Future Development in Biopesticide Research: Biotechnological Exploration*, https://doi.org/10.1007/978-981-16-3989-0_1

1.1 Introduction

The quest for food production to satisfy the world's ever-growing population remains a conscious preoccupation dating as far as 300 AD to the present day (Alexandratos and Bruinsma 2012). According to the United Nations' projection, the world population has seen an increase of 0.8 billion per decade with 2020 seeing a population of approximately 7.7 billion. Interestingly, statistics showed sufficient agro production to sustain the global demand but with insignificant input from most of the third-world countries (Fan and Rosegrant 2008). This leaves about a staggering 820 million global populace in ravaging hunger (WHO 2018). The global and regional growth of aggregate production would have fared better in a world without crop pests. Pests are any species of living agents that cause damage to crops and their stored products. Some of these agents include fungi, bacteria, nematodes, weeds, rodents, and insects. According to Pandya (2018), pests account for 30% loss of potential yield (with major loss from developing countries) and 14% damage (Jankielsohn 2018) in storage pests. Improving crop yield to an industrial-scale requires the deliberate application of conventional fertilizers and pesticides. The use of these synthetic chemicals, especially in the Green Revolution era, went along with attendant consequences such as poisoned foods, environmental degradation and health challenges. Later on, this scenario raised concern about sustainable development, considered as the judicious exploitation of the environment for the benefit of both the present and future generations (Burton 1987). This consciousness of sustainable development was first reflected in the "Silent Spring", a book by Rachel Carson in 1963 and through a series of lectures and conventions. This resulted in the 2030 Agenda for Sustainable Development Goal (SDGs) of the United Nations that was born in 2015. The SDGs spelt out a robust blueprint through which sustainable development can be achieved in all spheres of human endeavors.

The central message of sustainable development was to create a nexus between socially acceptable economic growth and environmental stewardship. However, during the green revolution, conventional agriculture, despite its success mounted huge pressure on the ecosystem through many facets including environmental pollution, land degradation, unsustainable use of natural resources, climate change, distortion of ecological services and biodiversity loss. Some of these negative impacts can be curtailed and controlled through changes in consumption patterns and the efficient use of natural resources. When these behavioral shifts, green technology and chemistry, and sustainability principles are factored into large-scale farming the three-dimensional concepts (Fig. 1.1) of sustainable development would be achieved. Within this framework, agriculture would be tied to achieve profitability, community well-being and environmental safety.

Thus, agriculture that is directed to achieve economic viability, environmental objectives and social acceptability can be regarded as sustainable agriculture. Sustainable agriculture is directly or indirectly connected to all the various variants of sustainable development, including the 17 SDGs and Green Chemistry (Perlatti et al. 2014; Ganasen and Velaichamy 2016; Saleh and Koller 2018). Green Chemistry

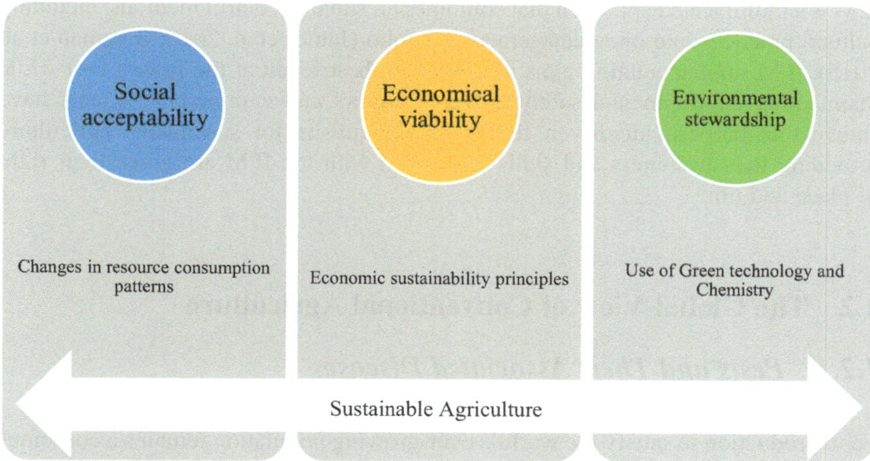

Fig. 1.1 Three dimensional concepts of sustainable development in agriculture

(processing, synthesis and use of innocuous chemicals) directly connects sustainable agriculture and the SDGs in eight areas based on the consumption of possible renewable chemicals and the associated green technologies. These eight goals are SDG15, SDG14, SDG12 and SDG6 (concerned with the environmental conservation and restoration which mostly require organic materials), SDG7 and SDG9 (concerned with green energy and technology respectively), and SDG1 and SDG2 (concerned with improving agro-outputs, which thus require biofertilizers and biopesticides).

Biopesticides are naturally occurring organisms and substances derived from plants and natural inorganic compounds that can control pests' populations by different mechanisms of action (Tijjani et al. 2016), excluding those that interfere with the nervous system of pests (Marrone 2019). Biopesticides are of three categories: microbial biopesticides (microorganisms and their products that have pest controlling influences or compounds), biochemical biopesticides (natural substances with an active agent that control pests by non-toxic mechanisms) and plant-incorporated protectants (transgenic plants) (Kumar 2012; Ibrahim and Shawer 2014; Leahy et al. 2014). These bio-based pesticides exert their effects through different modes of action which, are classified into five groups: metabolic poison, growth regulators, gut disruptors, neuromuscular toxins and non-specific multi-site inhibitors (Sparks and Nauen 2015). Moreover, in most cases, biopesticides arm multiple modes of action against targeted pests making it difficult for the pest to develop resistance as is common with synthetic pesticides (Hassan and Gokçe 2014). Due to their eco-friendliness and low toxicity properties, they do not harm non-targeted organisms including humans and the environment. They are also specific, easily biodegradable, pose no post-harvest contamination problem, and are suitable in an integrated pest management system (Marrone 2009). The effectiveness of biopesticides is made pronounced in integrated pest management (IPM).

IPM is a multifaceted approach that combines all suitable control methods, including cultural practices into one management portfolio (James et al. 2010; Barzman et al. 2015). IPM implementation aims to obtain the best result at the lowest cost while maintaining environmental safety. Several authors and commercial farmers have shown that biopesticide-driven IPM is a prerequisite for sustainable agriculture providing that awareness and skills associated with the IPM are given their right of place and time.

1.2 The Global-View of Conventional Agriculture

1.2.1 Pests and Their Associated Diseases

Food production to satisfy the world's ever-growing population remains a conscious preoccupation, right from the Tertullian era (300 AD) to this present age (Alexandratos and Bruinsma 2012). According to the United Nations projection, the global population will reach 9.15 billion in 2050. With a figure of 6.9 billion in 2010 and 2.25 billion in 1970, and presently the world population in 2020, is approximately 7.70 billion, there is seemingly a population expansion at a rate of 0.8 billion per decade. Current statistics hold that agro production's aggregate value is sufficient to sustain the global demand, but contribution from most third-world countries is abysmally low due to confluence of factors, including poverty, sectional and leadership failures to revamp agriculture (Fan and Rosegrant 2008). And the result is widespread hunger amongst a great number (hundreds of millions) of the world population. WHO (2018) puts the number of those who suffer hunger worldwide at 820 million people, which ironically, is similar to the target number of people, the programme "world without hunger" hopes to reach by 2030. The regional and global increase in aggregate production of agro-yields would have fared better in the situations where there are no crop pests.

Phytopathogenic bacteria may cause the following symptoms, but are not limited to, galls, leafspots, blights, overgrowths, wilts, specks, soft rots, chlorotic halos, cankers and scabs (Cooper and Gardener 2006). Relatedly, some of the diseases they cause can be described as crown gall, bacteria ring rot, fire blight, black rot of cabbage, bacterial soft rot, walnut bacterial blight, and brown rot (Sobiczewski 2008). Phytoplasma can cause pear decline and aster yellows (Mergenthaler et al. 2020). However, sometimes not the case, plant diseases caused by fungi are recognized from the particular organ of the plant they affect and the type of symptom elicited. Based on these, the following fungal diseases are distinguished as follows: damping-off disease, powdery mildew, downy mildew, vascular wilts, root and foot rots, rusts, galls, dieback and anthracnoses (Brown and Ogle 1997). Some notable plant viruses are tobacco mosaic virus, plum pox virus, yellow leaf curl virus, potato virus X, Africa cassava mosaic virus, brome mosaic virus, cauliflower mosaic virus, and potato virus Y (Scholthof et al. 2011). Nematodes are worm-like animals with some species known to parasitize plants. They are known to be the agents of root knots, yellow patches, yellow dwarfs, lesions, root-tip swelling, stem rot, flagging, cyst and stubby-root and corky ringspots (Lucas and Campbell 2012; Mduma et al. 2015). Mistletoe and dodder are well-recognized plant parasites. Insects are known to be the most significant pests. Approximately 0.5% of the insect population are crop pests (Jankielsohn 2018), although they also perform four major ecological services: nutrient cycling, decomposition, predation and pollination (Losey and Vaughan 2006). According to Gallai et al. (2009), insects' pollination accounts for 72% of crop reproduction worldwide with a 9.5% contribution to crop production yield. An active manifestation of pests is partly created by intentional human activities and clearing of vegetation to create space for crops and livestock production to ensure food security thereby compromising the ecosystem and its functions. An immediate cascade effect is witnessed in the distortion of biodiversity, causing insects to aggressively compete with humans in terms of space and nutrients. It is undeniable that the effects of pests are of economic significance from a global perspective. Summarily, the world has to contend with approximately 40 thousand pestilent species for the optimal production of food for the ever-increasing population of humans.

1.2.2 Global Economic Significance of Insect Pests

Top trans-continent insect pests cause huge global losses to crops. More than 180 host plants including cotton and chickpea are attacked by cotton bollworm (*Helicoverpa armigera*) with an economic loss of $2 billion on an annual basis (Tay et al. 2013) while onion thrips (*Trips tabaci*) ranked top as the most important pest of onion (Negash et al. 2020) and other plant hosts. More than 500 host plants belonging to 60 plant families suffer pestilent attack from tobacco whitefly (*Bemisia tabaci*) along with the potential of reducing crop yield up to 50% (Gangwar and Gangwar 2018). More than 200 plants, including tomato and common bean, are destroyed by *Tetranychus urticae* commonly known as the two-spotted spider mite which has resulted in a control cost of $400 million per year (Litskas et al. 2019). An annual budget of between $4 and $5 billion has been estimated to cover weekly insecticide application and yield lost to the insecticide-resistant diamondback moth (Zalucki et al. 2012) which destroys more than 15 genera of plants (Willis 2017), including *Brassica* (cabbage). *Spodoptera litura*, commonly known as taro caterpillar, has been reported to cause 0.85 million tonnes of loss per year in an arable field of 1.46 million hectares planted with soybean and cotton (Sharma et al. 2018). The polyphagous *S. litura* covers more than 120 species of plants as a pest (Bragard et al. 2019). The red flour beetle, a well-known secondary pest, feeds on stored food products such as dry fruits, cereals and cocoa beans.

Myzus persicae, the green peach aphid, is a resistant global pest and virus vector that feeds on more than 400 plant species (Silva et al. 2012). Their hosts are mostly essential crops such as oilseed rape, potato and tomato. They have the potential to reduce yield up to 30% in unprotected farmland (Alyokhin et al. 2020). A study conducted in 12 African countries demonstrated that in a year, losses incurred from maize cultivation and harvesting reach up to 4.1 to 17.7 million tonnes following an infestation of the fall armyworm *Spdoptera frugiperda* (Kassie et al. 2020). *S. frugiperda* is an invasive pest and can affect many crop types, especially maize and cotton (De Groote et al. 2020; Willis 2017). Thrips (*Frankliniella occidentalis*), Mediterranean fruit fly (*Ceratitis capitate*) and codling moth (*Cydia pomonell*) attack pepper, citrus and apple, respectively with substantial damage done to more than 177 plant genera (Abdullah et al. 2015; Willis 2017). The cowpea weevil is a pest that feeds on stored cowpea and legumes in the tropics with 10% to 50% storage loss (Tiroesele et al. 2015; Sanon et al. 2018). The infestation of cotton, maize and other plant species by the noctuid moth of the cotton leafworm (*Spodotera littoralis*) wildly occurs in Africa and Europe, thereby posing a threat to food security (Ahmed et al. 2019). Alfalfa and pea have been extensively attacked by *Acyrthosiphon pisum* (pea aphid) (Calevro et al. 2019). Citrus are attacked by the Asian citrus psyllid (*Dtaphorina citri*) (Monzo and Stansly 2017) and tomato leafminer (*Tuta absoluta*) (Biondi and Desneux 2019). Apart from the preference for the specific plants previously stated, these last three pestilent species have also affected more than 46 plant genera (Willis 2017).

1.2.3 Application of Synthetic Pesticides and Their Significance

Industrial application of pesticides, which was accelerated in 1940, aimed at reducing the impact of pests, and marked the coming of age of the practice of modern agriculture and disease control (Unsworth 2010). Broadly, pesticides can be classified as: synthetic or natural pesticides. Synthetic compounds used in modern agriculture as insecticides, herbicides, rodenticides, fungicides, and molluscicides are either from organochlorine, organophosphate, carbamate or pyrethroids (Mitra et al. 2011; Ndakidemi et al. 2016). Popular examples include Dichlorodiphenyltricholroethane (DDT), aldrin, toxaphene, endrin, chlordane, mirex, dieldrin, and heptachlor (Ritter et al. 1995). Pesticide's modes of action interfere with behavior, growth, reproduction and life cycle stages, development, and nervous system (Qi et al. 2001) of pests. Conventional pesticides have demonstrated broad-spectrum effects, quicker action, longer residual activity, convenience, and; highly effective against the target pests (Felsot and Rack 2007; Maneepitak and Cochard 2014). Pesticides improve the productivity of farm harvest, offer protection against crop losses and control vector-borne disease (Aktar et al. 2009). Furthermore, pesticides improve farm produce shelf life, marketability and profitability of these agricultural protects agricultural land and stored grain (Gill and Garg 2014; Kumar and Kalita 2017). Without crop protection, agricultural production losses would rise to between 48-83% (Glare et al. 2016). The global cost of pesticides has been estimated to be around $38 billion (Pan-Germany 2012). The use of insecticides, molluscicides and herbicides in Thailand has maintained their rice exporting prowess despite reducing the number of farmers cultivating the plant (Maneepitak and Cochard 2014). Conversely, a farming practice without pesticides would create lower productivity and a high food price (Damalas 2009).

Despite the aforementioned benefits of synthetic pesticides, they have disadvantages such as high cost/long duration of development (Sanganyado et al. 2015); crop contamination, pest resistance, reduction, and a threat to the bird populations especially by DDT (Mitra et al. 2011); the depopulation of biological control agents and pollinators (Ndakidemi et al. 2016). Depopulation of biological control usually results in an eruption of secondary pests while reduced pollinators cause low productivity. Agrochemicals have also been confirmed as having serious health implications for humans and the connected ecosystem (Carvalho 2017). Some of the identified health-related challenges caused by pesticides are irritation, enzyme inhibition, allergic sensitization, oxidative damage and neurotransmission inhibition (Hallenbeck and Cunningham-Burns 1985). Residues of some synthetic pesticides, for example, toxaphene and DDT, remain persistent in the soil for years, and are washed off into water bodies with the attendant effect of contaminating groundwater and aquatic biota (Carvalho 2017). Humans become intoxicated through bioaccumulating contaminants in food chains common in terrestrial and aquatic environments (Lushchak et al. 2018). Organochlorine and organophosphate compounds undergo the evaporation-condensation process and consequently are transported far and wide, with extensive effects felt at a considerable distance from the point of application. For instance, chlorpyrifos applied on a banana plantation in the intertropical region of Central America was detected in the ice pack in the Artic (Carvalho 2017), essentially making the use of pesticides a global challenge. Pesticides as environmental pollutants disrupt microbial biomass and biodiversity, cause soil fertility losses, and adversely affect vital biochemical and enzymatic activities in soil (Gill and Garg 2014) apart from causing unintended health challenges to humans. Thus, averting these negative consequences caused by synthetic pesticides has become a necessity, making the quest for alternatives to synthetic pesticides, an existential discussion in conventional agricultural practices. As such biopesticides are considered as a solution option to the challenge of pollution (Koul 2011). Bio-based pesticides have been favoured in recent times, with a 2% drop annually of synthetic pesticides against a 10% growth rate of biopesticides (Damalas and Koutroubas 2018). This trend reduced environmental pollution from synthetic pesticides, increased the production of wholesome crops, and improved biodiversity, thus addressing some of the dimensions of sustainable agriculture.

1.3 Sustainable Agriculture: Definition, Concept and Context

Sustainable agriculture employs the judicious and continuous use of critical resources in meeting both the needs of today but seeks to find a balance that does not compromise the need of future generation by integrating the understanding of the ecosystems in the exploitation of these natural resources (Burton 1987). This idea of practicing sustainability predates most of the ancient civilizations, including countries like Greece, Rome and China (Cato 1979; King 1911). However, the sixties marked the development of global consciousness with the much publicized book authored by Rachel Carson titled, The Silent Spring (Carson 1963), which emphasized the harm done to the environment by agriculture. Moreover, in the "Tragedy of the Commons", written by Garret Hardin in 1968 posited that the rational drive of selfish interest would end up compromising common interest and would finally exhaust available natural resources (Hardin 2009; Frischmann et al. 2019). The profound computational simulation by Meadows (1972) predicted economic and social collapse if a man fails to impose a limit to growth linked to the use of the inarguably limited natural resources. After 40 years, their predictions were confirmed in the visible signs of global pollution and its associated problems that threaten sustainable development. On the premise of depleting the natural resources reserve and pollution, the first UN Conference held in Stockholm in 1972 was to discuss human impact on the environment and its relatedness to economic development (Maurer and Bogner 2019). Additionally, it was necessary to find a common ground to inspire and guide the global population to preserve the human environment. Progressively, the World Conference in 1979 focused on the influence and assessment of anthropogenic and natural causes and the contributions to climate change with its implications to human society. From the foregoing it is becoming evident that our planet has limited non-renewable resources and progress is not synonymous with economic growth alone. Consequently, the human development index came into the picture that encompasses economic and social achievements.

The Brundtland ("Our Common Future") report (1987), provides, the simplest and most accepted meaning of sustainable development, as it stated that limitation is imposed on sustainable development by the collection of technology, social organization and the inefficient status of the biosphere to mitigate the impacts of human activities. Agenda 21, the draft document of the United Nation Conference of 1992 built a nexus between the socioeconomic sector and the environment. The document focused on explaining the deterioration of the environment and how it can be integrated into a sustainable development plans. Within this framework, sustainable development stands on the tripod of economics, social and the environment and if pursued with sincerity of purpose would certainly guarantee a more prosperous future through improved living standards and safe ecosystems. A chapter in the Agenda 21 document was dedicated to rural development and sustainable agriculture which require a major shift in agriculture, macroeconomic policies and the environment (Blandford 2011). Beyond this tripod and the five domain concepts (economic, socio-cultural, technology, environment and public policy) of Steward W. C (Marteel-Parrish and Newcity 2017), sustainable agriculture can also be looked at through the lens of the 12 Principle of Green Chemistry (Anastas and Warner 1998) and the 17 Sustainable Development Goals (United Nations). Conceptualizing sustainable agriculture on the tripod dimensions means agro-practice that connects environmental soundness (efficient use of finite resources, prevention of air, land and water contamination; and reduction of health hazards), economic viability (reliable and profitable production activities) and social acceptability (self-sufficiency, improved quality of life and equality). Given this concept, equal emphasis will be placed on each of the dimensions of the Tripod over a long period (Zhen et al. 2005).

The Green Chemistry concept was enunciated in the Pollution Prevention Act of 1990 and developed by the trio of Joseph Breen, Tracy Williams and Paul Anastas in 1998 (Tundo and Griguol 2018) and has gained prominence in theory, development and practice since then. The adjective "green" connotes benign solvent, catalyst/reagent and energy consumption components associated with a reaction (Ivankovic et al. 2017). This means adopting the safest innovative measures in line with some elements of chemical knowledge to protect the environment and its receptors including man. Green chemistry is considered as sustainable chemistry (Tundo and Griguol 2018) when conventional chemistry is implicated in the area of economic consideration, efficient use of materials and waste reduction (Manahan 2006). It covers areas such as efficient processes, renewable materials, green solvents and catalysts, and benign products (Song and Han 2015). As a multidimensional concept, green chemistry encompasses processing, synthesis, and the use of chemicals (solvents, reagents, catalysts, products and feedstock) that reduce risks to the environment (its receptors) and humans (Verma et al. 2018). Its goal is to produce reduced wastes by using safer and better chemicals synthesized in the safest and most efficient process. Green chemistry has 12 Principles (PGC) which cover four concepts: (1) the design of processes in which all raw materials are incorporated in the final product (2) the use of innocuous substances whenever possible (3) energy-efficient process design and (4) the choice of the best method of waste disposal (Ubuoh 2016; Nnaji and Igbuku 2019). Despite the lofty promises of green chemistry, it has its challenges embedded in time and cost (Ivankovic et al. 2017). Switching from the old and conventional process to green process is usually laced with a considerable time that covers for design, or redesigns as the case may be, for a new process. Robust design, required by green technology, requires a high cost of implementation. Lack of information and resources had also plagued the applications with negativity. For instance, ionic liquids seem to have a high prospect for green chemistry but if balanced against a strict definition of the 12 green chemistry principles, they are not considered green.

Green and sustainable agriculture holds a very strong position in the SDG agenda because it is directly and indirectly connected to the 17 SDGs such that no goal is left unlinked (Omilola and Robele 2017). Sustainable agriculture is considered as an agro-practice that promises long-term productivity with minimal harmful effects due to the use of biofertilizers, biopesticides and organic manures. Sustainable agriculture is the outcome of farming that integrates green chemistry and sustainability principles to serve the economic, social and environmental interest. It holds the central position within the most recognized and acceptable sustainable development recipe.

1.4 Sustainable Agriculture: Its Place in the Tripod Concept, SDGs and Green Chemistry

The Green Revolution intensified food production in Asia and other countries through technology, chemical inputs, irrigation, agricultural policies and strong institutional frameworks (Nelson et al. 2019). Interestingly, food prices decreased significantly, the income of rural dwellers increased, poverty declined by almost a double in Asia between 1970 and 1995, and conservation of marginal forest lands (Pinstrup-Andersen and Hazell 1985; Evenson and Gollin 2003; Pingali 2012; Stevenson et al. 2017). Simultaneously, with some important exceptions, countries in the African continent performed below their agricultural potentials due to insignificant employment of external inputs (Shimeles et al. 2018). The resulting poor agricultural productivity and growth coupled with high population increase necessitated these countries to be net importers of food. The Green Revolution made significant economic progress but also created critical challenges such as pesticide impacts, pollution, land degradation, loss of biodiversity, unsustainable use of water and other resources. Naturally, agriculture is the highest user of natural resources such as water, land or nutrients, although the extent of their uses depends on a particular farming system. As noted by Mancosu et al. (2015) major land portions are used for agriculture, and the sector consumes 70% of the water used globally. Consequently, agricultural practices put huge pressure on greenhouse gas emissions and climate change, contribute majorly to the distortion of ecological services and functions. Moreover, with, the associated temperature increase due to climate change, it is expected that agro-yield levels, particularly in the third-world countries, will be reduced (Huang et al. 2015; Sapkota et al. 2017; Perrone 2018). Some of these occurrences can be curtailed and controlled through deliberate actions, changes in consumption patterns, efficient use of both renewable and non-renewable resources, the involvement of green technology, strategic positioning of institutional frameworks and implementation of salubrious policies for better agricultural regimes. The achievement of economic and environmental objectives of agriculture within the concept of sustainability would certainly influence the social construct through human development indices such as quality education, better health, well-being, and social infrastructure. Thus, the tripod dimension of sustainable development will receive a significant boost from sustainable agriculture since its achievement seeks three things: economic profitability, social well-being and environmental stewardship.

The scope of the (SDGs) UN 2030 Agenda of 2015 integrates the three-dimensional angle of sustainable development as its subset. The various variants of sustainable developments including the SDGs are connected directly or indirectly to the many dimensions of sustainable agriculture. Sustainable agriculture can address hunger (SDG2), drastically reduce poverty (SDG1), increase productivity where the agents of change (SDG5 and SDG10) are invested on, stewardship of the natural resources is upheld (SDG6, SDG12, SDG14 and SDG15); innovation and technology play their significant roles (SDG9), and appropriate responses are taken to address climate change (SDG13). Further, sustainable agriculture has the potentials to stimulate economic growth and improve livelihoods (SDG8). These SDGs are also linked with sustainable agriculture in different decrees. The remainder that are linked to sustainable agriculture are SDG3 (through inclusive economic growth, nutritious products), SDG4 (through a capacity building), SDG7 (inadequate alternative energy), SDG11 (value addition in agro-system, ecosystem resilience), SDG16 (peace and justice in tandem with economic growth), and SDG17 (government involvement).

However, government and private institution involvement through partnerships, policies, regulations, laws, and investment has an overriding influence on sustainable agriculture. Figure 1.2 depicts how SDG1 and SDG2 (concerned directly with improved productivity of agro-outputs; with organic manure, biofertilizers and biopesticides), SDG6, SDG12, SDG14 and SDG15 (concerned with efficient use and restoration of natural resources; with organic materials) and SDG7 (concerned with the provision of green energy; using non-fossil sources) are directly linked to green chemistry based on the consumption of green chemicals and employment of associated technology (SDG9). Thus, sustainable agriculture is directly linked to eight of the 17 SDGs with green chemistry as the deciding factor. Within the context of sustainable agriculture, innocuous substances are not only meant to substitute conventional chemicals but can also reverse the consequences of synthetic chemicals.

Concerning the 12 Principle of green chemistry, an organic substance which is renewable and biodegradable will satisfy most of the conditions given in comparison to synthetic chemicals (Perlatti et al. 2014; Song and Han 2015; Ganasen and Velaichamy 2016; Gonzalez 2017; Saleh and Koller 2018). However, their production in commercial quantities is limited because of optimization bottlenecks (Hassan and Gokçe 2014). Apart from being renewable and biodegradable, these organic chemicals/substances should produce less residual chemicals, energy and solvent; be safe and their production process should be. One relevant group of compounds to green chemistry worthy of mentioning are biofertilisers, regarded as bioactive compounds derived from the activities of bacteria, fungi or algae. Biofertilisers are capable of causing plants to uptake nutrients by their interaction with the rhizosphere of the benefiting plant (Carvajal-Munoz and Carmona-Garcia 2012; Igiehon and Babalola 2017; Itelima et al. 2018). Apart from nutrient uptake, other benefits associated with biofertilisers are reduction in fertilizer usage, soil fertility improvement, tolerance to (a)biotic stresses, and improvement in crop yield (Kumar 2018). Most biofertilisers are either phosphorus solubilizing or nitrogen-fixing (Yimer and Abena 2019). Phosphorus solubilizing microorganisms release organic acids which dissolve phosphate and tricalcium phosphate bearing rock particles thereby

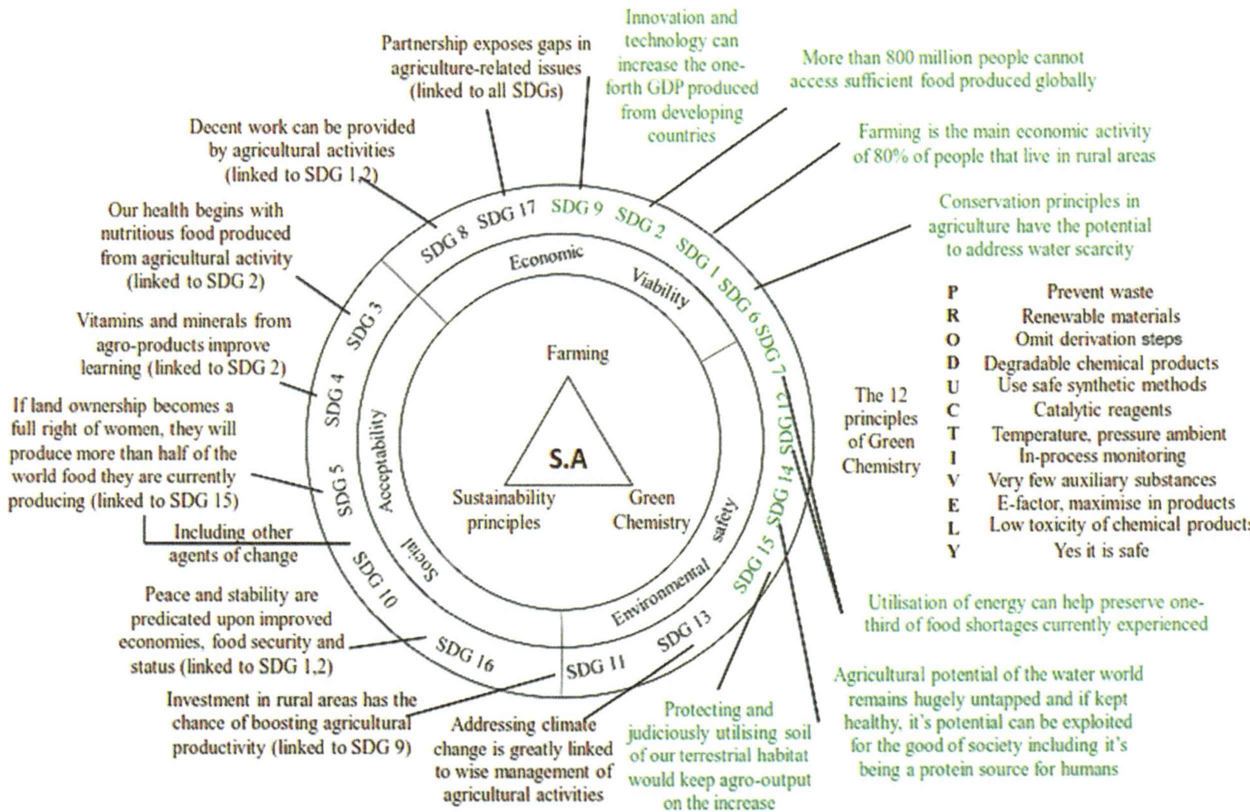

Fig. 1.2 Connection of sustainable agriculture and 8 SDGs and green Chemistry. (Adapted from Anastas and Warner 1998; FAO 2013)

facilitating plant uptake of phosphorus (Saeid et al. 2018). Phosphorus solubilizing microorganisms are not selective nor specific to particular plants but have broad-spectrum action (Bhattacharya 2019). The nitrogen-fixing candidates fix atmospheric nitrogen into forms which are biologically available to plants, easing uptake into cells. The microorganisms involved in this category are either symbiotic or free living. Examples of microorganism that fix nitrogen are *Azobacter* (for rice, vegetable, wheat), *Azospirillum* (for sugarcane, rice), *Azolla* (for rice), *Rhizobium* (for leguminous crops), and blue-green algae (for rice). There is quite a significant body of evidence that demonstrate the application of biofertilizers is efficient in different kinds of crops, including cotton, tomatoes, potatoes and others (Htwe et al. 2019; Khanna et al. 2019; Gortari et al. 2019). Some of these biofertilisers also exhibit biocontrol activities and in this context dually function as biopesticides (Gortari et al. 2019).

1.5 Biopesticides: Definition and Scope

Biopesticides are considered as a naturally occurring organisms or bio-based formulations that control pests through different mechanisms of action (Tijjani et al. 2016). They are products or byproducts derived from animals (nematode; *Heterorhabditis* spp.), insects (*Trichogramma* spp.) plant parts or extracts (example, a finely ground flower of *Chrysanthemum cinerariaefolium*) and microorganisms (*Bacillus thuringiensis*, *Verticillium*, *lecanii*, *Neodiprion sertifer*) (Pavela 2014; Rodgers 1993). However, there is a deviation in the generally accepted definition, as it is that natural products should be regarded as chemical pesticides, if they can have an impact on the pest nervous system (Marrone 2019). For example, rotenone, as a plant-based pesticide, has the potency of killing and controlling insect pests by exerting its toxic effect primarily on nerve and muscle cells (Singh 2014). Similarly, nicotine (a fast-acting nerve toxin), sabadilla (affects nerve cell membrane action), pyrethrins (interrupts the normal transmission of nerve impulses), and fluoroacetate (causes depletion of glutamic acid thereby affecting the nervous system) are all effective pesticides that interfere with the insect nervous system (Oguh et al. 2019). However, most literature classifies these compounds as botanical pesticides (Brudea et al. 2012; Chengala and Singh 2017; Bateman et al. 2018; Arshad et al. 2019). A chemical homologue of a biopesticide can also be regarded as a biopesticide (Oguh et al. 2019). Biopeticides intended for the control of herbivorous insects or pests can be regarded as phytosanitory biocontrol agent or bioproduct.

1.5.1 Biopesticides' Categories and Their Modes of Action

Biopesticides are either microbial, biochemical or plant-incorporated protectant (PIP) biopesticides. Their modes of action come under five groups: neuromuscular toxins, metabolic poisons, gut disruptors, growth regulators, and non-specific multi-site inhibitors based on the physiological processes they affect (Sparks and Nauen 2015). However, some of the modes of action are still not specifically clear, especially with biochemical biopesticides. Microbial biopesticides exert their control through antagonism, predation, parasitism, and antibiosis (Mishra et al. 2018) For a natural substance to be considered as a biochemical biopesticide, its mechanism of action must be nontoxic (Ivase et al. 2017; Inam-ul-Haq et al. 2019). Plant-incorporated protectants are dependent on the incorporated molecule which may be derived from microorganisms or plants.

1.5.1.1 Microbial Biopesticides

Microbial biopesticides could be bacteria, fungi, viruses, protozoa and nematodes, or compounds derived from these organisms that influence pest activities, through competition, pathogenicity or inhibitory toxins. These agents are broadly divided into multifactorial microbial generalists and hyperparasitic microbial specialists. The generalists control a wider range of pests whereas the specialists act against a particular pest. More than 3000 microbes have been recognized to cause diseases in insects implicating two major groups of nematodes (*Steinernema*; 55 species and Heterorhabditis; 12 species), more than 100 bacteria, 800 fungi, 1000 protozoa, and 1000 viruses (Sparks et al. 1999; Dowds and Peters 2002; Casadevall 2007; Mills and Kean 2010; Ravensberg 2011; Lewis and Clarke 2012; Singh et al. 2015; Nawaz et al. 2016; Marche et al. 2018; Ruiu 2018). Specific examples are *Bacillus thuringiensis*, *Paenibacillus* (bacteria), HearNPV (Baculovirus), *Metarhizium anisopliae*, *Verticillium* (fungi), *Heterorhabditis*, *Steinernema* (nematodes), *Nosema*, *Vairimorpha* (protozoa), *Chlorella*, *Anabaena* (microalgae; Costa et al. 2019). This category of biopesticides has the advantages of specificity (non-pathogenic to non-target), synergisms (can be used alongside synthetic pesticides), eco-friendliness (their residue has no negative impact on the ecosystem or eco-receptors), permanent effects (the microorganism becomes an integral component of the insect population or its habitat exhibiting the inhibitory effects) and growth improvement to plants (Nawaz et al. 2016). However, our understanding of microbial pesticides is hampered by challenges such as detailed scientific research, ecological study, and mass-production technologies (Haase et al. 2015). These challenges may differ from the known and common entomopathogenic microorganisms.

The bacteria *B. thuringiensis* is entomopathogenic, and produces Bt toxins. When insects ingest Bt toxins, the following sequence of events occurs: binding of the toxins to the midgut receptors, a pore-forming process is triggered, disruption of the intestinal barrier functions and finally infestation leading to the death of insects. A similar mechanism is confirmed in mosquito and blackfly control with *Lysinibacillus sphaericus* (formally *Bacillus sphaericus*) active agent. In this example, the complementary biosynthesis of crystal proteins (BinA and B) and Mtx (mosquitocidal toxin) act as the insecticidal toxins (Ruiu 2018). In instances of fungal infection, the host cuticle serves as a point of contact to fungi, and when the environmental conditions are favourable, fungal spores and conidia germinate. The enzymatic and mechanical actions enhance the penetration of the fungi into the host body. Consequently, the mycelia develop internally giving rise to different types of spores, conidia, metabolites, toxins and virulence factors (Ruiu 2018). Baculoviruses exert their effects via the production of crystalline occlusion bodies, possessing infectious particles, in the host cell. Once contaminated food is ingested, the occlusion bodies within the midgut release virions (occlusion derived viruses; ODVs) affecting the membranes of microvillar epithelial cells through the action of their envelope proteins (Townsend et al. 2010). The cadaver of the affected insects liquefies thereby dispersing the virus particle in the environment. Symbiotic nematodes, transport entomopathogenic bacteria to the internal host system via natural openings. The bacteria elicit their insecticidal toxins and virulence factors and metabolites that encourage the reproduction of the obligate nematodes.

1.5.1.2 Biochemical Biopesticide

Biochemical biopesticides are substances of natural origin with active agents to control pests by mechanisms that are not toxic to the host, the environment and humans (Kumar 2012; Leahy et al. 2014). By this definition, a natural chemical can be considered a biopesticide if it acts as an attractant, deterrents repellant, antifeedant, suffocant, confusants, arrestants, or desiccant (Stankovic et al. 2020). Being natural implies that such chemicals would be discrete or mixed bioactive substances from nature. However, a synthetic analogue that is identical to a natural compound, both structurally and functionally (exhibits the same mode of action). Certain factors have made some synthetic analogues of naturally occurring substances to dominate the commercial market (Dang et al. 2016). Although toxicity is a subjective term, a substance could be said to be nontoxic if direct lethality of the target host does not arise as a result of the chemical or biological interference of the substance active ingredients with the physiology of the target pest. This definition does not guarantee the absence of ill-fated biochemical and metabolic reactions in the target pest organism by the presumed nontoxic substance. Instead, the initiation of such ill-fated reactions is linked to one or more physical processes attributable to the substance. For instance, essential oil causes asphyxia (a physical process) which obstructs pest respiration leading to death. A substance still merits the nontoxic status if its active ingredients invoke biochemical reactions that interfere with the behavior or reproductive system of the target pests without resulting in death. A substance is environmentally safe if it is exogenous to that environment and has no impact on the physicochemical signature of the environment or affects the ecological services provided by that environment and causes no distortion or harm to ecological receptors including wildlife and humans. Chemicals that pass these criteria of naturalness, nontoxicity and eco-friendliness are semiochemicals (pheromones and allelochemicals), essential oil (from neem, sour orange), insect growth regulators (juvenile hormones, chitin synthesis inhibitors), plant growth-promoting regulators (Rhizobacteria) and natural minerals (diatomaceous earth, kaoline).

The semiochemical mode of action (MoA) is concerned with the disruption of hormones and neuropeptides associated with metamorphosis and insects' growth. The MoA of mineral-based insecticides (kaoline, insecticide soaps, diatomaceous earth) is mostly physical. The abrasive nature and sorption properties of diatomaceous earth, and the waxy layer of insects are damaged giving way for desiccation and eventual death (Nukenine et al. 2010; Sousa et al. 2013). Similarly, kaoline exerts its insecticidal effect through its sorption property, which causes desiccation in insects. Besides, surface activity, the coating property of kaoline can cause reduced sublethal effects, repellence and oviposition deterrence (Yee 2008). The mode of action of insecticidal soap is expressed through cuticle dissolution leading to suffocation and desiccation. Bioactive compounds in botanical extracts can cause inhibition of hyphal growth, structural modifications of mycelia, changes in the cell wall, partitioning of cell membranes, and separation of the cytoplasmic membrane in entomopathogenic fungi (Lengai and Muthomi 2018). Plant extracts apart from inducing behavioral changes (as it concerns feeding habit, oviposition and mating behaviour) in insect pests also inhibit insect reproduction, growth and development. Essential oils act as antifeedants, repellants, and oviposition deterrents. Besides, they possess active ingredients that make them larvicidal, ovicidal and insecticidal thereby displaying properties that interfere in all stages of insect metamorphosis (Sarma et al. 2019). The MoA of semiochemicals acts by inhibiting lipid biosynthesis resulting in a significant decrease in total lipids in immature insects (Linda et al. 2010), disruption and prevention of metamorphosis caused by the binding of juvenile hormone analogues to the receptor of juvenile hormone in insects (Jindra and Bittova 2020), and inhibition of moulting and chitin synthesis which determines growth and development of insects (Cohen 2001; Ullah et al. 2019).

1.5.1.3 Plant-Incorporated Protectants

A plant-incorporated protectant (PIP) is a biopesticide produced by a gene inserted into a plant through transgenesis (Ibrahim and Shawer 2014). PIP does not require killing the pest but renders the plant unsuitable for an attack. In some cases, the protected plant may act as a repellant or disrupt the normal physiology of the insect pests when insects ingest PIPs. Once the PIP is ingested it overcomes the digestive and physical barriers and then gets to the target site where it acts. The digestive system has been confirmed as a strong determinant of insect vulnerability and susceptibility therefore, gut function disruption has been a common theme in the development and discovery of PIPs (Nelson and Alves 2014). Insecticidal proteins, particularly Bt, are suitable for application in PIPs and are thus being explored in pest control (Koch et al. 2015). The insecticidal property of *Bacillus thuringiensis* was first discovered in 1902 against silkworm (*Bombyx mori*) and from then the search for Bt strains as an insect control agent has continued (Jisha et al. 2013). The insecticidal proteins from Bt are effective, diverse and specific thus they are widely used as a model in PIP biotechnology, however, Schwenk et al. (2020) has demonstrated that Bt shows non-negligible pathogenic potentials. The insecticidal crystal protein produced by Bt is known as Cry proteins (δ-endotoxins) but they are diverse thus, they exhibit insect selectivity. For instance, there are those that are selective for Lepidoptera, Coleoptera and those for Diptera (Maciel et al. 2014). Currently, no less than 70 classes (based on sequence homologies and target selectivity) of Cry proteins have been used to protect corn, cotton, potato, soybean and other crops (Pardo-Lopez et al. 2013). The Cry proteins are toxins produced during the sporulation period but toxins produced during the vegetative phase are called vegetative insecticidal proteins (Vips) and are commonly used in PIPs. More than 50 Vips proteins, including Vips 1, Vips 2 and Vips 3, have been reported (Shingote et al. 2013; Chakroun et al. 2016; Sopko et al. 2019). Other insecticidal proteins from other bacteria proved to be effective in transgenic control are toxic complex (Tc) proteins expressed by *Photprhabdus* and *Xenorhabdus* (Shingote et al. 2013). Also, plants possess transgenic enzyme inhibitors that have been explored in PIP technology, for example, α-amylase inhibitors (Franco et al. 2002). Mir1-CP protease from maize, enhancing protease from Baculovirus has also shown potency in protecting plants via the PIP technology (Mohan et al. 2006; Wei et al. 2018). Besides, double-stranded ribonucleic acids (dsRNAs) are commonly used as approved PIPs (Parker and Sander 2017) due to the rapid progress in ascertaining RNAi biological processes (Liu et al. 2020). The dsRNA triggers host-induced gene silencing and protein synthesis inhibition which improves endogenous gene expression in plants while causing increased pest mortality within the plants (Raruang et al. 2020).

The Bt mode of action could be explained based on the correlation of the Cry protein ingestion and insect susceptibility. Once the Cry protein reaches the midgut after ingestion it attacks the "brush border" epithelium with the attendant manifestation of feeding cessation (Lee et al. 2003) With the right concentration of the toxin, ATPases concerned with active transport, become inhibited, followed by modulation of endogenous potassium channels and pore formation that occasionally leads to uncontrolled ionic flux, the collapse of normal cellular function and death (Knaak et al. 2010). As noted earlier, the Cry proteins exist in different classes and structures with structure-dependent toxicities specific to particular insect orders. For instance, Cry 3 and Cry 1 proteins are toxic to Coleoptera and Lepidoptera respectively (Chakroun et al. 2016). In the case of the dsRNA, after ingestion, it becomes biochemically cleaved by dicer into small molecules of interfering RNA (siRNA of ca. 20 nucleotides) after getting in contact with a target cell in the gut of insects (Sopko et al. 2019). The machinery of the RNA interference (RNAi) aids siRNA targeting mRNA for destruction (Zhou and Rana 2013). The destruction of the targeted mRNA inhibits its translation into proteins, essential for the insect pest thus leading to reduced growth or death (Rodrigues and Figueira 2016). The dsRNA PIP, the regulatory approved in this class was used against corn rootworm by inhibiting the synthesis of SnF7 proteins necessary for vacuolar sorting protein (Ramaseshadri et al. 2013; USEPA 2017). The prospects of RNAi PIP are higher

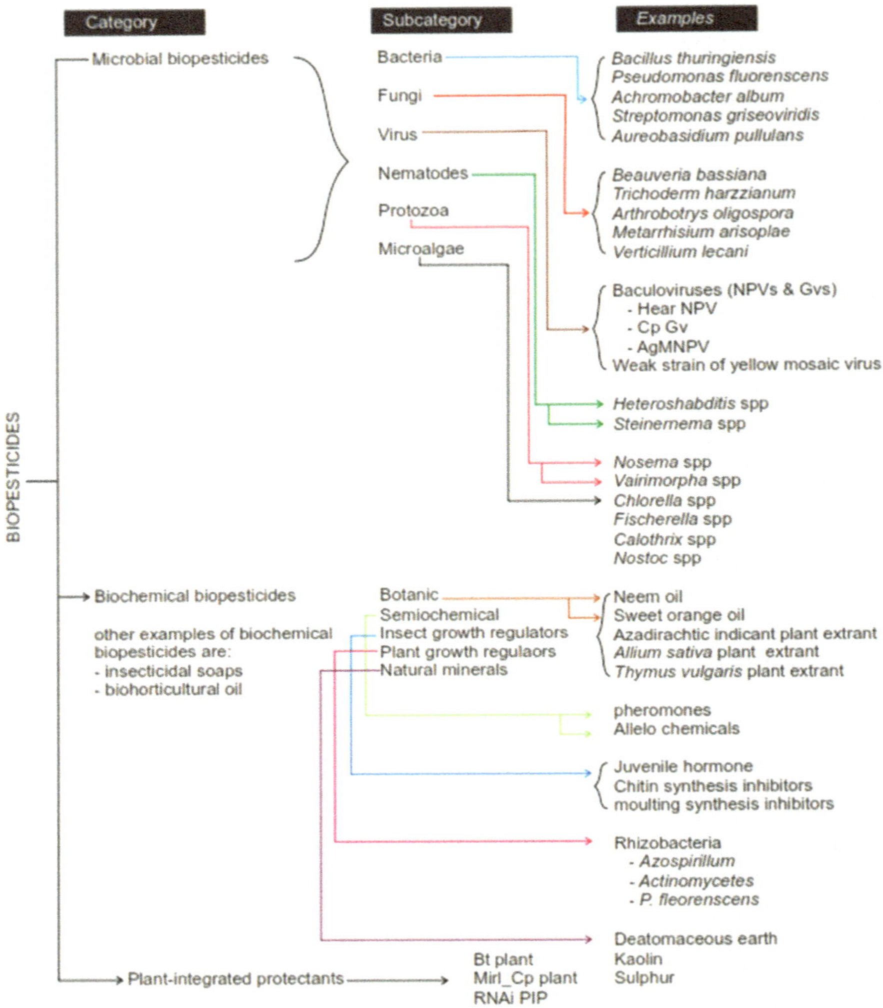

Fig. 1.3 The three different categories of biopesticides and some selected examples

in insects with long dsRNA because of the higher number of siRNAs. The three different categories of biopesticides are illustrated in Fig. 1.3.

1.5.2 Production, Commercialization and Market Prospect of Biopesticides

Biopesticide production starts with bioprospecting from the natural environment, screening (*in vitro*), *in vivo* experiments, purification, formulation and registration (Fig. 1.4) (Lengai and Muthomi 2018). Botanical pesticides are obtained from plants in the natural environment while microbial bioprospecting can be obtained from

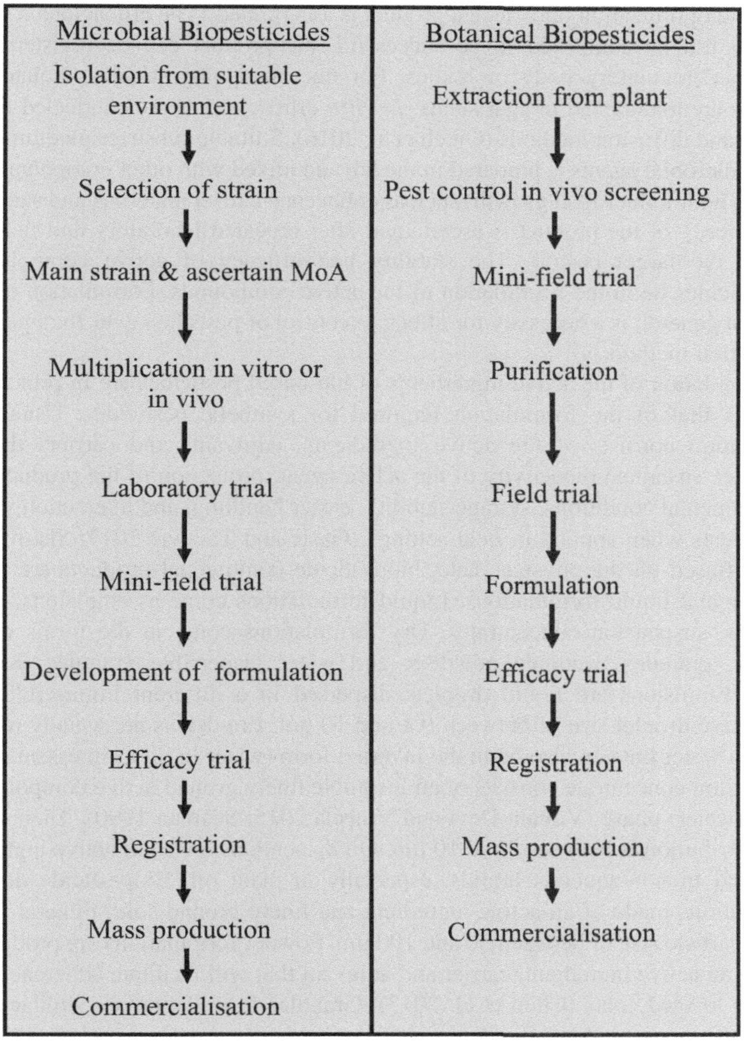

Fig. 1.4 Commercial production chart of microbial and biochemical biopesticides. (Adapted from Lengai and Muthomi 2018)

compost, manure or rhizosphere. Botanicals are first cleaned of impurities, extracted and the extracts are then screened *in vitro* against pest of interest using different methods (Altemimi et al. (2017). From the screened plants, the most effective plant can be evaluated through field-trials, and from these plant extract the most active constituents can be identified and ascertained for optimum formulation (Khot et al. 2012) The active compounds are usually identified with either or in a combination of different spectrometric techniques including high-performance liquid chromatography (HPLC). The active ingredients are further combined with carriers, surfactants, emulsifiers and other components, followed by intensive *in vitro* and laboratory trials and optimization until such a product is ascertained to be efficacious. Once the efficacy trial is confirmed to be successful, the product can be registered with concerned regulatory body or bodies. For microbial biopesticides, isolated pure cultures are maintained in agar slants. *In vitro* efficacy trials are conducted through culture and diffusion methods (Oikeh et al. 2016). Suitable substrate medium for the active microbial agents is prepared in the lab and mixed with other components (that will not inhibit microbial growth) such as enhancers, carrier materials and stabilizers. The efficacy of the product is ascertained after repeated laboratory and field trials against the target pest(s). The stability and efficacy of active compounds of biopesticides lie in the formulation of the active compounds. Formulation of pesticides, in general, is a necessity for effective control of pests however for a particular application method.

Formulation of the active ingredients of bio-based pesticides are in principle the same as that of the formulation required for synthetic pesticides. Usually, the formulation consists of the active ingredients, adjuvants and carriers that will guarantee sustained bioactivity of the active agent, protection of the products from environmental conditions, storage stability, easier handling, and interaction with the target pests when applied in field settings (Gasic and Tanovic 2013; Sharma et al. 2019). Based on the physical state, biopesticide commercial products are divided into dry and liquid formulations. Liquid formulations come as emulsions, oil dispersions, suspension concentrates. Dry formulations come in the forms of dust, powder, granules, wettable powders, and water dispersible granules (Knowles 2005). Emulsions are liquid droplets dispersed in a different immiscible liquid with phase droplet size of between 0.1 and 10 μm. Emulsions are usually regarded as oil in water but can also be in the inverted form (water in oil as in essential oils). Suspension concentrate consists of an insoluble finely ground active compound in a liquid (water) phase (Vimala-Devi and Vineela 2015; Seaman 1990). Their particle size distribution is between 1and 10 μm. Oil dispersions are solid active ingredients dispersed in non-aqueous liquids, especially in plant oil. Biopesticide dusts are formulations made of an active ingredient and finely ground solid mineral powder with a particle size of between 50 and 100 μm. Powder formulations are produced by mixing an active ingredient, carrier and adjuvant that will facilitate adherence of the product to seed coats (Chen et al. 2013). Granules formulation are similar to dust formulations except that the granular particles are heavier and larger (100–1000 μm). Wettable powders are formulations made up of finely active ingredients (5 μm), surfactants, dispersing agents and inert fillers which are used after suspension in aqueous medium (Tadros 2005). Water dispersible granules are a powder formulation to be used when dispersed in water.

Currently, biopesticides command a 5% pesticide market, with a corresponding value of approximately $3billion globally (Damalas and Koutroubas 2018) and are increasing at an annual growth rate of between 10-15% (Marrone 2014). The likely pressures behind this increase are demands for organic vegetables, tree crops and vineyards; and the need for improved food safety. Thus far, the average product number of registered biopesticide products in India is 970; in the US they are 452 and 86 in Europe (Hassan and Gokçe 2014; Mishra et al. 2020). Microbial biopesticides share 63% of the global biopesticide market (Hassan and Gokçe 2014), with *Bacillus thuringiensis* derived products accounting for 90% (Kumar and Singh 2015). Other microbial biopesticides that have made significant in-roads in market penetration are *Trichoderma gamsii*, *Trichoderma harzianum*, and *Beauveria bassiana* (Mishra et al. 2020). Botanical biopesticides, on the other hand, are dominated by the essential oil, pyrethrins, rotenone, and azadirachtin.

Tables 1.1, 1.2 and 1.3 display a few selected commercial products and their manufacturers. It was projected that biopesticides would have equal utility in plant protection by 2050 in comparison to the role currently played by chemical pesticides. This would become a reality if Africa and Southeast Asia, make a sincere and concerted effort in bringing about changes in farming practices (Olson 2015) as well as investing in research in this direction. However, biopesticide production and consumption is hindered and will continue to be so unless all stakeholder, including, end-users, marketers, regulators and researchers involved in the development and commercialization chain, align to the same objectives. For instance, marketers often disagree with researchers and regulators to such an extent that the end-users get dissuaded about the potency of the final products (Kumar and Singh 2014). Furthermore, researchers are discouraged due to the daunting procedure of documentation, submission procedures and registration required for a product to enter the market. Often, the process takes several years, coupled with the high cost involved in the registration of new pest controlling agents. These factors, as well as others that are unique to various regions of the world, contribute to higher cost of biopesticide products. Moreover, the social acceptability of biopesticides compared to the conventional and long-accepted synthetic chemical pesticides, as well as the end-users' perception of biopesticide "slow action" are strong aggravating factors. To maximally exploit the prospects of biopesticides, regulatory authorities should have a fast and flexible but not-quality-compromising requirements that consider the morally relevant differences that mitigate their use compared to synthetic pesticides especially in the registration process. Additionally, special concessions should be provided for organizations, institutions and individuals to encourage them during biopesticide production and registration. It is expected that more efficacious biopesticides would be produced in the future.

Table 1.1 Commercial examples of biopesticides (adapted from Mondal and Parween 2000; Kabaluk et al. 2010)

Product name	Active ingredient	Plant species	Company
Botanical insecticides			
Prentox Pyrethrum	Pyrethrins	*Chrysanthenum cinerariaefolium*	Prentiss
Py-rin Growere	Pyethrins	*Chrysanthenum cinerariaefolium*	Wilbur-Ellis
Pycon	Pyrethrins	*Chrysanthenum cinerariaefolium*	Agropharm
Premim Pyganic	Pyrethrins	*Chrysanthenum cinerariaefolium*	MGK
Neemx 90 EC	Azadirachtin	*Azadirachta indica*	Thermo Trilogy
NeemAzal	Azadirachtin	*Azadirachta indica*	Trifolio-M
Trineem	Azadirachtin	*Azadirachta indica*	Tagros
Vironone	Rotenone	*Derris* spp.	Vipesco
Rotenone-Copper	Rotenone	*Lonchocarpus* spp.	Bonide
PBNox	Rotenone	*Tephrosia* spp	Penick
Stalwart	Nicotine	*Nicotiana tabacum*	United Phosphorus Ltd
Nicotine 40%	Nicotine	*Nicotiana tabacum*	Dow AgroScience
Natur Gro R-50	Ryania	*Ryania speciose*	AgriSystems International
Veratran D	Sabadilla	*Schoenocaulon officinale*	Dunhill Chemical
Nemguard	Garlic extract	*Allium sativa*	ECOSpray Ltd
Botanical fungicides and bactericides			
Bioxeda	Clove oil	*Syzygium aromaticum*	Xeda International
Iodux 40	Laminarin	*Laminaria digitata*	Geomar
Vertigo	Cinnamaldehyde	*Cassiatora* spp.	Monterey
Milsana	Milsana	*Reynoutria sachalinensis*	Geomar
Botanical herbicides			
Interceptor	Pine oil	*Pinus* spp.	Certified Organics Ltd
Barrier H	Citronella oil	*Cymbopogon* spp.	Barrier Biotech Ltd
Hinder	Pelagonic acid	*Geraniaceae* family members	Amvac
IGR commercial products			
Applaud	Buprofezin	Protects rice against homopteran pests	Nihon Nohyaku
Logic	Fenoxycarb	Protects apples from worms	Ciba-Geigy
Atabron	Chlorfluazuron	Protects vegetables, cottons against lepidopteran pests	Trigard

Table 1.2 Commercial products of different microbial pesticides

Microbial biopesticide	Insect.	Fungi.	Bacteri.	Others	Commercial product	Company	Reference
Bt and sub species	*				Dipel	Valent BioScience	Li et al. (2007)
		*			BT Sulfur 15–50	Loveland Products Inc	www.greenbook.com
			*		Defender	Arbico Organiics	Mishra et al. (2015)
				1	Thurisav-3	Agrochem	Kabaluk et al. (2010)
Bacillus subtilis				1	Sting	Stanes Company	Berlitz et al. (2014)
Pseudomonas fluorenscens				1	Bio-Cure-B	Stanes Sompany	Khalil et al. (2012)
Aureobasidium pullulans			*		Blossom Protect	Bio-Ferm	Rosello et al. (2013)
Salmonella enteriditis subsp. danysz				3	Biorat G	LabioFam	Kabaluk et al. (2010)
Streptomyces griseoviridis K61		*			Mycostop	Arbico Organics	Bailey (2010)
Lactobacillus spp				2	Organo-Sol	Organica	Kremer (2019)
Beauveria bassiana	*				Mycotrol	Mycotech Corp.	Steinkraus and Tugwell (1997)
Trichoderma harzzianum		*		1	Plant Guard	Ajay Bio Tech	Radwan et al. (2018)
Sclerotina minor IMI 344141				2	Sarritor	Redox Industry Limited	Abu-Dieyeh and Watson (2007)
Baculovirus	*				Cyd-X	Certis USA	Quarles (2013)
AgNPV	*				Baculovirus Nitral		Haase et al. (2015)
Yellow Mosaic Virus (weak strain)				4	AgroGuard-Z		Rajput et al. (2020)
Nematode	*				Nemabact		Ilyashenka and Ivaniuk (2008)

Insect. insecticide, *Fungi.* fungicide, *Bacteri.* bactericide, *1* nematicides, *2* herbicides, *3* rodenticides, *4* viruside
* Applicable

Table 1.3 Latest commercial phytosanitory bioproducts, their target pest and protected plants

S/ No	Product (manufacturers)	Active ingredients	Disease/pest controlled (host)	Remarks	References
1	Grandevo DF 2 (Marrone Bio Innovations)	*Chrombacterium subtsugae* sp. PRAA4-1	*Tetranychus urticae* (Tomato + more than 150 plant species)	The bioproduct significantly reduces fecundity of adult pest and nymphal mortality	Golec et al. (2020)
2	Agree WG (Cetis USA)	*Bacillus thuringiensis*	Lepidopterous larva	Controls both resistant and non-resistant diamondback moth larvae	Sterk et al. (2020)
3	Nogall (Bio-Care Technology)	*Agrobacterium radiobacter* sp. K1026	Crown gall (Grapes plus thousands of plant species)	Strain K1026 produces toxic compounds that inhibits other *Agrobacterium* spp. that cause the disease	Kerr and Bullard (2020)
4	Biogard (CBC and Intrachem Bio)	*Bacillus* thuringiensis	Glassy clover land snail (Cotton field)	The product causes luminal secretion and hemocyte infiltration causing molluscicidal activity in the *Munacha. cartusiana* snail pest	Abd El-Atti et al. (2020)
5	Tricotop (Biotop)	*Trichoderma* spp.	Fungal pathogens of vegetables	The product is a cold adaptive biocontrol agent active with 0C	Morel et al. (2020)
6	Zequanox (Marrone Bio Innovations)	Dead cells of *Pseudomonas fluorescens*	Mussels	Mussels ingest the product as food and once consumed, the mussel stomach lining becomes ruptured with consequent death	scr.zacks.com (2020)
7	Aflasafe SN01 (BAMTAARE-IITA)	*Aspergillus flavus* genotype native to Senegal	Aflotoxin (Ground nut, maize)	Effective bioagent for aflatoxin mitigation	Senghor et al. (2020)
8	BotaniGard ES (BioWorks)	*Beauveria bassiana*	Aphids (Banana plus host of other plant species)	BotaniGard act by contact enabling the active agent to grow on the	Prince and Chandler (2020)

(continued)

Table 1.3 (continued)

S/No	Product (manufacturers)	Active ingredients	Disease/pest controlled (host)	Remarks	References
				cuticle of the insect pests	
9		Protease inhibitors (PIs)	Diverse pests	PIs are legumes' proteins that inhibit protease activity of phytopathogens	Rodríguez-Sifuentes et al. (2020)
10	Spray-induced gene silencing (SIGS)	Interference RNA (RNAi)	Diverse pests	dsRNA is an emerging biopesticides in which RNAi influence the degradation of target pest mRNA causing gene silencing and subsequent synthesis of vital protein	Zhang et al. (2020) and Biedenkopf et al. (2020)
11	Metarril E9 (Koppert)	*Metarhizium anisopliae*	Asian longhorned beetle	The active agent produced more conidia on the surface of pest cadavars and on a wider thermotolerance with optimum ranges of 25-30 °C.	Clifton et al. (2020)
12	Nemastim (Pheronym)	Nematodes	Insect pests	The product results in a 5X insect kill in comparison to EPN	Kaplan (2020)
13		*Beauveria bassiana*	wheat weevils (Grain silos)	The active agent is an highly resistant microsclerotia used against insect in grain silos	Trejo et al. (2020)
14	Not provided	*Bacillus thuringiensis* sp. JXBT-0296	Root and butt rot of Cassia nodosa (+ hundred species of trees)	Strain JXBT act as an excellent insecticide of Ganoderma lucidum (moths) that causes the root and butt rot disease	Fenshan (2020)
15	[a]Antifungal	Crucifalexins	Grey mould (Wine grapes +	Potent antifungals synthesized from	

(continued)

Table 1.3 (continued)

S/No	Product (manufacturers)	Active ingredients	Disease/pest controlled (host)	Remarks	References
			more than 200 plant species)	brassinin through engineered metabolic pathways	Calgaro-Kozina et al. (2020)
16	[a]EPN	*Heterorhabdits* spp	Pod borer (Beans)	Efficacy comparable to commercial H. indica	Thakur et al. (2020) and Vashisth et al. (2019)
17	[a]EO	Polyphenolic extracts	Mediterranean Fruit fly	Active ingredient extracted from olive mill wastewater	Di Ilio and Cristofaro (2020)
18	[a]Fumigant	Monoterpenes	*Drosophila suzukii* (soft-skinned fruit crops)	Monoterpenes interact with Drosophila suzukii type 1 tyramine receptor thus having an inhibitory effect on the target pest	Finetti et al. (2020)
19		*Aspergillus flavus* F3	*Elasmolomus pallens* (Groundnut)	The strain F3 is active against *E. pallens*, a post-harvest pest of groundnut	Umaru and Simarani (2020)

Note: 1–11 (Recently registered biopesticides or existing biopesticides whose proof of efficacy has been recently proven); 12–14 (Patented biopesticides); 15–19 (Recent research backing a positive proof of concept)
[a]Potential formulation

1.5.3 Biopesticide Prospects and Limitations

A close examination of biopesticides and conventional synthetic pesticides demonstrates parallel similarities, particularly in potency. Conventional pesticides have a broad-spectrum effect, are quicker in action, have longer residual activity, are highly effective against target pests, and are convenient (Felsot and Rack 2007; McCoy and Frank 2020). The negative effects of synthetic pesticides have prompted the placement restrictions on a significant number of them. Approximately 1000 active ingredients of conventional pesticides were authorized in 2001, but there has been a decline, to as low as 250 in 2009 (Jensen 2015) and the entrance of new chemical pesticides has also reduced from 70 in 2000 to 28 in 2012 (McDougall 2013). This new perspective has given impetus to the increased demand for alternative pesticides that must counteract the negative effects that are leading to the slow demise of questionable practice of chemical pesticide applications. Biopesticide usage tends to have little or no harmful effects on nontarget organisms, humans, and the environment due to its specificity and low toxicity compared to synthetic pesticides. It has been hypothesized that in the near future biopesticides can replace synthetic pesticides without significantly affecting crop yield (Leng et al. 2011; Arora et al. 2016; Mishra et al. 2020). At present an operational compromise appears to be in the utilization of biopesticides in conjunction with synthetic pesticides, which has also proven effective both quantitatively and qualitatively (Ujagir and Byrne 2009). It is not an ideal solution but has its merits in the reduction of pollution and deleterious effects that would otherwise be the case with sole use of these synthetic pesticides. One of the outstanding benefits of the use of biopesticides is their eco-friendliness. They are easily biodegradable and produce minimal residues thus, their presence in the air, water and terrestrial ecosystems is absent or minimal. The high specificity of biopesticides guarantees that they only harm the target pests and encourage the proliferation of beneficial organisms such as pollinators, predators and parasitoids for the overall benefits to protected crops. Further, biopesticides have proven to be effective against insect pests that have developed resistance to conventional pesticides. By extension, the insurgence of secondary pests will be limited if not completely obliterated. In this sense, the continued use of conventional pesticides would mean spending more for the poor results, increasing environmental impacts and health challenges.

Furthermore, biopesticides would greatly reduce the effect of bioaccumulation of toxic compounds in the food chain that are likely affecting humans, especially infants and adults (Kumar 2012). The use of crudely extracted plant insecticides, has been demonstrated to be more economically attractive and successful in agro-dependent rural areas in controlling insect pests in comparison to synthetic compounds (Tulipa and De 2019). For example, *Tephrosia vogelii* is a pesticidal plant used by 80% of farmers in southeastern Africa. Commercial cultivation relies heavily on pesticidal plants (Stevenson et al. 2017). Thus, cultivation of pesticidal plants can provide an opportunity for acquiring additional income through entrepreneurial ventures. These advantages of using biopesticides (see Table 1.4) ranging from environmental stewardship to healthy foods and feeds for animal consumption and the lessening of health issues have not been fully achieved due to certain challenges. These challenges are responsible for limiting the full adoption of biopesticides as pest and disease control options. One such constraints in relation to the application is dose determination of the active ingredients from biopesticide sources under real-life conditions (Shiberu and Getu 2016). This limitation is compounded because of the bioactive compounds' concentration, as it is influenced by the environment under which these pesticidal plants grow and also by their varieties. The diversity of these pesticidal plants results in differences in their responses to pathogens and herbivores (Ghorbani et al. 2005; Sales et al. 2016). Moreover, dose inaccuracy may be affected by the method of extraction, which may reduce the concentration or processing methods (Sesan et al. 2015). The disparity of efficacy determination during laboratory tests versus field tests is also a source of limitation. Additionally, the rapid development process and production of pesticides in different regions around the world implies that there are few, if any, standard

Table 1.4 Merits and demerits of biopesticide (Adapted from Hassan and Gokçe 2014; Marrone 2009)

Factor	Advantages	Disadvantages
Eco-friendliness	Biopesticides have low toxicity against humans and the environment	Due to their low toxicity, biopesticides exhibit slow action against target pest
Environmental persistence	Biopesticides when applied leave little or no residues thus have no pre-harvest interval. This factor is key in export crops	
Multi-mode of action	Due to their multiple mode of action, pest hardly develop resistance or cross-resistance to biopesticides	
Biodegradability	Biopesticides are characteristically low volatile compounds does pose risks to the environment and its receptors	Poor stability is often between 2–4 days which necessitate frequent and repeated application for effective eradication
Specificity	Biopesticides have little or no adverse effect on non-target organisms	Biopesticides limit actions against broad range of pests thereby would require diverse plant protection strategies
Safety profile	Safety during application, makes it convenient for workers to complete agro-assignment on timely basis, including harvest operation	
Suitability in IPM	Biopesticides has the potential of surpassing conventional control agents when used in IPM programme and reduce the use of classical pesticides	
Usefulness of co-wastes	Wastes from biopesticides production are used as fertilizers	
Improving productivity	Use of biopesticides lead to increased yield and in complete sense defines organic farming producing and wholesome and toxin-free food and crops	
Cost	Cost to develop biopesticides is cost effective	Cost of production of a certified biopesticide product is comparatively higher
Standardization		Standardization of the quality of biopesticides remains a limitation

preparation methods and guidelines for the determination of field efficacy of most of the active components of these new or modified biopesticides. The susceptibility of biopesticides to several environmental conditions including moisture and temperature tends to reduce the product shelf-life. This is made worse in uncontrolled environments such as those found in the field (Koul 2011). Microbial biopesticide formulations especially those of fungi and viruses display limited activity spectra against pest complexes; and their effectiveness is a function of the dynamic interaction between the environment, pathogens and the hosts (Ansari et al. 2012).

Some of the recognized limitations for microbial biopesticides are the need for highly virulent strains, complex handling requirements, intricate life cycles of candidates, and the phenomenon of slow biochemical activity, often a consequence of microbial lag phase and time needed for acclimatization and production of enzymes in response to identified threat (Glare et al. 2016). The combination of these limitations results in variable effects, which are considered significant challenges to research and development. In addition, there is a significant cost during product development. For example, cost incurred from substantial product feasibility studies, data gathering, and chemical analyses, such as toxicity testing, characterization of active ingredients and formulation as well as label articulation and packaging, are all tedious processes required from a regulatory standpoint for product acceptability and registration. These processes often serve as a deterrent for most innovators and researchers, preventing the product from reaching commercialization, due to resource limitations (Stoneman 2010). Consideration of the initial substantial investment requirements versus turnover and profit margins for potential biopesticide start-up companies, particularly with limited product lines, provides little encouragement or incentives to investors. In most instances, these fledging companies must also include capital finances for the construction of new facilities and initial production costs. However, a possible solution to the latter, especially with a single product line, will be from government investments into facilities and infrastructural development which these researchers may employ in product development and commercial production, but are charged a fee, which significantly reduces the financial burden. Additionally, Kumar and Singh (2014) highlight the resistance to broad applications of biopesticides by farmers in developing countries and the lack of trust, which they attribute to insufficient information and a lack of awareness of the efficacy of these biopesticides and the misunderstanding, especially, with microbial pesticides within the value chain. It is also important to note, that currently available biopesticides have been shown to have increased effectiveness when complemented with varieties of pest control methods using the integrated pest management approach (Grasswitz 2019).

1.6 Biopesticides as a Component in Integrated Pest Management (IPM)

1.6.1 Definition and Purpose of IPM

One unavoidable long-term outcome of using pesticides is the inevitable development of resistance and the outbreaks of secondary pests. Moreover, the understanding that there are often a myriad of pests that can attack a cultivation site has meant that any approach that will offer a lasting solution must use several approaches to tackle each of the problems. Apart from being science-based and sustainable, it is also a decision-making process that fosters the integration of cultural, physical, biological and chemical tools to identify and reduce the risk associated with pests for economic viability, social acceptability and environmental safety (USDA-ARS 2018). Some practitioners see IPM as alternating chemical compounds with different mechanisms of action groups to establish pest control efficacy and reduce pesticide resistance (Dara 2019). Elimination of pests, minimal consumption of chemical, promotion of environmental stewardship and human health preservation are overarching objectives with the IPM approach. There are at least 77 variances to the definition of IPM, most of which are based on operational interpretation, strategies and objectives, yet there is a core philosophy within the practice of IPM that is constant to most of its advocates (Dufour 2001). The implementation of IPM is aimed at obtaining the best result at the lowest cost, generating the least hazard to humans and the environment and avoiding the development of resistant pest strains. Moreover, IPM generates positive externalities, including lasting effects on farmers' health (Naranjo et al. 2015). Reports from diverse literary resources have indicated that the implementation of IPM requires, to varying degrees, education, tools, consumer preference, regulation, governance, moral values, socio-cultural and economic conditions, environmental awareness and retail marketing (Parsa et al. 2014; Jayasooriya and Aheeyar 2016; Rezaei et al. 2019). The intellectual buy-in of these factors in the implementation of IPM must be robust and represent a modern model of pest control management systems. This modern model of IPM underscores the science, art and enterprise components of sustainable crop production with four major components: knowledge and resources; planning and organization; communication and pest management (Dara 2019).

1.6.2 The Key Components of IPM

Knowledge and resources, planning and organization, and communication are different elements that are crucial to the IPM scheme. Knowledge of pest biology and ecology, their potential risk to plants, various control strategies and options, and their suitability are key for farmers to make informed decisions regarding pest control (Asante et al. 2001). A lack of sufficient knowledge remains one contributing key factor limiting IPM implementation including the meaning and understanding of IPM (Adam et al. 2010). The identification of an effective control strategy also requires knowledge about the specific stages in the insect pest life cycle that are most damaging to the plant, their niches, the nature of the damage and their economic significance, seasonal population trends and vulnerability of each stage to control options (Lefebvre et al. 2015). The effectiveness of control options in IPM is critical because one effective option for a particular circumstance may not be effective in another different situation. Planting date adjustment is an effective control strategy for pests known to follow seasonal patterns, and during the favourable season, it can be controlled with natural enemies (Heimpel and Cock 2018). While entomopathogenic nematodes have been confirmed to be effective against soil pests, viruses and bacteria can be effective against insect pests using chewing mouthparts and fungi against different kinds of pests (Zalom et al. 2018). The planning and organization component of IPM is concerned with data collection, raw data processing and informed decision making. It is important to know that regular monitoring of fields for pest spread precedes data collection. As a basic step of crop protection, early detection of pests of economic interest can allow for curbing them with minimal cost and less intricate control tactics and prevent intensification of damage. Drones in recent times have been employed in detecting and locating areas exposed to (a)biotic stressors through aerial imagery (Santesteban et al. 2017; Dara 2019). The partnership between plant science industries and IPM practitioners would have a great influence on these aspects through the integration of IPM principles and awareness of product development strategies and business plans thereby facilitating the offering of ideal marketing materials and sales services for the safest pest control products.

Profound recordkeeping that factor cultural practices, pest identity, their damages, seasonal fluctuations, the influence of environmental factors, and effective treatment strategies will help build institutional knowledge on improving crop production management (Stenberg 2017). When vital information concerning pest management is communicated among growers and within groups through traditional and modern communication channels, this would not only spread appropriate knowledge but can also help the whole circle of crop production and sustainable crop production. Most information shared through outreaches is obtained through research and outreach, considered to be an integral part of IPM. Research in IPM aims at detecting and anticipating pests, their characteristics, associated problems, factors that are favourable to injurious pests and development of strategies that can lead to the development of prophylaxis and treatment of possible (a)biotic stressors (USEPA 2017). A study conducted by Parsa et al. (2014) identified inadequate technical support and training as critical barriers to IPM implementation. This assertion was based on information extracted from IPM practitioners and professionals around 96 countries. Extension services and science-based solutions were shown to play a positive role in the IPM success of fruits and vegetables in New Zealand (Cameron 2007). Information gathered through scientific findings when disseminated through effective outreach to the end receivers involving extension educators and researchers would certainly play an important role in IPM implementation and management.

1.6.3 Pest Management Approach and Its Principles

Pest management terminology evolved from integrated control (a combination of biological and chemical control) when it became obvious that integrated control can accommodate more than chemical and biological controls (Ruiu 2015; Hajek and Eilenberg 2018). With time pest management became a preferable term to pest control even though the contextual description of both may be regarded as similar. The IPM principles are geared towards the prevention of possible pest problems. Some of the recommended practices may not be effective in all situations thus, practitioners and professionals must choose one or more strategies appropriate for their given situation to manage pest populations to a tolerable low-level with preferably, minimum economic damage (Dara 2019). The additive effect of the different control options (cultural control, host plant resistance, biological control, behavioral control, physical control, microbial control and chemical control) which in themselves can be employed individually to provide a certain degree of control level, can provide significant results in reducing pest damage (FAO 2013; Grasswitz 2019). However, it is important to consider the IPM as a systemic approach in reducing pests below their damaging economic threshold (population density cut-off above which control action should be initiated). The adoption of IPM based on general principles, rather than reliance on a single pest control method, entails sustainable pest management integrated with a variety of farming situations (Dara 2019). The holistic IPM approach will be better served through the adoption of eight implementation principles shown in Table 1.5 which follows a logical order with Principles 4–7 involving the application of biopesticides. Tables 1.6 and 1.7 display the roles played by the different categories of biopesticides.

1.6.4 Biopesticides in IPM Implementation

When prevention and suppression stages of IPM fail and are confirmed by monitoring, a decision should be promptly made to put in place a non-chemical option. It is noteworthy that PIP-cultivars, an example of biopesticides, have been advised for this stage of IPM implementation (Hodson and Lampinen 2018). The first non-chemical option to take is a mechanical approach, which may include a sticky trap (pheromone traps) alongside augmentation (with natural enemies) biological control. These natural enemies are either predators (insects that feed on pests; e.g. lady beetle, ground beetle, lacewings, centipedes) or parasitoids (wasps, *Trichogramma*) (Naranjo et al. 2015). A strategy in which microbial biopesticides are integrated into IPM to boost the effect of augmented natural enemies is known as biointensive pest management (BIPM). However, if, BIPM is not able to reduce the population density or the effect of the pest, it is advised that least-toxic pesticides be used. This category of pesticides includes insecticidal soaps (active against whiteflies, aphids, and other soft-bodied insects), horticultural oils (smother aphids, scales, mites), botanicals (neem oil, garlic oil, excluding pyrethrum), and insect growth regulators (chitin synthesis inhibitors, moulting synthesis disruptors). Insect growth regulators are also grouped under least-toxic pesticides but they are not allowed to be used in organic farming (Rahman et al. 2016). Although biopesticide usage as plant protection compounds is intensified in Principles 4 to 7, their utility is also reflected in all the different stages of the IPM pyramid shown in Fig. 1.5. For instance, the pheromone bait trap is used in monitoring the population density of

Table 1.5 The eight principles of IPM adapted from ANNEX III of Framework Directive 2009/128/EC (Barzman et al. 2015)

Principle 1: prevention and suppression	Principle 2: monitoring	Principle 3: decision-making	Principle 4: non-chemical methods	Principle 5: pesticide selection	Principle 6: reduced pesticide use	Principle 7: anti-resistance strategies	Principle 8: evaluation
Potential injurious organisms should be targeted, prevented and suppressed through combination of: • Crop rotation and intercropping • Adequate use of cultivation techniques • Use of resistant cultivars and certified seeds and planting resources where appropriate • Relying on balanced nutrient supply and germ-free/optimal water management • Preventing	Where possible use adequate techniques and tools to monitor pests, including field observation, warnings, early diagnosis system, forecasting and the making use of reliable advice from experts	Based on the outcome of Principle 2, practitioners decide on when and what plant protection measures to take. Threshold values from reliable research are key component for decision-making in the area of timing, need for, and methods of pest control	If satisfactory results can be possible, non-chemical methods (physical, mechanical and biological methods) should be used instead of chemical pesticides	When pesticides are applied, they should be specific and possess the list side effects so that the environment and non-target organisms are not harmed including humans	As low as reasonable possible dose of effective pesticides should be used to control pest of interest. Alongside, reduced application frequency or partial application in line with allowable level of risk in crops; and limit risk for resistance development of pests	Apply the strategy of rotating (or multiple) pesticides to protect crops or plant where the risk of resistance against a particular treatment option is evident	The success of the applied plant protection option should be checked based on the monitored pests and the records on the use of pesticides

(continued)

Table 1.5 (continued)

Principle 1: prevention and suppression	Principle 2: monitoring	Principle 3: decision-making	Principle 4: non-chemical methods	Principle 5: pesticide selection	Principle 6: reduced pesticide use	Principle 7: anti-resistance strategies	Principle 8: evaluation
spread of pests through sanitation • Protecting/enhancing beneficial organisms • Preserving biodiversity near farmland							

Table 1.6 Field application of the different categories of biopesticides

Category and sub-category	Pest being controlled and plant being protected	Applicable principle(s)	Reference
Plant-incorporated protectant (PIP)			
Transgenic plants	Pea weevil larvae, peach potato aphid, *Heliothis virescens*, *Agrotis* spp, *Spodoptera frugiperda,* Corn, cotton, potatoes, tobaco	Principle 1: Prevention and suppression	Stevens et al. (2012)
Microbial biopesticide			
Bacterial: Bt formulations	Lepidoptera, coleopterans, dipterans, mosquito larva *P. xylostella*, *H. armigera*, *S. litura*, Cotton, cruciferous vegetables, corn, tomato, cabbage	Principle 4: Non-chemical methods	Ansari et al. (2012)
Fungal:	*Beauveria bassiana,* Whiteflies, termites, beetles, mosquitoes, *Cydia pomonnella*, *Leptinotarsa decenmlineata*, Rice, maize crop, potato plant, banana tree, sugarcane	Principle 4: Non-chemical methods	Bhattacharya (2019)
Nematodes: *Steinernema* spp.	White grub, mole crickets, *P. xylostella*, *Delia radicum*, *P. japonicum*, *A. aegyptii*, Cabbage	Principle 4: Non-chemical methods	Ansari et al. (2012)
Protozoa: Microsporida	Lepidoptera, Orthoptera, corn borer, hoppers	Principle 4: Non-chemical methods	
Viral: Baculoviruses	Codling moth, tobacco budworm, alfalfa looper, gypsy moth, *S. exigua*, *P. xylostella*, *H. armigera*, *S. litura*, Apple, cotton, alfalfa, corn, sorghum, tomatoes, plum	Principle 4: Non-chemical methods	Granados and Williams (1986) and Rai et al. (2001)
Biochemicals			
Natural minerals	German Cockroach, pulse beetles, rice weevil, onion thrips, *Sitophilus* spp., *Ceratitis capitate*, *D. suzukii*, *R. dominica*, Wheat, maize plant, rice, grains	Principle 6 and 7	Pierattini et al. (2019)
	Plant extract/oil Two-spotted mite, cabbage root flies, cowpea weevil, *Sitophilus oryzae*, *Tribolium castanum*, *Aedes albopictus*, Rice grain, cowpea, tobacco plant, potato	Principle 6 and 7	Grdisa and Grsic (2013) and Erdogan et al. (2012)
	Semiochemicals Silkworm moth, gypsy moth, caterpillars, bark beetle, aphids, *Rhagoletis* spp, *Helicoverpa armigera*, *Poillia japonica*, Tomato, maize plant, *Allium porrum*, pine trees	Principle 6 and 7	Gebreziher (2018)

Table 1.7 Strategic application of semiochemicals

Control strategy	Principle	Purpose or important information	Reference
Monitoring	Employs kairomone- and pheromone-baited traps	Detection of pest invasion, Population density and fluctuation, Detection of first peak flight activity	El-Ghany (2019) and Piccardi (1980)
Mass tapping	Host attractant mechanism in a baited traps	Catching and removal of attracted insect from the population	Pinero and Dudenhoeffer (2018)
Attract and kill	Use of attractants on a substrate to attract pest and kill later on	Used in both field and stored products	Sonenshine (2004) and Nandagopal et al. (2008)
Matting disruption	Synthetic sex pheromones are employed to cause: False trail following, Camouflage, Desensitization and sensory imbalance	Disruption of chemical communication is the weapon of use in this strategy	El-shafie and Faleiro (2017) and Benelli et al. (2019)
Push-pull strategy	Deterring of insects from protected crops with simultaneous attraction of the pests in order areas and killed subsequently	Use of alarm pheromones as deterrent	Heuskin et al. (2011) and Kumari and Kaushik (2016)
Biological control	Use of kairomonal substances to attract natural enemies: predators and parasitoids	HIPVs, OIPVs and insect host semiochemicals	Renou and Guerrero (2000) and Mensah and Moore (2011)
Mono-control strategy			
Arrestment	Semiochemical induces formation of clusters	Matting enhancement and host scouting success	Sonenshine (2004)
Attractant	Luring of insects by semiochemicals and HIPVs to sources that containing inhibiting or killing agent	Matting enhancement, Location of habitat and hosts	El-Shafie and Faleiro (2017)
Antifeedants	Disallowing insects from entering a microhabitat for feeding on a particular surface	Provide surplus food for unaffected insects	Deutsch and Guédot (2017)
Confusant	Silencing of one semiochemical by another due to higher concentration	Spacing in reproduction	Knipling (1976)

insect pests (FAO 2013). Garlic oil as a repellant to certain insects serves a purpose under physical means of controlling pests. Likewise, insecticidal soap affecting soft-bodied insects (Curkovic 2016) is a typical example of a mechanical mode of action. Diatomaceous earth and kaolin also serve the same physical control. As a principle, toxic pesticides are used as the last resort and are unavoidable in certain critical situations. Normally, WHO classes III and IV are preferred due to the lesser danger

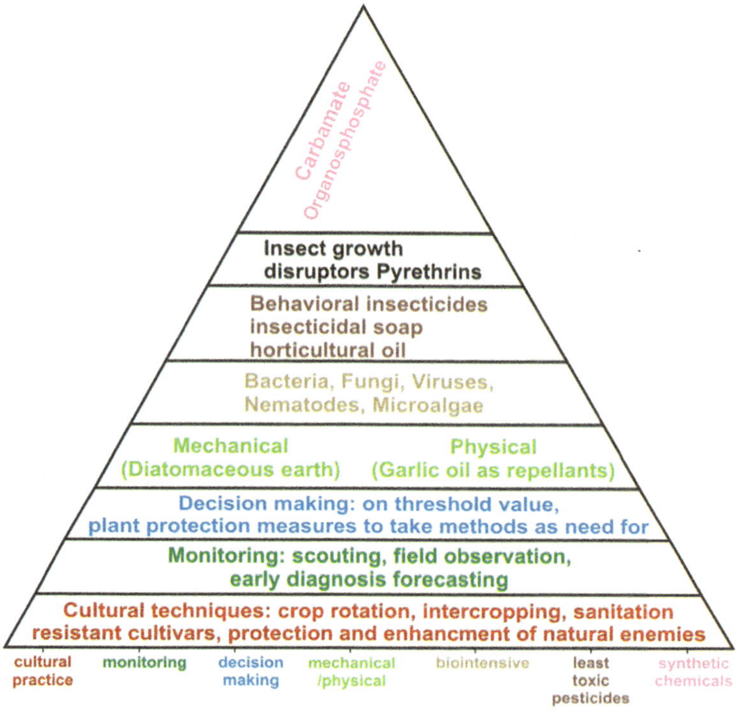

Fig. 1.5 Pyramid of biopesticide driven integrated pest management. (Adapted from James et al. 2010)

they pose in comparison to Classes I and II. Classes III and IV which are used in some IPM implementations are pyrethroids, carbamates and organophosphate in increasing order of toxicity (Chandler et al. 2011).

The optimal performance of biopesticides in the IPM portfolio scheme is contingent upon compatibility between agents, delivery precision, application frequency and timing (Chandler et al. 2011). Equally, the effectiveness of IPM implementation lies to some extent on result evaluation. This can be done by determining whether IPM results were achieved, and, if there are no other possibilities of improvements. A programme of an IPM can be evaluated through:

1. Taking cognizance of any changes and prophylactic measures that avoid future problems
2. Changing economic thresholds with respect to experience

Thresholds are levels of pest populations that require taking control action to prevent health, economic or esthetic losses representing different action thresholds. When an action threshold is set based on economic consideration, such threshold is regarded as an economic threshold (ET). Most thresholds in the IPM programme are set on ET.

1. Tracking the benefits and costs of an IPM programme
2. Visual monitoring and counting of pests and nontarget organisms during treatment and before treatment
3. Treatment records: methods, cost, rate, dates, time, etc.
4. Feedback from site users and clients
5. Increased productivity, healthy environment and improved biodiversity
6. Where need be, implement improvement actions after deep knowledge of what treatment to take

1.7 Current Research and Future Direction

Biopesticides have the potential to completely replace synthetic chemicals used in pest control. Consequently, it has attracted global attention in recent times for several positive reasons highlighted previously in this chapter. Apart from being green molecules, more active substances are discovered through research thereby raising the stake of their potential as innovative pest control agents. Currently, there are more than 175 registered bio-based pesticides globally and the commonly used biopesticides are from neem and Bt derivatives (Moosavi and Zare 2016; Raza et al. 2019; Samada et al. 2020). Bt pesticides are the most popular and diversified biopesticides. They are components of most PIPs, biochemical and microbial biopesticides. Most recent statistics holds that 75% of biopesticides used consist of Bt-based products (Samada et al. 2020). Neem has proven to be the most widely used botanical biopesticide (Rodgers 1993; Leng et al. 2011; Pavela 2014; George et al. 2014; Pavela and Benelli 2016; Huang et al. 2020). Moreover, neem has demonstrated multiple modes of action by acting as an antifeedant, sterilant, ovicidal and insecticide. However, owing to its instability in environmental conditions, such as sunlight, its effectiveness is short-lived. Pesticidal plants are preferentially used in less developed countries to control pests. However, the determination of the active ingredients of these pesticidal plants remains credible research gap. Efforts are ongoing to improve the characterization of effective phytochemicals and their concentration in finished products but precision and standardization are still current issues. Interest should be devoted to phytochemicals and their production to control pests. This however, will require knowledge of natural product chemistry.

The instability of neem products under ultraviolet light is also worthy of investigation in the pest control systems. Neem efficacy can be enhanced through the use of synergists such as piperonyl butoxide (Dar et al. 2014). Additionally, the new branch of science, nanotechnology could be used to address the limitations experienced with pesticide formulations especially concerning storage stability, release rate and effectiveness as it relates to biopesticides (Perlatti et al. 2013; Chaudhary et al. 2017). Nanopesticides in agriculture have been demonstrated to play roles in controlled release, enhanced dosage reduction, genetic material delivery into crops and detection of pathogens (Manchikanti 2019). Although, some misgivings have been expressed for nanopesticide applications, pertaining to resistance by the target pests, deleterious impact on humans, interference with genetic codes among others (Ganguli 2019). Emerging technologies such as fusion protein and recombinant DNA are currently used to enhance the efficacy of biopesticides. Moreover, omics technologies are expected to be applied in the environment of Bt and its Cry protein delivery to target pests as well as in the elucidation of novel toxin discovery. This is likely to be achieved with optimization studies that exploit both recombinant DNA techniques and proteomics (Nakasu et al. 2014). Strain selection during bioprocessing of active microbial agents can be aided using sequencing technology to directly target genes found to be associated with insecticidal traits (Glare et al. 2016). Baculovirus application in many cropping systems has shown its effectiveness and reliability especially in the control of soybean fields. Thus, on this premise, a projection was made for the application of Baculovirus to protect two million hectares of a farm from the velvet bean caterpillar cost-effectively (Sun 2015). Moreover, the success in the use of Baculovirus can be improved through *in vitro* culturing processes, changes in formulations and genetic engineering. It was further proposed that such an approach will not only necessitate the use of foreign genes but this may improve reaction times (Szewczyk et al. 2006).

An important prerequisite for the success of commonplace applications of biopesticides is the research breakthrough that ensures efficiency in the production, formulation and delivery of the formulated products; these are essentially critical to biopesticide commercialization. In order to facilitate the market penetration of biopesticides, there is a need for cooperation among researchers, venture capitalists, investors, producing companies and farmers with greater considerations placed on long term gains (Rezaei et al. 2019). Public-funded agro-programmes must also prioritize assistance in research and development of these biopesticides. Additionally, the amelioration of the bottleneck of regulatory protocols, will guarantee business feasibility and affordability that will encourage bioentrepreneurs towards these bio-based commercial products. This will play a significant role in driving the agenda of economic sustainability. One major hurdle that is standing on the way of an adequate supply of biopesticides is regulatory protocols that will guarantee the affordability of bio-based commercial products (Rao et al. 2007). It is acknowledged that there is a need for regulatory processes in developing countries; however, there is a need that these tools of governance must not become a hindrance to credible advances that will improve the environmental management provided by these eco-friendly products.

Researchers should also consider ways of using available technologies at the production scale-up with objectives focused on cost reduction, efficiency and reliability of commercial biopesticides. This will positively affect the efficacy and reliability of biopesticides when integrated into a flexible IPM programme. The latter will ensure the protection of the ecosystem through the use of minimal synthetic pesticides at the most reasonable cost. A pest management system in which IPM is connected to computer-based system support can offer improved stored grain protection (Samada et al. 2020). It should be acknowledged that strict compliance with biological control has more chances to fail and thus should be supported with other variants of pest control options. In line with the concept of the "green consumerism" biopesticides are exclusively used in organic farming to produce chemical-free foods and crops (El-Shafie and Faleiro 2017). Upholding organic farming as a tradition will improve and conserve valuable natural resources such as nutrient loss, topsoil, erosion, compaction and improvement of higher economic value for farm crops (Umar 2013). The full adoption of biopesticides and their variant forms including their roles in IPM requires awareness and training among the different stakeholders to sustainably control insect pests and vectors.

1.8 Conclusion

Conventional agriculture has been relied upon to satisfy the food needs of the global population which is growing at a rate of 0.8 billion per decade. The application of synthetic pesticides has contributed immensely to mitigating the negative effects of more than 40,000 extant crop pests leading to the high productivity of farm yields. On an average basis, pests are responsible for 30% crop yield loss and 14% damage to stored food products. These pesticides cause health challenges to humans, environmental pollution, development of pest resistance, and narrowing of biodiversity among others. For these reasons the need to significantly reduce the continuous reliance on synthetic pest controlling agents is pertinent. This can be achieved through the adoption of bio-based pesticides. The use of alternative pesticides in crop cultivation reduces environmental pollution and solves pest resistance problems and improves biodiversity, including natural enemies. Crop cultivation practices that ensure environmental conservation, economic viability and social acceptability will result in sustainable agriculture. Sustainable agriculture is directly or indirectly linked to the 17 SDGs. The 12 Green Chemistry Principles have shown that sustainable agriculture is directly connected to eight out of the 17 SDGs. As green compounds, biopesticides have shown to substitute classical pesticides with a possible increase in crop productivity. In order to achieve an optimum increase in crops' productivity, biopesticides are used in an integrated pest management (IPM) scheme, which goes in tandem to achieve minimal use of chemical pesticides at the lowest cost. With the right education, skill, research on how to improve shelf life and stability problems, and partnerships among stakeholders, biopesticide-driven IPM can make chemical pesticide-free agriculture a reality and create a nexus between socially acceptable economic viability and environmental safety.

References

Abd El-Atti MS, Khalil AM, Elsheakh AA et al (2020) Biological control of Monacha cartusiana "glassy clover land snails" by microbial biopesticides Biozed and Biogard, using bait technique. Biocatal Agric Biotechnol 25:101572

Abdullah ZS, Greenfield BP, Ficken KJ et al (2015) A new attractant for monitoring western flower thrips, Frankliniella occidentalis in protected crops. SpringerPlus *4*(1):1–9

Abu-Dieyeh MH, Watson AK (2007) Efficacy of Sclerotinia minor for dandelion control: effect of dandelion accession, age and grass competition. Weed Res 47(1):63–72

Adam BD, Siaplay PW, Flinn BW et al (2010) Factors influencing economic profitability of sampling-based integrated pest management of wheat in country elevators. J Stored Prod Res 46:186–196

Ahmed KS, Mikhail WZ, Sobhy HM et al (2019) Impact of nanosilver-profenofos on cotton leafworm, *Spodoptera littoralis* (Boisd.) larvae. Bull Natl Res Centre 43(1):46

Aktar W, Sengupta D, Chowdhury A (2009) Impact of pesticides use in agriculture: their benefits and hazards. Interdiscip Toxicol 2(1):1–12

Alexandratos N, Bruinsma J (2012) World agriculture towards 2030/2050: the 2012 revision. ESA Working No. 12-03. FAO, Rome

Altemimi A, Lakhssassi N, Baharlouei A et al (2017) Phytochemicals: extraction, isolation, and identification of bioactive compounds from plant extracts. Plants 6(4):42

Alyokhin A, Nault B, Brown B (2020) Soil conservation practices for insect pest management in highly disturbed agroecosystems—a review. Entomol Exp Appl 168(1):7–27

Anastas PT, Warner JC (1998) Green Chemistry: theory and practice. Oxford University Press, New York

Ansari MS, Ahmad N, Hasan F (2012) Potential of biopesticides in sustainable agriculture. In: Environmental protection strategies for sustainable development. Springer, Dordrecht, pp 529–595

Arora NK, Verma M, Prakash J et al (2016) Regulation of biopesticides: global concerns and policies. In: Bioformulations. For sustainable agriculture. Springer, New Delhi, pp 283–299

Arshad M, Ullah MI, Çagatay NS et al (2019) Field evaluation of water plant extracts on sucking insect pests and their associated predators in transgenic Bt cotton. Egypt J Biol Pest Control 29 (1):1–6

Asante SK, Tamo M, Jackai LEN (2001) Integrated management of cowpea insect pests using elite cultivars, date of planting and minimum insecticide application. Afr Crop Sci J 9:655–665

Bailey PR (2010) U.S. Patent Application No. 12/739,924

Barzman M, Barberi P, Birch ANE et al (2015) Eight principles of integrated pest management. Agron Sustain Dev 35(4):1199–1215

Bateman ML, Day RK, Luke B et al (2018) Assessment of potential biopesticide options for managing fall armyworm (Spodoptera frugiperda) in Africa. J Appl Entomol 142(9):805–819

Benelli G, Lucchi A, Thompson D, Ioriatti C (2019) Sex pheromone aerosol devices for matting disruption: challenges for a brighter future. Insects 10(10):308

Berlitz DL, Knaak N, Cassal MC et al (2014) *Bacillus* and biopesticides in control of phytonematodes. In: Basic and applied aspects of biopesticides. Springer, New Delhi, pp 3–16

Bhattacharya A (2019) Changing environmental condition and phosphorus-use efficiency in plants. In: Changing climate and resource use efficiency in plants. Academic Press, Cambridge, MA

Biedenkopf D, Will T, Knauer T et al (2020) Systemic spreading of exogenous applied RNA biopesticides in the crop plant Hordeum vulgare. ExRNA 2(1):1–10

Biondi A, Desneux N (2019) Special issue on Tuta absoluta: recent advances in management methods against the background of an ongoing worldwide invasion. J Pestic Sci 92:1313–1315

Blandford D (2011) The contribution of agriculture to green growth. Report to the OECD

Bragard C, Dehnen-Schmutz K, Di Serio F et al (2019) Pest categorisation of *Spodoptera litura*. EFSA J 17(7):1–11

Brown J, Ogle H (1997) Fungal diseases and their control. Plant Pathog Plant Dis Aust *1* (4):443–467

Brudea V, Rîşca I, Enea C et al (2012) Efficacy of some biopesticides and plant secondary metabolites against fall webworm *Hyphantria Cunea* Drury (F. Arctiidae-Lepidoptera) in the lab conditions. Cercetari Agron Moldova 45(1):73–80

Burton I (1987) Report on reports: our common future: the world commission on environment and development. Environ Sci Policy Sustain Dev 29(5):25–29
Calevro F, Tagu D, Callaerts P (2019) *Acyrthosiphon pisum*. Trends Genet 35(10):781
Calgaro-Kozina A, Vuu KM, Keasling JD et al (2020) Engineering plant synthetic pathways for the biosynthesis of novel antifungals. ACS Central Sci 6(8):1394–1400
Cameron PJ (2007) Factors influencing the development of integrated pest management (IPM) in selected vegetable crops: a review. N Z J Crop Hortic Sci 35(3):365–384
Carson R (1963) Silent Spring. Penguin Books, Harmondsworth
Carvajal-Munoz JS, Carmona-Garcia CE (2012) Benefits and limitations of biofertilization in agricultural practices. Livest Res Rural Dev 24(3):1–8
Carvalho FP (2017) Pesticides, environment, and food safety. Food Energy Secur 6(2):48–60
Casadevall A (2007) Determinants of virulence in the pathogenic fungi. Fungal Biol Rev 21 (4):130–132
Cato MP (1979) Di Agri Cultura. In: Hooper WD, Ash HB (eds) Translation of on agriculture, Revised edn. Harvard University Press, Cambridge, MA
Chakroun M, Banyuls N, Bel Y et al (2016) Bacterial vegetative insecticidal proteins (Vip) from entomopathogenic bacteria. Microbiol Mol Biol Rev 80(2):329–350
Chandler D, Bailey AS, Tatchell GM et al (2011) The development, regulation and use of biopesticides for integrated pest management. Philo T R Soc B Biol Sci 366(1573):1987–1998
Chaudhary S, Kanwar RK, Sehgal A et al (2017) Progress on *Azadirachta indica* based biopesticides in replacing synthetic toxic pesticides. Front Plant Sci 8:610
Chen KN, Chen CY, Lyn YC et al (2013) Formulation of a novel antagonistic bacterium based biopesticide for fungal disease control using microencapsulation techniques. J Agric Sci 5 (3):153–163
Chengala L, Singh N (2017) Botanical pesticides—a major alternative to chemical pesticides: a review. Int J Life Sci 5(4):722–729
Clifton EH, Jaronski ST, Hajek AE (2020) Virulence of commercialized fungal entomopathogens against Asian longhorned beetle (Coleoptera: Cerambycidae). J Insect Sci 20(2):1
Cohen E (2001) Chitin synthesis and inhibition: a revisit. Pest Manag Sci 57(10):946–950
Cooper J, Gardener M (2006) Bacterial plant pathogens and symptomology. Ed. Tom Schultz, WSU County Extension, SJI
Costa JAV, Freitas BCB, Cruz CG et al (2019) Potential of microalgae as biopesticides to contribute to sustainable agriculture and environmental development. J Environ Sci Health Pt B 54 (5):366–375
Curkovic T (2016) Detergents and soaps as tools for IPM in agriculture. Integrated pest management (IPM). Environ Sound Pest Manag 5:12–23
Damalas CA (2009) Understanding benefits and risks of pesticide use. Sci Res Essays 4 (10):945–949
Damalas CA, Koutroubas SD (2018) Current status and recent developments in biopesticide use. Agriculture 8:13
Dang CH, Nguyen CH, Im C et al (2016) Synthesis and application of pheromones for integrated pest management in Vietnam. Integrated Pest Management (IPM). Environ Sound Pest Manag 5:103–127
Dar SA, Dar NA, Bhat MA et al (2014) Prospects, utilization and challenges of botanical pesticides in sustainable agriculture. Int J Mol Biol Biochem 2:1–14
Dara SK (2019) The new integrated pest management paradigm for the modern age. J Int Pest Manag 10(1):12
De Groote H, Kimenju SC, Munyua B et al (2020) Spread and impact of fall armyworm (Spodoptera frugiperda JE Smith) in maize production areas of Kenya. Agric Ecosyst Environ 292:106804
Deutsch A, Guédot C (2017) Utilizing insects' sense of smell for pest management. University of Wisconsin—Extension, Cooperative Extension

Di Ilio V, Cristofaro M (2020) Polyphenolic extracts from the olive mill wastewater as a source of biopesticides and their effects on the life cycle of the Mediterranean fruit fly Ceratitis capitata (Diptera, Tephriditae). Int J Trop Insect Sci 12:1–8
Dowds BCA, Peters A (2002) Virulence mechanisms. In: Gaugler R (ed) Entomopathogenic nematology. CAB International, New York, pp 79–98
Dufour R (2001) Biointensive integration pest management (IPM). National Centre for Appropriate Technology/Appropriate Technology Transfer for Rural Arears (AATRA). https://www.attra.ncat.org/attar-pub/ipm.html
El-Ghany A (2019) Semiochemicals for controlling insect pest. J Plant Prot Res 59(1):1–11
El-Shafie HAF, Faleiro JR (2017) Semiochemicals and their potential use in pest management. In: Biological control of pest and vector insects, vol 12, pp 1–22
Erdogan P, Yildirim A, Sever B (2012) Investigations on the effects of five different plant extracts on the two-spotted mite Tetranychus urticae Koch (Arachnida: Tetranychidae). Psyche 10:1155–1159
Evenson RE, Gollin D (2003) Assessing the impact of the Green Revolution, 1960 to 2000. *Science* 300(5620):758–762
Fan SA, Rosegrant MW (2008) Investing in agriculture to overcome the world food crisis and reduce poverty and hunger. IFPRI (International Food Policy Research Institute), Washington, DC
FAO (2013) International code of conduct on pesticide management—guidance on pest and pesticide. Manag Policy Dev 1:34–36
Felsot AS, Rack KD (2007) Chemical pest control technology: benefits, disadvantages, and continuing roles in crop production systems. American Chemical Society, Washington, DC
Fenshan W (2020) *Bacillus thuringiensis* JXBT-0296 with insecticidal activity on *Ganoderma lucidum* chi moth and application thereof. China National Intellectual Property Administration CN111676173A. https://patents.google.com/patent/CN111676173A/en?q=biopesticides&before=priority:20201231&after=priority:20200101&oq=biopesticides+2020. Accessed 2 Nov 2020
Finetti L, Ferrari F, Calo G et al (2020) Modulation of Drosophila suzukii type 1 tyramine receptor (DsTAR1) by monoterpenes: a potential new target for next generation biopesticides. Pest Biochem Physiol 165:104549
Franco OL, Rigden DJ, Melo FR et al (2002) Plant α-amylase inhibitors and their interaction with insect α-amylases: structure, function and potential for crop protection. Eur J Biochem 269 (2):397–412
Frischmann BM, Marciano A, Ramello GB (2019) Retrospectives: tragedy of the commons after 50 years. J Econ Perspect 33(4):211–228
Gallai N, Salles JM, Settele J et al (2009) Economic valuation of the vulnerability of world agriculture confronted with pollinator decline. Ecol Econ 68:810–821
Ganasen S, Velaichamy V (2016) Innovations in green chemistry towards sustainable development. OIDA Int J Sustain Dev 9(11):11–14
Ganguli P (2019) Patenting issues in the development of nanobiopesticides. In: Nano-biopesticides today and future perspectives. Academic Press, New York, pp 367–395
Gangwar RK, Gangwar C (2018) Lifecycle, distribution, nature of damage and economic importance of whitefly, *Bemisia tabaci* (Gennadius). Acta Sci Agric 2(4):36–39
Gasic S, Tanovic B (2013) Biopesticide formulations, possibility of application and future trends. Pestic Ifitomed 28(2):97–102
Gebreziher HG (2018) The role of herbivore-induced plant volatiles (HIPVs) as indirect plant defense mechanism in a diverse plant and herbivore species; a review. Int J Agric Environ Food Sci 2(4):139–147
George DR, Finn RD, Graham KM et al (2014) Present and future potential of plant-derived products to control arthropods of veterinary and medical significance. Parasit Vectors 7(1):1–12

Ghorbani R, Wilcockson S, Leifert C (2005) alternative treatments for late blight control in organic potato: antagonistic micro-organisms and compost extracts for activity against *Phytophthora infestans*. Potato Res 48:181–189

Gill HK, Garg H (2014) Pesticide: environmental impacts and management strategies. Pestic Toxic Aspects 8:187

Glare TR, Gwynn RL, Moran-Diez ME (2016) Development of biopesticides and future opportunities. In: Microbial-based biopesticides. Humana Press, New York, pp 211–221

Golec JR, Hoge B, Walgenbach JF (2020) Effect of biopesticides on different Tetranychus urticae Koch (Acari: Tetranychidae) life stages. Crop Prot 128:105015

Gonzalez MA (2017) Introduction to Green and Sustainable Chemistry. Elsevier, New York

Gortari F, Nowosad MIP, Laczeski ME et al (2019) Biofertilizers and biocontrollers as an alternative to the use of chemical fertilizers and fungicides in the propagation of yerba mate by mini-cuttings. Rev Árvore 43(4)

Granados RR, Williams KA (1986) in vivo infection and replication of baculoviruses. In: Granados RR, Federici BA (eds) The biology of baculoviruses. CRC, Boca Raton, pp 89–108

Grasswitz TR (2019) Integrated pest management (IPM) for small-scale farms in developed economies: challenges and opportunities. Insects 10(6):179

Grdisa M, Grsic K (2013) Botanical insecticides in plant protection. *Agric Conspectus Sci* 78 (2):85–93

Haase S, Sciocco-Cap A, Romanowski V (2015) Baculovirus insecticides in Latin America: historical overview, current status and future perspectives. Viruses 7(5):2230–2267

Hajek AE, Eilenberg J (2018) Natural enemies: an introduction to biological control. Cambridge University Press, Cambridge

Hallenbeck WH, Cunningham-Burns KM (1985) Pestcides and human health. Springer-Verlag, New York

Hardin G (2009) The tragedy of the commons. J Nat Resour Policy Res 1(3):243–253

Hassan E, Gokçe A (2014) Production and consumption of biopesticides. In: Advances in plant biopesticides. Springer, New Delhi, pp 361–379

Heimpel GE, Cock MJW (2018) Shifting paradigms in the history of classical biological control. BioControl 63:27–37

Heuskin S, Verheggen FJ, Haubruge E et al (2011) The use of semiochemical slow-release devices in integrated pest management strategies. Biotechnol Agron Soc Environ 25:459–470

Hodson AK, Lampinen BD (2018) Effects of cultivar and leaf traits on the abundance of Pacific spider mites in almond orchards. Arthropod Plant Interact 13:453–463

Htwe AZ, Moh SM, Soe KM et al (2019) Effects of biofertilizer produced from Bradyrhizobium and Streptomyces griseoflavus on plant growth, nodulation, nitrogen fixation, nutrient uptake, and seed yield of mung bean, cowpea, and soybean. Agronomy 9(2):77

Huang J, Tichit M, Poulot M et al (2015) Comparative review of multifunctionality and ecosystem services in sustainable agriculture. J Environ Manage 149:138–147

Huang J, Rucker A, Schmidt A et al (2020) Production of constitutive and induced secondary metabolites is coordinated with growth and storage in Norway spruce saplings. Tree Physiol 40:928–942

Ibrahim RA, Shawer DM (2014) Transgenic Bt-plants and the future of crop protection (an overview). Int J Agric Food Res 3(1):14–50

Igiehon NO, Babalola OO (2017) Biofertilizers and sustainable agriculture: exploring arbuscular mycorrhizal fungi. Appl Microbiol Biotechnol 101(12):4871–4881

Ilyashenka D, Ivaniuk V (2008) Potato stem nematode in Belarus. Zemdirbyste Agric 95:74–81

Inam-ul-Haq M, Hyder S, Nisa T et al (2019) Overview of biopesticides in Pakistan. In: Plant growth promoting rhizobacteria (PGPR): prospects for sustainable agriculture. Springer, Singapore, pp 255–268

Itelima JU, Bang WJ, Onyimba IA et al (2018) A review: biofertilizer; a key player in enhancing soil fertility and crop productivity. J Microbiol Biotechnol Rep 2(1):22–28

Ivankovic A, Dronjic A, Bevanda AM et al (2017) Review of 12 principles of green chemistry in practice. Int J Sustain Green Energy 6(3):39–48

Ivase TJP, Nyakuma BB, Ogenyi BU et al (2017) Current status, challenges, and prospects of biopesticide utilization in Nigeria. Acta Univ Sapientiae Agric Environ 9(1):95–106

James B, Atcha-Ahowe C, Godonou I et al (2010) Integrated pest management in vegetable production: a guide for extension workers in West Africa. IITA, Ibadan

Jankielsohn A (2018) The importance of insects in agricultural ecosystems. Adv Entomol 6 (2):62–73

Jayasooriya HJC, Aheeyar MMM (2016) Adoption and factors affecting on adoption of integrated pest management among vegetable farmers in Sri Lanka. Proc Food Sci 6:208–212

Jensen JE (2015) Perspectives on the implementation of IPM in EU—the advisory perspective. In: IPM innovation in Europe, Poznan, Poland January 15–17, 2015

Jindra M, Bittova L (2020) The juvenile hormone receptor as a target of juvenoid "insect growth regulators". Arch Insect Biochem Physiol 103(3):e21615

Jisha VN, Smitha RB, Benjamin S (2013) An overview on the crystal toxins from Bacillus thuringiensis. Adv Microbiol 3(5):462

Kabaluk T, Svircev A, Goettel M et al (2010) Use and regulation of microbial pesticides in representative jurisdictions worldwide. IOBC Global, 108

Kaplan F (2020) Nematode pheromone extract dramatically increases efficacy of nematode dispersal for crop biocontrol. United States Patent and Trademark Office US 10,736,326 B2. https://www.pheronym.com/2020/08/25/pheronym-granted-patent-for-ag-bio-remediation-breakthrough/. Accessed 2 Nov 2020

Kassie M, Wossen T, De Groote H et al (2020) Economic impacts of fall armyworm and its management strategies: evidence from southern Ethiopia. Eur Rev Agric Econ 47:1473–1501

Kerr A, Bullard G (2020) Biocontrol of crown gall by *Rhizobium rhizogenes*: challenges in biopesticide commercialisation. Agronomy 10(8):1126

Khalil MS, Kenawy A, Gohrab MA et al (2012) Impact of microbial agents on *Meloidogyne incognita* management and morphogenesis of tomato. J Biopest 5(1):28

Khanna R, Pawar J, Gupta S et al (2019) Efficiency of biofertilizers in increasing the production potential of cereals and pulses: a review. J Pharmacog Phytochem 8(2):183–188

Khot LR, Sankaran S, Maja JM et al (2012) Applications of nanomaterials in agricultural production and crop protection: a review. Crop Prot 35:64–70

King FH (1911) Farmers of forty centuries, or, Permanent agriculture in China, Korea and Japan. Mrs. FH King

Knaak N, Franz AR, Santos GF et al (2010) Histopathology and the lethal effect of Cry proteins and strains of *Bacillus thuringiensis* Berliner in *Spodoptera frugiperda*. Braz J Biol 70(3):677–684

Knipling EF (1976) Role of phermones and kairmones for insect suppression systems and their possible health and environmental impacts. Environ Health Perspect 14:145–152

Knowles A (2005) New developments in crop protection product formulation. Agrow Reports UK: T and F Informa, London, pp 153–156

Koch MS, Ward JM, Levine SL et al (2015) The food and environmental safety of Bt crops. Front Plant Sci 6:283

Koul O (2011) Microbial biopesticides: opportunities and challenges. CAB reviews: perspectives in agriculture. Vet Sci Nutr Nat Resour 6:056

Kremer RJ (2019) Bioherbicides and nanotechnology: current status and future trends. In: Nanobiopesticides today and future perspectives. Academic Press, New York, pp 353–366

Kumar S (2012) Biopesticides: a need for food and environmental safety. J Biofertil Biopest 3 (4):1–3

Kumar VV (2018) Biofertilizers and biopesticides in sustainable agriculture. In: Role of rhizospheric microbes in soil. Springer, Singapore, pp 377–398

Kumar D, Kalita P (2017) Reducing postharvest losses during storage of grain crops to strengthen food security in developing countries. Foods 6(1):8

Kumar S, Singh A (2014) Biopesticides for integrated crop management: environmental and regulatory aspects. J Biofertil Biopest 5:e121
Kumar S, Singh A (2015) Biopesticides: present status and the future prospects. J Fertil Pest 2:1–2
Kumari A, Kaushik N (2016) Oviposition deterrents in herbivorous insects and their potential use in integrated pest management. Indian J Exp Biol 54(3):163–174
Leahy J, Mendelsohn M, Kough J et al (2014) Biopesticide oversight and registration at the US Environmental Protection Agency. In: Biopesticides: state of the art and future opportunities. American Chemical Society, Washington, DC, pp 3–18
Lee MK, Walters FS, Hart H et al (2003) The mode of action of the *Bacillus thuringiensis* vegetative insecticidal protein Vip3A differs from that of Cry1Ab δ-endotoxin. Appl Environ Microbiol 69(8):4648–4657
Lefebvre MS, Langrell RH, Gomez-y-Paloma S (2015) Incentives and policies for integrated pest management in Europe: a review. Agron Sustain Dev 35:27–45
Leng P, Zhang Z, Pan G et al (2011) Applications and development trends in biopesticides. Afr J Biotechnol 10(86):19864–19873
Lengai GM, Muthomi JW (2018) Biopesticides and their role in sustainable agricultural production. J Biosci Med 6(6):7
Lewis EE, Clarke DJ (2012) Nematode parasites and entomopathogens. In: Vega F, Kaya H (eds) Insect pathology, 2nd edn. Academic Press, London, pp 395–424
Li H, Buschman LL, Huang F et al (2007) DiPel-selected Ostrinia nubilalis larvae are not resistant to transgenic corn expressing Bacillus thuringiensis CrylAb. J Econ Entomol 100(6):1862–1870
Linda PC, Kim YS, Tong L (2010) Mechanism for the inhibition of the carboxyltransferase domain of acetyl-coenzyme A carboxylase by pinoxaden. Proc Natl Acad Sci 107(51):22072–22077
Litskas VD, Migeon A, Navajas M et al (2019) Impacts of climate change on tomato, a notorious pest and its natural enemy: small scale agriculture at higher risk. Environ Res Lett 14(8):084041
Liu S, Jaouannet M, Dempsey DA et al (2020) Research review paper RNA-based technologies for insect control in plant protection. Biotechnol Adv 39:107463
Losey JE, Vaughan M (2006) The economic value of ecological services provided by insects. BioScience 56(4):311–323
Lucas GB, Campbell L (2012) Introduction to plant diseases: identification and management. Springer Science & Business Media, Berlin
Lushchak VI, Matviishyn TM, Husak VV et al (2018) Pesticide toxicity: a mechanistic approach. EXCLI J 17:1101
Maciel H, Zingaretti S, Maciel G (2014) Comparative analysis of the amino acid sequence of the Cry protein of codify the toxic protein to insects of the orders Lepidoptera, Diptera and Lepidoptera/Diptera. BMC Proc 8:120–P120
Manahan SE (2006) Green chemistry and the ten commandments of sustainability. ChemChar Research, Columbia, MO
Manchikanti P (2019) Bioavailability and environmental safety of nanobiopesticides. In: Nano-biopesticides today and future perspectives. Academic Press, New York, pp 207–222
Mancosu N, Snyder RL, Kyriakakis G, Spano D (2015) Water scarcity and future challenges for food production. Water 7(3):975–992
Maneepitak S, Cochard R (2014) Uses, toxicity levels, and environmental impacts of synthetic and natural pesticides in rice fields—a survey in Central Thailand. Int J Biodivers Sci Ecosyst Serv Manag 10(2):144–156
Marche MG, Camiolo S, Porceddu A et al (2018) Survey of Brevibacillus laterosporus insecticidal protein genes and virulence factors. J Invertebr Pathol 155:38–43
Marrone PG (2009) Barriers to adoption of biological control agents and biological pesticides. Integrated pest management. Cambridge University Press, Cambridge, pp 163–178
Marrone PG (2014) The market and potential for biopesticides. In: Gross AD, Coats JR, Duke SO, Seiber JN (eds) Biopesticides: state of the art and future opportunities. American Chemical Society, Washington, DC, pp 245–258

Marrone PG (2019) Pesticidal natural products—status and future potential. Pest Manag Sci 75 (9):2325–2340

Marteel-Parrish A, Newcity KM (2017) Highlights of the impacts of green and sustainable chemistry on industry, Academia and Society in the USA. Johnson Matthey Technol Rev 61 (3):207–221

Maurer M, Bogner FX (2019) How freshmen perceive environmental education (EE) and education for sustainable development (ESD). PLoS One 14(1):e0208910

McCoy T, Frank D (2020) Organic vs. conventional (synthetic) pesticides: advantages and disadvantages. Virginia Tech

McDougall P (2013) R&D trends for chemical crop protection products and the position of the European Market. A consultancy study undertaken for ECPA. www.ecpa.eu/information-page/regulatoryaffairs/publications-regulatory-affairs

Mduma HS, Mkindi M, Karani A et al (2015) Major signs and symptoms caused by biotic and abiotic agents on plants in the tropical Africa. Int J Sci Res 6:750–759

Meadows DH (1972) The limits to growth: a report for the Club of Rome's project on the predicament of mankind. Universe Books, New York

Mensah RK, Moore C (2011) Exploitation of semiochemicals for the management of pest and beneficial insects with special emphasis on cotton cropping systems in Australia: a review. J Biol Control 25(4):253–269

Mergenthaler E, Fodor J, Kiss E et al (2020) Biological and molecular evidence for the transmission of aster yellows phytoplasma to French marigold (Tagetes patula) by the flatid planthopper Metcalfa pruinosa. Ann Appl Biol 176(3):249–256

Mills NJ, Kean JM (2010) Behavioral studies, molecular approaches, and modeling: methodological contributions to biological control success. Biol Cont 52(3):255–262

Mishra J, Tewari S, Singh S et al (2015) Biopesticides: where we stand? In: Plant microbes symbiosis: applied facets. Springer, New Delhi, pp 37–75

Mishra RK, Bohra A, Kamaal N et al (2018) Utilization of biopesticides as sustainable solutions for management of pests in legume crops: achievements and prospects. Egypt J Biol Pest Cont 28 (1):3

Mishra J, Dutta V, Arora NK (2020) Biopesticides in India: technology and sustainability linkages. 3 Biotech 10:1–12

Mitra A, Chatterjee C, Mandal FB (2011) Synthetic chemical pesticides and their effects on birds. Res J Environ Toxicol 5(2):81–96

Mohan S, Ma PWK, Pechan T et al (2006) Degradation of the S. frugiperda peritrophic matrix by an inducible maize cysteine protease. J Insect Physiol *52*(1):21–28

Mondal KAMSH, Parween S (2000) Insect growth regulators and their potential in the management of stored-product insect pests. Integr Pest Manag Rev *5*(4):255–295

Monzo C, Stansly PA (2017) Economic injury levels for Asian citrus psyllid control in process oranges from mature trees with high incidence of huanglongbing. PLoS One 12(4):e0175333

Moosavi MR, Zare R (2016) Present status and the future prospects of microbial biopesticides in Iran. In: Agriculturally important microorganisms. Springer, Singapore, pp 293–305

Morel MA, Tapia-Vázquez I, Castro-Sowinski S et al (2020) Fighting plant pathogens with cold-active microorganisms: biopesticide development and agriculture intensification in cold climates. Appl Microbiol Biotechnol 5:1–14

Nakasu EY, Williamson SM, Edwards MG et al (2014) Novel biopesticide based on a spider venom peptide shows no adverse effects on honeybees. Proc R Soc B Biol Sci 281(1787):20140619

Nandagopal V, Prakash A, Rao J (2008) Know the pheromones: basics and its application. *J Biopest 1*(2):210–215

Naranjo SE, Ellsworth PC, Frisvold GB (2015) Economic value of biological control in integrated pest management of managed plant systems. Annu Rev Entomol 60:1–13

Nawaz M, Mabubu JI, Hua H (2016) Current status and advancement of biopesticides: microbial and botanical pesticides. J Entomol Zool Stud 4(2):241–246

Ndakidemi B, Mtei K, Ndakidemi PA (2016) Impacts of synthetic and botanical pesticides on beneficial insects. Agric Sci 7(06):364

Negash B, Azerefegn F, Ayalew G (2020) Insecticide resistance management against thrips (Thysanoptera: Thripidae) on onion in the central Rift Valley of Ethiopia. Int J Trop Insect Sci 5:1–9

Nelson ME, Alves AP (2014) Plant incorporated protectants and insect resistance. In: Insect resistance management. Academic Press, New York, pp 99–147

Nelson ARLE, Ravichandran K, Antony U (2019) The impact of the Green Revolution on indigenous crops of India. J Ethnic Foods 6(1):8

Nnaji JC, Igbuku A (2019) Improving the eco-innovation status of Africa using sustainable chemistry principles. Eur J Sustain Dev Res 3(1):em0068

Nukenine EK, Goudoungou JW, Adler C et al (2010) Efficacy of diatomaceous earth and botanical powders against the maize weevil, Sitophilus zeamais Motschulsky (Coleoptera: Curculionidae) on maize. *Julius-Kühn-Archiv* 425:881

Oguh CE, Okpaka CO, Ubani CS et al (2019) Natural pesticides (Biopesticides) and uses in pest management—a critical review. Asian J Biotechnol Genet Eng 17:1–18

Oikeh EI, Omoregie ES, Oviasogie FE et al (2016) Phytochemical, antimicrobial, and antioxidant activities of different citrus juice concentrates. Food Sci Nutr 4(1):103–109

Olson S (2015) An analysis of the biopesticide market now and where it is going. Outlooks Pest Manag 26(5):203–206

Omilola B, Robele S (2017) The central position of agriculture within the 2030 Agenda for Sustain Development, vol 1683. International Food Policy Research Institute, Washington, DC

Pandya IY (2018) Pesticides and their applications in agriculture. Asian J Appl Sci Technol 2 (2):894–900

Pan-Germany (2012) Pesticide and health hazards. Facts and figures. www.pan germany.org/download/Vergift_EN-201112-web.pdf. Accessed 14 Oct 2013

Pardo-Lopez L, Soberon M, Bravo A (2013) *Bacillus thuringiensis* insecticidal three-domain Cry toxins: mode of action, insect resistance and consequences for crop protection. FEMS Microbiol Rev 37(1):3–22

Parker KM, Sander M (2017) Environmental fate of insecticidal plant-incorporated protectants from genetically modified crops: knowledge gaps and research opportunities. Environ Sci Technol 51:12049–12057

Parsa S, Morse A, Bonifacio TCB et al (2014) Obstacles to integrated pest management adoption in developing countries. Proc Natl Acad Sci U S A 11:3889–3894

Pavela R (2014) Limitation of plant biopesticides. In: Advances in plant biopesticides. Springer, New Delhi, pp 347–359

Pavela R, Benelli G (2016) Essential oils as ecofriendly biopesticides? Challenges and constraints. Trends Plant Sci 21(12):1000–1007

Perlatti B, de Souza Bergo PL, Fernandes da Silva MF (2013) Polymeric nanoparticle-based insecticides: a controlled release purpose for agrochemicals. In: Trdan S (ed) Insecticides-development of safer and more effective technologies. InTech, Rijeka, pp 523–550

Perlatti B, Forim MR, Zuin VG (2014) Green chemistry, sustainable agriculture and processing systems: a Brazilian overview. Chem Biol Technol Agric 1(1):5

Perrone T (2018) How agriculture and climate change are related: causes and effects. Accessed 6, 2019.

Piccardi P (1980) Insect sex-communication and prospects for pheromones in pest management. Italian J Zool 47(3–4):397–408

Pierattini EC, Bedini S, Venturi F et al (2019) Sensory quality of essential oils and their synergistic effect with diatomaceous earth, for the control of stored grains insects. Insects 10(4):114

Pinero JC, Dudenhoeffer AP (2018) Mass trapping designs for organic control of the Japanese beetle, *Popillia japonica* (Coleoptera: Scarabaeidae). Pest Manag Sci 74(7):1687–1693

Pingali PL (2012) Green revolution: impacts, limits, and the path ahead. Proc Natl Acad Sci 109 (31):12302–12308

Pinstrup-Andersen P, Hazell PB (1985) The impact of the Green Revolution and prospects for the future. Food Rev Int 1(1):1–25

Prince G, Chandler D (2020) Susceptibility of *Myzus persicae*, *Brevicoryne brassicae* and *Nasonovia ribisnigri* to fungal biopesticides in laboratory and field experiments. Insects 11 (1):55

Qi B, Gordon G, Gimme W (2001) Effects of neem-fed prey on the predacious insects Harmonia conformis (Boisduval) (Coleoptera: Coccinellidae) and Mallada signatus (Schneider) (Neuroptera: Chrysopidae). Biol Cont 22(2):185–190

Quarles W (2013) New biopesticides for IPM and organic production. IPM Pract 33:1–20

Radwan EM, El-Malla MA, Fouda MA et al (2018) Appraisal of positive pesticides influence on pink bollworm larvae, *Pectinophora gossypiella* (Saunders). Egypt Acad J Biol Sci F Toxicol Pest Control 10(1):37–47

Rahman S, Biswas SK, Barman NC et al (2016) Plant extract as selective pesticide for integrated pest management. Biotechnol Res 2(1):6–10

Rai HS, Verma ML, Gupta MP et al (2001) Efficacy of foliar application of Nuclear Polyhedrosis Virus against Pod Borer, *Helicoverpa armigera* Hubn in Chickpea. Symp Biocontrol Based Pest Management for Quality Plant Protection in the Current Millennium. PAU, Ludhiana, July 18–19, 2001

Rajput VS, Jhala J, Acharya VS (2020) Biopesticides and their mode of action against insect pests: A review. Int J Chem Stud 8(2):2856–2862

Ramaseshadri P, Segers G, Flannagan R et al (2013) Physiological and cellular responses caused by RNAi- mediated suppression of Snf7 Orthologue in Western Corn Rootworm (Diabrotica virgifera virgifera) larvae. PLoS One 8(1):e54270

Rao RGV, Rupela OP, Rao VR et al (2007) Role of biopesticides in crop protection: present status and future prospects. Indian J Plant Prot 35(1):1–9

Raruang Y, Omolehin O, Hu D et al (2020) Host induced gene silencing targeting Aspergillus flavus aflM reduced aflatoxin contamination in transgenic maize under field conditions. Front Microbiol 11:754. https://doi.org/10.3389/fmicb.2020.00754

Ravensberg WJ (2011) A roadmap to the successful development and commercialization of microbial pest control products for control of arthropods, vol 10. Springer Science & Business Media, Berlin

Raza G, Makhdoom R, Zaheer H (2019) Genetic improvement of Bacillus thuringiensis against the cotton bollworm, *Earias vitella* (Fab.) (Lepidoptera: Noctuidae), to improve the cotton yield in Pakistan. Egypt J Biol Pest Cont 29(1):1–6

Renou M, Guerrero A (2000) Insect parapheromones in olfaction research and semiochemical-based pest control strategies. Annu Rev Entomol 45(1):605–630

Rezaei R, Safa L, Damalas CA et al (2019) Drivers of farmers' intention to use integrated pest management: integrating theory of planned behavior and norm activation model. J Environ Manage 236:328–339

Ritter L, Solomon KR, Forget J et al (1995) A review of selected persistent organic pollutants. In: International Programme on Chemical Safety (IPCS). PCS/95.39, vol 65. World Health Organization, Geneva, p 66

Rodgers PB (1993) Potential of biopesticides in agriculture. Pest Sci 39(2):117–129

Rodrigues TB, Figueira A (2016) Management of insect pest by RNAi—a new tool for crop protection. In: RNA interference. IntechOpen, London

Rodríguez-Sifuentes L, Marszalek JE, Chuck-Hernández C et al (2020) Legumes protease inhibitors as biopesticides and their defense mechanisms against biotic factors. Int J Mol Sci 21 (9):3322

Rosello G, Bonaterra A, Frances J et al (2013) Biological control of fire blight ofapple and pear with antagonistic *Lactobacillus plantarum*. Eur J Plant Pathol 137:621–633

Ruiu L (2015) Insect pathogenic bacteria in integrated pest management. Insects 6(2):352–367

Ruiu L (2018) Microbial biopesticides in agroecosystems. Agronomy 8(11):235

Saeid A, Prochownik E, Dobrowolska-Iwanek J (2018) Phosphorus solubilization by *Bacillus* species. Molecules 23(11):2897
Saleh HEDM, Koller M (2018) Introductory chapter: principles of green chemistry. In: Green Chemistry. IntechOpen, London
Sales MDC, Costa HB, Fernandes PMB (2016) Antifungal activity of plant extracts with potential to control plant pathogens in pineapple. Asian Pac J Trop Biomed 1:26–31
Samada LH, Usman SF, Tambunan SF (2020) OnLine J Biol Sci 20(2):66–76
Sanganyado E, Munjanja B, Nyakubaya VT (2015) Biopesticide residue in water. In: Biopesticides handbook, vol 14. CRC, Boca Raton, pp 93–117
Sanon A, Zakaria I, Clementine LDB (2018) Potential of botanicals to control Callosobruchus maculatus (Col.: Chrysomelidae, Bruchinae), a major pest of stored cowpeas in Burkina Faso: a review. Int J Insect Sci 10:1179
Santesteban LG, Di Gennaro SF, Herrero-Langreo A, Miranda C, Royo JB, Matese A (2017) High-resolution UAV-based thermal imaging to estimate the instantaneous and seasonal variability of plant water status within a vineyard. Agric Water Manag 183:49–59
Sarma R, Adhikari K, Mahanta S et al (2019) Combinations of plant essential oil based terpene compounds as larvicidal and adulticidal agent against Aedes aegypti (Diptera: Culicidae). Sci Rep 9(1):1–12
Scholthof KBG, Adkins S, Czosnek H et al (2011) Top 10 plant viruses in molecular plant pathology. Mol Plant Pathol 12(9):938–954
Schwenk V, Riegg J, Lacroix M et al (2020) Enteropathogenic potential of Bacillus thuringiensis isolates from soil, animals, food and biopesticides. Foods 9(10):1484
Seaman D (1990) Trends in the formulation of pesticides: an overview. Pest Sci 29(4):437
Senghor LA, Ortega-Beltran A, Atehnkeng J et al (2020) The atoxigenic biocontrol product Aflasafe SN01 is a valuable tool to mitigate aflatoxin contamination of both maize and groundnut cultivated in Senegal. Plant Dis 104(2):510–520
Sesan TE, Enache E, Iacomi M et al (2015) Antifungal 33XXX against Botrytis cinerea Pers. in the Blackcurrant Crop (Ribes nigrum L). Acta Scim Polonorum Technologia Alimentaria 1:29–43
Sharma KR, Raju SVS, Jaiswal DK, Thakur S (2018) Biopesticides: an effective tool for insect pest management and current scenario in India. Indian J Agric Allied Sci 4: 59, 62
Sharma A, Jaronski S, Reddy GVP (2019) Impact of granular carriers to improve the efficacy of entomopathogenic fungi against wireworms in spring wheat. J Pestic Sci. https://doi.org/10.1007/s10340-019-01161-1
Shiberu T, Getu E (2016) Assessment of selected botanical extracts against Liriomyza Species (Diptera: Agromyzidae) on tomato under glasshouse condition. Int J Fauna Biol Stud 1:87–90
Shimeles A, Verdier-Chouchane A, Boly A (2018) Introduction: understanding the challenges of the agricultural sector in Sub-Saharan Africa. In: Shimeles A, Verdier-Chouchane A, Boly A (eds) Building a resilient and sustainable agriculture in sub-saharan Africa. Palgrave Macmillan, Cham
Shingote PR, Moharil MP, Dhumale DR et al (2013) Distribution of vip genes, protein profiling and determination of entomopathogenic potential of local isolates of *Bacillus thuringiensis*. Bt Res 4 (1)
Silva AX, Bacigalupe LD, Luna-Rudloff M et al (2012) Insecticide resistance mechanisms in the green peach aphid Myzus persicae (Hemiptera: Aphididae) II: costs and benefits. PLoS One 7 (6):e36810
Singh D (ed) (2014) Advances in plant biopesticides. Springer, Berlin
Singh B, Arora R, Gosal SS (2015) Biological and molecular approaches in pest management. Scientific Publishers
Sobiczewski P (2008) Bacterial diseases of plants: Epidemiology, diagnostics and control. Zemdirbyste 95:151–157
Sonenshine DE (2004) Pheromones and other semiochemicals of ticks and their use in tick control. Parasitology 129(S1):S405

Song J, Han B (2015) Green chemistry: a tool for the sustainable development of the chemical industry. Natl Sci Rev 2(3):255–256

Sopko MS, Narva KE, Bowling AJ et al (2019) Modification of Vip3Ab1 C-terminus confers broadened plant protection from lepidopteran pests. Toxins 11(6):316

Sousa AH, Faroni LR, Andrade GS et al (2013) Bioactivity of diatomaceous earth to Sitophilus zeamais (Coleoptera: Curculionidae) in different application conditions. Rev Bras Engenharia Agríc Ambiental 17(9):982–986

Sparks TC, Nauen R (2015) IRAC: Mode of action classification and insecticide resistance management. Pestic Biochem Physiol 121:122–128

Sparks TC, Thompson GD, Kirst HA et al (1999) In: Hall FR, Menn JJ (eds) Biopesticides: use and delivery. Humana Press, Totowa, NJ, pp 171–188

Stankovic S, Kostic M, Kostic I et al (2020) Practical approaches to pest control: the use of natural compounds. In: Pests-classification, management and practical approaches. IntechOpen, London

Steinkraus DC, Tugwell NP (1997) *Beauveria bassiana* (Deuteromycotina: Moniliales) effects on *Lygus lineolaris* (Hemiptera: Miridae). J Entomol Sci 32(1):79–90

Stenberg JA (2017) A conceptual framework for integrated pest management. Trends Plant Sci 22:749–769

Sterk G, Hannegraaf J, Kolokytha P (2020) Effects of Bacillus thuringiensis subsp. aizawai GC91 (Agree WG) on R&D colonies of the buff-tailed bumblebee *Bombus terrestris* (2.5 part 2). Julius-Kühn-Archiv 465:53–57

Stevens J, Dunse K, Fox J et al (2012) Biotechnological approaches for the control of insect pests in crop plants. In: Soundararajan RP (ed) Pesticides-advances in chemical and botanical pesticides. InTech, Rijeka, pp 269–308

Stevenson PC, Isman MB, Belmain SR (2017) Pesticidal plants in Africa: a global vision of new biological control products from local uses. Ind Crop Prod 110:2–9

Stoneman B (2010) Challenges to commercialization of biopesticides. In: Proceedings microbial biocontrol of arthropods, weeds and plant pathogens: risks, benefits and challenges. National Conservation Training Center, Shepherdstown, WV

Sun X (2015) History and current status of development and use of viral insecticides in China. Viruses 7:306–319

Szewczyk B, Hoyos-Carvajal L, Paluszek M et al (2006) Baculoviruses—reemerging biopesticides. Biotechnol Adv 24:143–160

Tadros F (2005) Applied surfactants, principles and applications. Wiley-VCH Verlag GmbH. KGaA, Weinheim, pp 187–256

Tay WT, Soria MF, Walsh T et al (2013) A brave new world for an old world pest: *Helicoverpa armigera* (Lepidoptera: Noctuidae) in Brazil. PLoS One 8(11):e80134

Thakur N, Kaur S, Tomar P et al (2020) Microbial biopesticides: current status and advancement for sustainable agriculture and environment. In: New and future developments in microbial biotechnology and bioengineering. Elsevier, Amsterdam, pp 243–282

Tijjani A, Bashir KA, Mohammed I et al (2016) Biopesticides for pests control: a review. J Biopest Agric 3(1):6–13

Tiroesele B, Thomas K, Seketeme S (2015) Control of cowpea weevil, Callosobruchus maculatus (F.) (Coleoptera: Bruchidae), using natural plant products. Insects 6(1):77–84

Townsend RJ, Nelson TL, Jackson TA (2010) Beauveria brongniartii a potential biocontrol agent for use against manuka beetle larvae damaging dairy pastures on Cape Foulwind. N Z Plant Protec 63:224–228

Trejo SA, Reinoso S, Vargas WA, et al (2020) Bioinsecticide formulation comprising highly resistant beauveria bassiana microsclerotia and method of preparation thereof. United States Patent and Trademark Office. US20200281215A1. https://patents.google.com/patent/US20200281215A1/en?q=fungi+biopesticides&before=priority:20191231&after=priority:20190101&oq=fungi+biopesticides+2019. Accessed 2 Nov 2020

Tulipa D, De LC (2019) Biopesticides: a viable tool for organic farming. Int J Microbiol Res 11 (7):1660–1664
Tundo P, Griguol E (2018) Green chemistry for sustainable development. Chem Int 40(1):18–24
Ubuoh EA (2016) Green Chemistry: a panacea for environmental sustainability agriculture in global perspective. Global J Pure Appl Chem Res 4(1):21–29
Ujagir R, Byrne OM (2009) 18 Insect pests and their management. Lentil 13:282
Ullah F, Gul H, Yousaf HK et al (2019) Impact of low lethal concentrations of buprofezin on biological traits and expression profile of chitin synthase 1 gene (CHS1) in melon aphid, *Aphis gossypii*. Sci Rep 9(1):1–13
Umar MA (2013) Organic farming in Nigeria: problems and future prospects. Global Adv Res J Agric Sci 2:256–262
Umaru FF, Simarani K (2020) Evaluation of the potential of fungal biopesticides for the biological control of the seed bug, *Elasmolomus pallens* (Dallas) (Hemiptera: Rhyparochromidae). Insects 11(5):277
Unsworth J (2010) History of pesticide use. International Union of Pure and Applied Chemistry, Research Triangle Park, NC
USDA-ARS (United States Department of Agriculture-Agricultural Research Service) (2018) A national road map for integrated pest management. https://www.ars.usda.gov/ARSUserFiles/OPMP/IPM%20Road%20.Map%20FINAL.pdf
USEPA (2017) Registers innovative tool to control corn rootworm. https://www.epa.gov/newsreleases/epa-registers-innovative-tool-control-corn-rootworm. Accessed 28 June 2017
Vashisth S, Chandel YS, Chandel RS (2019) Comparative efficacy of indigenous heterorhabditid nematodes from north western Himalaya and Heterorhabditis indica (Poinar, Karunakar & David) against the larvae of *Helicoverpa armigera* (Hubner). Int J Pest Manag 65(1):16–22
Verma S, Goyal S, Shingla S (2018) Green Chemistry: a new approach to the synthesis, processing and application of chemical substances. Int J Biotech Bioeng 4:89–95
Vetter SH, Sapkota TB, Hillier J et al (2017) Greenhouse gas emissions from agricultural food production to supply Indian diets: Implications for climate change mitigation. Agric Ecosyst Environ 237:234–241
Vimala-Devi PS, Vineela V (2015) Suspension concentrate formulation of Bacillus thuringiensis var. kurstaki for efective management of Helicoverpa armigera on sunfower (Helianthus annuus). Biocontrol Sci Technol 25(3):329–336
Wei JZ, O'Rear J, Schellenberger U, Rosen BA, Park YJ, McDonald MJ, Zhu G, Xie W, Kassa A, Procyk L, Perez Ortega C (2018) A selective insecticidal protein from Pseudomonas mosselii for corn rootworm control. Plant Biotechnol J 16(2):649–659
Willis KJ (2017) State of the world's plants report-2017. Royal Botanic Gardens. https://stateoftheworldsplants.com/2017/plant-health.html
World Health Organization (2018) The state of food security and nutrition in the world 2018: building climate resilience for food security and nutrition. Food & Agriculture Organization
Yee WL (2008) Effects of several newer insecticides and kaolin on oviposition and adult mortality in western cherry fruit fly (Diptera: Tephritidae). J Entomol Sci 43(2):177–190
Yimer D, Abena T (2019) Components, mechanisms of action, success under greenhouse and field condition, market availability, formulation and inoculants development on biofertilizer. Biomed J Sci and Tech Res 12:9366–9371
Zacks Small-Cap Research (2020). https://scr.zacks.com/. Accessed 1 Nov 2020
Zalom FG, Bolda MP, Dara SK, et al (2018) UC IPM pest management guidelines: strawberry (insects and mites). University of California Statewide IPM Program, Oakland, CA

Zalucki MP, Shabbir A, Silva R et al (2012) Estimating the economic cost of one of the world's major insect pests, *Plutella xylostella* (Lepidoptera: Plutellidae): just how long is a piece of string? J Econ Entomol 105(4):1115–1129

Zhang K, Wei J, Huff Hartz KE et al (2020) Analysis of RNA interference (RNAi) biopesticides: double-stranded RNA (dsRNA) extraction from agricultural soils and quantification by RT-qPCR. Environ Sci Technol 54(8):4893–4902

Zhen L, Routray JK, Zoebisch MA, Chen G, Xie G, Cheng S (2005) Three dimensions of sustainability of farming practices in the North China Plain: a case study from Ningjin County of Shandong Province, PR China. Agric Ecosyst Environ 105(3):507–522

Zhou R, Rana TM (2013) RNA-based mechanisms regulating host–virus interactions. Immunol Rev 253(1):97–111. https://www.greenbook.net/loveland-products-inc/bt-sulfur-15-50-dust

Chapter 2
Entomopathogenic Fungi: Current Status and Prospects

Ana Carla da Silva Santos, Rosineide da Silva Lopes, Luciana Gonçalves de Oliveira, Athaline Gonçalves Diniz, Muhammad Shakeel, Elza Áurea de Luna Alves Lima, Antonio Félix da Costa, and Vera Lucia de Menezes Lima

Abstract Entomopathogenic fungi are microorganisms capable of infecting and killing arthropods and therefore have a great potential in pest management. As the extensive use of synthetic pesticides has led to increased resistance in insects, decreased natural enemies, and had negative impacts on environmental and human health, the search for eco-friendly control agents is urgent. Entomopathogenic fungi are promising alternatives in this regard and are attracting global attention, with increasing efforts and financial investments being made for the development, commercialization and use of fungus-based control products. Despite scientific and technological advances, there is still a need for studies to expand the number of species applicable in pest management and improve their performance in the field. There is also a need to increase user awareness regarding their correct use with the aim to establish their widespread adoption and market potential. This chapter covers the main taxonomic groups that comprise entomopathogenic fungi, their modes of action to establish insect infection and spread, and the insect's defense mechanisms against these fungi. Furthermore, techniques of fungal isolation, selection, and production are discussed. The usage status, challenges, and prospects of

A. C. da S. Santos · A. G. Diniz · E. Á. de L. A. Lima
Departamento de Micologia Centro de Biociências, Universidade Federal de Pernambuco, Recife, Pernambuco, Brazil

R. da S. Lopes · L. G. de Oliveira · A. F. da Costa
Instituto Agronômico de Pernambuco, Recife, PE, Brazil

M. Shakeel
Key Laboratory of Bio-Pesticide Innovation and Application of Guangdong Province, College of Plant Protection, South China Agricultural University, Guangzhou, China

V. L. de M. Lima (✉)
Departamento de Bioquímica Centro de Biociências, Universidade Federal de Pernambuco, Recife, Pernambuco, Brazil
e-mail: lima.vera.ufpe@gmail.com

S. De Mandal et al. (eds.), *New and Future Development in Biopesticide Research: Biotechnological Exploration*, https://doi.org/10.1007/978-981-16-3989-0_2

mycoinsecticides are also addressed, highlighting their application potential for sustainable agricultural production.

Keywords Biological control · Fungal entomopathogens · Biopesticides · Pest control · Mycoinsecticides

2.1 Introduction

Entomopathogenic fungi are heterotrophic, eukaryotic, unicellular, or pluricellular insect pathogens which exhibit asexual and/or sexual reproduction and are classified in phylogenetically diverse taxa (Mora et al. 2017). These organisms can infect and kill arthropods, using them as hosts to develop at least part of their life cycle, and exhibit different degrees of virulence that vary according to the fungus and host from extreme specialists with narrow host ranges to generalists with broad host ranges (Vega et al. 2012). Narrow host range is commonly associated with obligate pathogens that depend entirely on the host for obtaining nutrients, while generalists are usually considered facultative pathogens that can survive on nutrients not obtained from the living host and play other ecological roles, besides being entomopathogenic (Vega et al. 2009; Goettel et al. 2010).

The number of insect-associated fungal species is estimated between 20,000 and 50,000. However, the real diversity of these species and their ecological significance is still poorly understood (Blackwell 2011). Fungus-insect associations include pathogenic associations that are of great importance for insect population stability in natural ecosystems. As entomopathogenic fungi are the causative agents responsible for most insect diseases, they have a great potential for use in insect pest management programs, reducing insect populations to levels that do not cause economic damage to crops. They can also be used in association with, or substitution for, chemical products (Alves 1998; Mascarin and Quintela 2013; Maina et al. 2018; Lopes et al. 2020). Some entomopathogenic fungi species also have the potential to control disease vectors (Tanzini et al. 2001; Scholte et al. 2004; Delgado and Murcia 2011).

Due to problems related to the extensive use of synthetic chemical pesticides, such as environmental impact, toxic effects to non-target organisms including humans and insecticidal resistance, ecofriendly products are claimed to provide a safer and more sustainable form of pest management (Naqqash et al. 2016; Sandhu et al. 2017; Baron et al. 2019; Silva et al. 2020). In response to this demand, interest in expanding the use of entomopathogenic fungi in insect control has been growing. Investments in research and technology have enabled the development of increasingly better mycoinsecticides in terms of performance and cost (Santoro et al. 2005; Marrone 2019). However, many advances are still required if the use of fungi in pest management is to become more widespread. This goes beyond enhancing the shelf life, effectiveness, and persistence of these agents, as it is necessary to discuss the issues of registration and regulation of these products, as well as special efforts that should be directed towards raising awareness among producers about their importance and correct use.

This chapter deals with entomopathogenic fungi, including the main taxonomic groups that comprise these organisms, their modes of action towards establishing insect infection and further spread, as well as the insect's defense mechanisms against these fungi. We also address isolation techniques, selection, production, and insect immunity. The status of the use, commercialization, possible constraints and prospects of mycoinsecticides are also discussed, highlighting the importance of fungal entomopathogens in developing more sustainable methods of agricultural production.

2.2 Entomopathogenic Fungi

The classification of the kingdom Fungi has been frequently updated due to the inclusion of DNA sequence data in studies. A recent proposal for the classification of the kingdom Fungi proposed nine subkingdoms and 18 phyla, based on the phylogeny and time of divergence of certain taxa (Tedersoo et al. 2018). Entomopathogenic fungi do not form a monophyletic group and can be found in several major fungal lineages some rich in entomopathogenic fungi, while others have few known insect pathogens. For example, the phyla Ascomycota and Entomophthoromycota include the vast majority of species pathogenic to insects, infecting hosts of 13 and 10 orders of insects, respectively. On the other hand, few entomopathogens are known in the phyla Chytridiomycota, Blastocladiomycota, Kickxellomycota and Basidiomycota, with members of each of these phyla infecting at most three orders of insects, as far as is known (Araújo and Hughes 2016; Mora et al. 2017).

2.2.1 *Chytridiomycota*

The phylum Chytridiomycota consists of a group with mobile cells named zoospores which have a single directed smooth flagellum at least once reflecting their aquatic life cycle (Barr et al. 1987). These zoospores respond to chemical gradients allowing them to locate their hosts actively, and can encyst when the environment becomes too cold, hot or dry, preventing water loss or cell collapse (Gleason and Lilje 2009). Most chytrids are saprobes, inhabiting mainly wet soils and freshwaters, and a minority of species inhabit marine environments (Araújo and Hughes 2016). However, most species are parasites of plants, animals, and other fungi (Sparrow 1960; Martin 1978; Karling 1981; Dewel et al. 1985). Infections caused by chytrids in insects appear to be rare when compared to other groups of fungi (Araújo and Hughes 2016). Most chytrids known for infecting insects were contained in Blastocladiales, genus *Coelomomyces* (Gul et al. 2014), but this group was recently transferred to the phylum Blastocladiomycota (Tedersoo et al. 2018), as will be discussed below. Few entomopathogens remain in the phylum Chytridiomycota (Araújo and Hughes 2016).

2.2.2 *Blastocladiomycota*

Blastocladiomycota is the only phylum of the subkingdom Blastocladiomyceta (Tedersoo et al. 2018) and the only fungal group characterized by an alternation of haploid and diploid generations (James et al. 2014). Formerly, it was considered an order belonging to the phylum Chytridiomycota (Kirk et al. 2008; Wijayawardene et al. 2018). Blastocladiomycota comprises zoosporic fungi that produce spores with a single smooth flagellum inserted posteriorly (Hibbett et al. 2007; Jerônimo et al. 2015). Members of this phylum are considered cosmopolitan, are present in different aquatic and terrestrial ecosystems, saprobes and/or parasites, in association with algae, macrophytes, invertebrates, amphibians, fungi, and oomycetes (Shearer et al. 2007). Species of *Catenaria* Sorokin (*C. spinosa* Martin, *C. anguillulae* Sorokin), *Coelomomyces* Keilin (*C. utahensis* (Romney) Couch & Nielsen) and *Polycaryum* Stempell (*P. laeve* Stempell) are common parasites in copepods and some insects, acting in the control of these populations (Gleason et al. 2010). The *Coelomomyces* and *Coelomycidium* genera, which comprise dozens of insect pathogenic species (Barr 2001), are reported mainly in hosts of the orders Hemiptera and Diptera (Samson et al. 1988; Gul et al. 2014).

2.2.3 *Kickxellomycota*

Kickxellomycota consists of a group of filamentous fungi characterized by the formation of regularly septate hyphae and septa with median, disciform cavities containing plugs. Asexual reproduction is by arthrospore, merosporangia (one- or two-spored), or trichospores; and sexual reproduction by zygospores (globose, broadly fusiform, hemifusiform, or long-cylindrical and coiled). These fungi are saprobes, mycoparasites or obligate symbionts (Tedersoo et al. 2018). They are commonly found in the mid- or hind-gut of various arthropods, for example, in the gut of aquatic insect larvae (Harpellales), isopods and springtails (Asellariales) (Benny et al. 2014; Tretter et al. 2014). Besides, their small thalli, they cannot be cultured in the laboratory without their hosts, making it difficult to study them. There are some Kickxellomycota insect parasitic species, for example, *Smittium* species (Harpellales). The parasitism in these species seems to have evolved secondarily from their commensalism (Naranjo-Ortiz and Gabaldón 2019).

2.2.4 *Entomophthoromycota*

The phylum Entomophthoromycota comprises terrestrial fungi (non-flagellate) that form coenocytic hyphae. Primary sporulation is by the production of forcibly discharged primary conidia which usually form one or more types of secondary conidia on unfavorable substrates. Secondary conidia can be actively or passively discharged and may or may not have the same shape as the primary conidium. Homothallic zygospores or azygospores are also produced and function as resting spores to promote survival during unfavourable environmental conditions (Humber 2012a; Benny et al. 2014). These fungi can be found in practically all temperate and tropical regions of the world (Benny et al. 2014).

Entomophthoromycota includes about 280 species. Almost all are pathogens of arthropods (Boomsma et al. 2014), but soil- and litter-borne saprobes and some saprobes or specialized phytopathogens are also known in the group (Humber 2012b; Gryganskyi et al. 2013). Some species (e.g. *Completoria* spp.) have a different nutritional mode as parasites of fern gametophytes or desmids, tardigrades, and nematodes. In contrast, others become opportunistic vertebrate pathogens or saprobes (Boomsma et al. 2014). Entomophthoromycota probably evolved first as specialized saprobes before shifting to obligate parasitism. The earliest divergence within Entomophthoromycotina appears to be associated with arthropod exuviae and corpses. The majority of the Entomophthoromycotina members can parasitize insects to some degree (Naranjo-Ortiz and Gabaldón 2019). The aphids (Hemiptera: Aphididae) and a variety of lepidopterans (mostly moths from many families), adult flies, and grasshoppers and locusts (Orthoptera: Acrididae) are most affected by pathogens of this group. A smaller set of genera and species attacks soil invertebrates such as nematodes (Benny et al. 2014). *Conidiobolus* genus include members that are not obligate parasites and are able to degrade chitin and infect living insects (Naranjo-Ortiz and Gabaldón 2019).

2.2.5 *Basidiomycota*

Together with the Ascomycota and Entorrhizomycota phyla, Basidiomycota forms the Subkingdom Dikarya (Tedersoo et al. 2018), the so-called superior fungi. Basidiomycota species are characterized by the formation of basidiospores (sexual spores) outside on basidium cells in the sexual life stage (Webster et al. 1984; Pringle et al. 2005; He et al. 2019). Another important and exclusive trait for the group is the clamp connection formed during karyogamy (division of the nuclei in the tip of growing hyphae), which helps to ensure the dikaryotic condition (i.e., two nuclei in the same cell) (Alexopoulos et al. 1996). Basidiomycota comprises an extraordinary diversity of fungi, including mushrooms, boletes, puffballs, earthstars, toadstools, and some microfungi, such as rusts, smuts, and yeasts. Among the members of Basidiomycota are great decomposers of wood and leaf litter, important for carbon recycling, food sources for humans and other animals as well as being pathogens of plants, and animals, etc. (Kendrick 2017; Zhao et al. 2017).

Few Basidiomycota genera are known as insect pathogens which infect termite eggs (i.e., *Fibulorhizoctonia* - Atheliales), and mainly scale insects (i.e., *Septobasidium* and *Uredinella* - Septobasidiales) (Araújo and Hughes 2016). The Septobasidiales attack exclusively scale insects (Hemiptera—Diaspididae) (Evans

1988). The order includes two genera of entomopathogens: *Uredinella*, that attacks single insects and *Septobasidium* that attacks whole colonies of up to 250 individuals. Another Basidiomycota genus, *Fibulorhizoctonia* has been described as an opportunistic pathogen of termite eggs found living inside the termite's nest, among their eggs and has been proven to act as a pathogen on a number of occasions (Matsuura et al. 2000). The fact that so few insect pathogenic lineages are known in a group as large as Basidiomycota suggests a strong direction on their adaptive radiation for other ecological functionalities than parasitizing insects (Boomsma et al. 2014).

2.2.6 Ascomycota

The phylum Ascomycota is the largest group within the Kingdom Fungi in terms of the number of species described (Kirk et al. 2008). According Tedersoo et al. (2018), these phyla are classified into three subphyla—Taphrinomycotina, Saccharomycotina, and Pezizomycotina. Taphrinomycotina encompasses saprobe fungi or parasites of vertebrates and plants and does not include members known to be pathogenic to insects (Hibbett et al. 2007). Many yeasts of the subphyla Saccharomycotina are associated with insects for dispersal, while the Pezizomycotina subphyla, the largest, most morphologically and ecologically complex group, is rich in insect-associated members (Humber 2008; Blackwell 2011).

In Pezizomycotina, insect pathogenic lifestyles evolved several times independently. The order Hypocreales (Sordariomycetes) is particularly rich in insect pathogens (Naranjo-Ortiz and Gabaldón 2019). Hypocreales comprises the best known entomopathogens among the Ascomycota, distributed mainly into families Clavicipitaceae, Cordycipitaceae, and Ophiocordycipitaceae. The Clavicipitaceae includes the well-known entomopathogens *Aschersonia*, *Hypocrella*, *Regiocrella*, and *Metarhizium* (Sung et al. 2007). Cordycipitaceae contains *the genera Beauveria*, *Cordyceps (=Isaria)*, *Akanthomyces (=Lecanicillium)*, among others (Sung et al. 2007; Kepler et al. 2017). Some species of these fungi can influence the host nervous system and modify their behavior to spread the fungal spores (Araújo and Hughes 2016; Naranjo-Ortiz and Gabaldón 2019). Besides being the best known, entomopathogens belonging to Hypocreales are the most common active ingredients in commercial biopesticides based on fungi (de Faria and Wraight 2007; Lacey et al. 2015).

In addition to the class Sordariomycetes, other lineages of Pezizomycotina, including basal lineages, have also evolved as entomopathogens: for example, species in the Dothideomycetes and Eurotiomycetes (Boomsma et al. 2014). Dothideomycetes includes scale insect pathogens, such as species of the genus *Podonectria* (Rossman 1978). In Eurotiomycetes, members of the order Onygenales are obligate parasites of larval honeybees (*Apis mellifera*) and members of the order Eurotiales are facultative pathogens without much host specialization. Examples of opportunistic entomopathogens in Eurotiales are *Aspergillus* and *Penicillium* species, notorious for their production of toxic compounds, which may represent a strategy for poisoning host tissues for later saprophytic exploitation (Sosa-Gomez et al. 2010; Boomsma et al. 2014).

Laboulbeniomycetes is an interesting group of insect parasites within the Pezizomycotina, though they are not considered pathogens (Vega et al. 2012). These fungi generally use a specialized haustorium to adhere to the cuticle of insects, mites, and some Diplopoda, mainly on antennae or mouthparts. These fungi have been reported associated with insects of the orders Blattodea, Coleoptera, Dermaptera, Diptera, Hemiptera, Hymenoptera, Mallophaga, Orthoptera, and Thysanoptera (Haelewaters et al. 2015). Species of Laboulbeniales are obligate haustorial ectoparasites, whose two-celled ascospores germinate on the host surface, producing a haustorium from an attached cell, and a perithecium from the outer cell of insects and a few other arthropods. Interestingly, no mycelium or asexual state is produced. A second-order, Pyxidiophorales, is a parasitic fungus rarely associated with arthropod hosts. These fungi are mycelial, and usually is observed as asexual states preceding the development of the sexual state (Vega et al. 2012; Naranjo-Ortiz and Gabaldón 2019).

2.3 Biology and Pathogenesis

Although variations of the infectious forms of fungi on insects exist, entomopathogenic fungi encounter similar challenges in initiating and establishing infection on their hosts. These challenges are overcome in steps, starting with the release of a large number of spores with sticky surfaces to maximize the fungal propagule's chances of meeting and adhering to the host cuticle. Successful adhesion is of great importance in the fungal infection process, mediated by non-specific mechanisms such as the hydrophobic cell wall of the conidia. Proteins such as hydrophobins and adhesins enable adhesion to the insect's exoskeleton, by recognizing and fixing spores to the host surface (Wang and Leger 2007; Greenfield et al. 2014; Skinner et al. 2014). Environmental conditions such as solar radiation, temperature, and humidity can influence the process of adhesion to the insect's surface (Chandler 2017).

Under appropriate environmental conditions the propagules germinate on contact with the insect's integument or less commonly, survive oral ingestion (Mora et al. 2017) or even inspiration (Altinok et al. 2019). Infection by ingestion is common to microorganisms such as viruses and bacteria but is rare for entomopathogenic fungi. However, although fungi can occassionally use oral and respiratory infections as an alternative to penetrate the host's cuticle (Schabel 1976; Wei et al. 2017; Biswas et al. 2018; Rafaluk-Mohr et al. 2018), these alternatives can lead to greater effectiveness in the infection of arthropods with antifungal compounds in their cuticle (Pedrini et al. 2010; Da Silva et al. 2015). However, little is known about the mechanism of oral infection. In some cases, ingested spores appear to adhere to parts of the oral cavity more than to the digestive tract. Spore germination is not observed in the intestine. Carefully planned studies are necessary to clarify the physiological and molecular mechanisms activated in spores after ingestion (Mannino et al. 2019).

Fungi of the main entomopathogenic groups infect insects through the tegument. This important aspect of their biology enables them to infect different stages of insect development, including eggs and pupae and therefore puts fungi at an advantage compared to other microorganisms that can only infect their hosts by ingestion (Goettel et al. 2010; Chandler 2017). For the tegument penetration process to occur, the spore attached to the cuticle should germinate, forming a germ tube which can differentiate into an appressorium (Vega et al. 2012). Fungi that do not produce appressoria penetrate the cuticle with the germ tube (Hajek and Delalibera 2010). For example, *Akanthomyces lecanii* (Zimm.) Spatafora, Kepler & B. Shrestha (= *Verticillium lecanii* (Zimm.) Viégas) can penetrate the host cuticle only with the germ tube, while *Beauveria bassiana* (Bals.-Criv.) Vuill. and *Metarhizium anisopliae* (Metschn.) Sorokīn produce appressoria (Bhattacharyya et al. 2004).

Carbohydrates present in the cuticle activate germination (Sánchez-Pérez et al. 2014) under favorable conditions of temperature and humidity. Certain physical, chemical, and nutritional conditions in the cuticle also must be met (Mora et al. 2017). For most fungal species, the optimum temperature for spore germination ranges from 20 to 30 °C. However, some spores with particular characteristics germinate outside this range (Skinner et al. 2014). Germination usually requires a relative humidity above 90% at this stage of the infection (Sánchez-Pérez Andersen et al. 2006). The presence of water results in conidia imbibition by osmosis. Because of the hydration, the internal hydrostatic pressure increases creating a thin-walled germ tube (Sánchez-Pérez Andersen et al. 2006). Cuticle penetration occurs through mechanical pressure combined with the action of enzymes such as lipases, lipoxygenases, chitinases and proteases. In addition to being important virulence factors for the entomopathogenic fungi, these enzymes are potentially applicable in insect control (Sánchez-Pérez et al. 2014). The presence of multiple cuticle penetration sites allows faster colonization of the insect by the pathogen, avoiding infection by opportunistic organisms (Altre and Vandenberg 2001).

Many fungi undergo a transition from hyphae to yeast-like blastospores after entering the host and reaching the hemocoel, increasing the nutrient acquisition rates (Mora et al. 2017). Once in the hemocele, blastospores or secondary hyphae spread and invade muscle tissues, adipose bodies, Malpighi tubes, mitochondria and hemocytes. At this stage, the production of secondary metabolites by certain fungi interrupts the physiological responses of the insect's immune system and helps to sedate or kill host cells, facilitating resource acquisition by the fungus (Donzelli and Krasnoff 2016). A balance between the growth of the fungus and the production of toxins must occur to avoid high levels of toxins causing the death of the host before the fungus has accumulated enough cell mass to develop the necessary structures required for its further dispersion (Boomsma et al. 2014). Some metabolites produced by entomopathogenic fungi act as toxins and can cause behavioral alteration such as lack of coordination, seizures, paralysis, loss of appetite, among others. Due to their toxicity, these metabolites have the potential to be developed as insecticides (Sánchez-Pérez et al. 2014; Silva et al. 2015).

The host body is destroyed due to mechanical damage to internal organs, as well as by the development of hyphae and fluid absorption by the fungus (Mascarin and Jaronski 2016; Fan et al. 2017). Most entomopathogenic fungi cause the death of their hosts before spore production, and only some of them sporulate in the living bodies of their hosts. Fungi associated with insects can produce spores of sexual and asexual origin, usually in elevated structures which favor their dispersion to new hosts. For example, *Metarhizium* species produce asexual spores directly on conidiophores on insect cadaver, and some species of *Akanthomyces* and *Cordyceps* form conidia on synnemata. The production of sexual spores may occur in structures immersed in a stroma; for example, *Ophiocordyceps nutans*, *Cordyceps ninchukispora*, *Cordyceps militaris* and *Elaphocordyceps paradoxa*, which develop ascus and ascospores within perithecia immersed in a stroma (Vega et al. 2012).

After the death of the insect, which usually occurs between 2 and 15 days post-infection, depending on the strain and species of the fungus, as well as the characteristics of the host (Boucias and Pendland 1998), the fungus emerges from the insect through intersegmental areas or holes such as the mouth and anus. Then the production of conidia and other propagules from the host cadaver become fungal dissemination foci that enable dispersal under prevailing environmental conditions (Skinner et al. 2014). In the search for a new host and continuation of the cycle, entomopathogenic fungi have developed different dissemination strategies. Rain, wind, and invertebrates help in dispersing spores (Gul et al. 2014). The infection at various stages of a host's life may be responsible for spreading the fungus in an insect population (Dromph 2001). Infected insects used as food by uninfected insects end up becoming a source of the spread of the fungus (Pell et al. 2010; Gul et al. 2014). In addition, some species can manipulate the host's behavior as an adaptation to increase transmission against susceptible new hosts (Chandler 2017).

2.4 Insect Immunity to Entomopathogenic Fungi

In nature, with time, all multicellular organisms have developed diversified defense mechanisms to combat infectious pathogens. Higher vertebrates counter the pathogen challenge with their adaptive immune system, whereas invertebrates such as insects rely on the sophisticated innate immune system for defense against invading pathogens (Shakeel et al. 2018). This system involves humoral and cellular responses. The former is best-demonstrated by clotting, melanization with prophenoloxidase, and the production of antimicrobial peptides, whereas the latter is mediated by haemocytes and comprises encapsulation, nodulation, and phagocytosis (Shakeel et al. 2017). The innate immune system recognizes the invading microorganisms by pattern recognition receptors such as β-1,3-glucan recognition proteins, peptidoglycan recognition proteins, C-type lectins, and scavenger receptors (Hultmark 2003). This signal recognition is then amplified via serine proteases following signaling pathway activation (Osta et al. 2004). Finally, the effector molecules are induced to combat invading microorganisms.

On the other hand, entomopathogenic fungi have developed multiple strategies to circumvent insects' defense systems (Butt 2002). For successful invasion into the insect body, entomopathogenic fungi use a set of enzymes to breach the cuticle. Once inside the insect, entomopathogenic fungi release secondary metabolites to suppress its immune system (Vey et al. 2002).

2.4.1 The Humoral Immune Response

Antimicrobial peptides (AMPs), mainly cationic peptides, play a vital role in insects' humoral immunity by killing invading microbes such as fungi, bacteria, and viruses (Qu and Wang 2018). Upon microbial infection, a series of these peptides is produced by the fat body and subsequently released into the hemolymph to combat invading microorganisms (Nehme et al. 2007). Gram-positive bacteria and fungi activate the Toll signaling pathway, while Gram-negative bacteria trigger the IMD pathway (Buchon et al. 2014). Insects produce specific AMPs as a result of pathway activation, such as defensin which is activated against Gram-positive bacteria whereas drosomycin and Metchnikowin are expressed in response to fungal infections in *Drosophila* (Fehlbaum et al. 1994). AMPs such as diptericin, cecropins, drosocin, and attacin are induced when *Drosophila* is challenged by Gram-negative bacteria (Lemaitre and Hoffmann 2007). Until now, as compared to antibacterial peptides, only a few antifungal peptides have been well characterized, including termicin from termites (Da Silva et al. 2003), heliomycin from the tobacco budworm *Heliothis virescens* (Lamberty et al. 1999), and gallerimycin from the greater wax moth *Galleria mellonella* larvae (Schuhmann et al. 2003). Furthermore, JAK/STAT and JNK signaling pathways are also triggered following fungal infection that results in the induction of immune gene expression in *Drosophila*'s fat body (Lu and Leger 2016).

2.4.2 The Cellular Immune Response

There is a wealth of literature on humoral immune responses, and the underlying mechanisms are better understood than insects' cell-mediated reactions. However, with the help of genetic techniques a significant advance has occurred in our understanding of the mechanisms regulating cellular defenses (Qu and Wang 2018). The cell-mediated immune response relies on hemocytes present in the body of insects. Hemocytes are categorized into three types (plasmatocytes, crystal cells, and lamellocytes) based on morphology and functional features (Lanot et al. 2001; Evans and Banerjee 2003). Hemocytes perform several important functions in the cell's immune response, including nodulation, encapsulation and phagocytosis

(Strand 2008). Phagocytosis, an evolutionary conserved immune response, is a process whereby cells recognize, bind, and ingest large particles (Rosales 2005). In insects, a subset of hemocytes perform phagocytosis in the hemolymph (Browne et al. 2013). Plasmatocytes and granulocytes are described as the major phagocytic cells in many insects (Rosales 2011). In *Drosophila*, plasmatocytes correspond to 90–95% of all hemocytes (Qu and Wang 2018). The phagocytosis process takes place directly via receptors on the cell surface or indirectly through opsonins that cover the particle to facilitate its detection by phagocytic receptors (Lu and Leger 2016). Among the receptors involved in phagocytosis, Eater and Dscam are the best understood. Eater mutant flies show impaired phagocytosis and reduced survival following bacteria infection (Kocks et al. 2005). However, the Eater mutant activity against resisting fungi has not been reported yet (Lu and Leger 2016). Dscam, another well studied receptor, compromises bacterial and plasmodium resistance demonstrating a broad spectrum defense (Dong et al. 2006).

2.4.3 Fungal Countermeasures to Insect Immunity

As mentioned earlier, once inside the insect, entomopathogenic fungi are confronted with an inhospitable environment, full of efficient immune responses by the host. However, in the arms race between pathogen and host, entomopathogenic fungi have also evolved several strategies to counter the host's defense system (Xu et al. 2017). There are several documented instances of entomopathogenic fungi successfully infecting their hosts by manipulating the host's immune defences. An example is *M. anisopliae* which evades the host immune system by masking the cell wall during colonization of hemocoel (Wang and Leger 2006). *Isaria fumosorosea* produces enzymes such as lipase, chitosanase, and chitinase for successful penetration through the host's cuticle (Ali et al. 2010). Moreover, *M. anisopliae* releases secondary metabolites, destruxins, to weaken the host's immune responses, such as in *Drosophila*, where the expression of AMP genes is down-regulated following destruxin injection (Pal et al. 2007). Similarly, *Beauveria bassiana* produces a quinone derivative, oosporein, to inhibit prophenoloxidase activity and downregulate the expression of gallerimycin, an antifungal peptide found in *Galleria mellonealla* (Feng et al. 2015). Furthermore, few strains of *Beauveria* and *Metarhizium* show resistance to Drosophila's Toll pathway products such as the antifungal peptide drosomycin (Lu et al. 2015; Tzou et al. 2002). *B. bassiana* interacts with the gut microbiota to cause the mosquito's rapid death via downregulating AMPs and dual oxidase expression in the mosquito midgut (Wei et al. 2017).

2.5 Isolation Techniques

The vast majority of entomopathogenic fungi are isolated from arthropods' bodies. However, these fungi can also be isolated from soil, organic waste, plants, and water (Vega et al. 2009; Behie and Bidochka 2014). Most entomopathogenic fungi can grow in artificial media *in vitro*. This growth capacity has led to the development of specific techniques to select particular groups of microorganisms (Meyling 2007).

One way to obtain entomopathogenic fungi is by collecting dead or live insects, which in the laboratory undergo a process of superficial disinfection in 70% alcohol, sodium hypochlorite, and sequential rinses in sterile distilled water. After drying on filter paper, they are transferred to Petri dishes containing culture medium plus antibiotic (Inglis et al. 2012; Pestano et al. 2017). In the case of insects ollected with evident mycelial growth, isolation can be done by transferring the fungus' vegetative or reproductive structures directly into the culture medium, using the streaking technique (Shin et al. 2009). In the case of cadavers that do not have evident fungal structures, these can be placed in a humid chamber to facilitate the exit of hyphae and fungal conidia through the exoskeleton of the insect (Inglis et al. 2012).

Homogenization of insect corpses is another technique used to isolate entomopathogenic fungi. Insect corpses are superficially disinfected and homogenized by different methods. The insects can be immersed in a phosphate-saline buffer and crushed in a mortar with a pestle. The material obtained is diluted and sown in a suitable medium. Mechanical crushers and blenders are used to homogenize larger insects with a hard exoskeleton; however, the use of this equipment can lead to cross-contamination of the samples (Inglis et al. 2012; Ozdal et al. 2012).

Soil is considered a favorable environment for entomopathogenic fungi, since it is the habitat of potential arthropod hosts and provides protection to mycelium and fungal spores against solar radiation, excessive heat, cold and water (Vega et al. 2009). One of the methodologys used for the isolation of fungi from the soil is the plating of serial dilutions in specific selective or semi-selective media, containing chemical compounds and antibiotics. Tetracycline, streptomycin and chloramphenicol are examples of antibiotics that prevent or reduce the growth of contaminating bacteria and fungi (Doberski and Tribe 1980; Beilharz et al. 1982; Liu et al. 1993; Fernandes et al. 2010; Castro et al. 2016). Chloramphenicol inhibits the growth of a broad spectrum of Gram-negative and Gram-positive bacteria, including anaerobic bacteria (Falagas et al. 2008). Thiabendazole is a fungicidal and parasiticidal chemical used in agriculture, which has been indicated as an appropriate antifungal for isolated entomopathogenic fungi in selective media (Luz et al. 2007; Rocha and Luz 2009).

Soil-occurring entomopathogenic fungi can also be isolated by spreading soil particles directly into the selective medium. Another strategy used to isolate fungi from soil particles involves the incorporation of the particles into an agar medium (Warcup 1950). In this technique, 5 to 15 g of soil are distributed in a Petri dish and 15 mL of molten agar medium (50 °C) are deposited on the soil particles.

Subsequently, this material is incubated, and after growth, the fungal isolates are transferred to the appropriate medium (Inglis et al. 2012).

Of the specific methods for isolating entomopathogenic fungi from soil samples, insect bait is the most used (Vega et al. 2012). This method was initially developed to obtain entomopathogenic nematodes from soil, and was later adapted as a method for the isolation of entomopathogenic fungi (Zimmermann 1986, 2007). The use of insect baits can be considered a selective method, since it selects certain species of entomopathogenic fungi (Vega et al. 2012) and appears to be more sensitive than plating on media (Keller et al. 2003). Insect baiting by the larva of the wax moth *Galleria mellonella* Linnaeus (*Lepidoptera: Pyralidae*) has been used to isolate a wide spectrum of entomopathogenic fungi present in the soil (Keller et al. 2003; Sharma et al. 2018). The use of different baits may result in obtaining different groups of entomopathogens (Goble et al. 2010).

Entomopathogenic fungi can also be obtained from plant tissues such as leaves, stems, roots, or fruits. For isolation, it is necessary that the plant tissue has been disinfected by immersion in 1 to 5% sodium hypochlorite and/or alcohol (70%), followed by sequential rinses in sterile distilled water. It is preferable to use intact seedlings or excised whole leaves, as long as the petiole is not immersed in the disinfestation solutions (Ownley et al. 2008). After disinfestation, the plant tissues are aseptically cut into pieces and seeded in suitable agar media. Ten to 14 days after isolation the leaf discs are examined visually to check the presence of endophytic fungi, which must grow on the inner plant tissue of superficially sterilized leaf discs (Humber 1997). It is important to note that disinfectant solutions may not guarantee 100% disinfestation since polar disinfectants are less effective in plant tissues with large amounts of cuticular waxes. Thus, the use of two or more disinfectants combined with a surfactant can be useful (Inglis et al. 2012).

The isolation of entomopathogenic fungi from aquatic environments is not so common. There is no isolation technique specifically described for obtaining aquatic entomopathogenic fungi. The most commonly observed examples in the literature are works in which fungi were obtained from insects of aquatic environments in their adult phases, such as some of the genera *Culex*, *Aedes,* and *Anopheles*. The reports of fungal infection observed in these studies are mostly from Entomophthora species (Weiser and Batko 1966; Eilenberg 2000; Pell et al. 2001; Scholte et al. 2004; Evans et al. 2018).

All the isolation techniques described have advantages and disadvantages. Thus, for selecting an isolation method, it is important to note the type of study to be carried out and the characteristics of the environment from which the fungi will be isolated.

2.6 Selection and Production

As entomopathogenic fungi comprise the most widely used group of biocontrol agents worldwide, improving their production techniques is crucial to increasing productivity and reducing costs while maintaining fungal viability and virulence and consequently improving their overall effectiveness in pest management (Santoro et al. 2005; Jaronski and Mascarin 2017). Furthermore, prior selection of strains with pest control potential has been very important for the success of biological control programs. As fungal entomopathogens have a diversity of hosts and present significant interspecific genetic variability, they are quite conducive to this process. To this end, research has been carried out with several species to obtain more accurate information about their effectiveness, both in their action on the target pest and in propagule production (Lacey et al. 2015; Jaronski and Mascarin 2017).

2.6.1 Selecting Fungi for Pest Control

The selection of an entomopathogenic fungus is the most important step in choosing a bioinsecticide. The process involves the determination of fungal virulence, assessment of reproductive aspects, and production in an artificial medium (Ambethgar 2009; Lopes et al. 2011). Many studies have been conducted to select virulent strains for pest control. For example, the potential of 27 entomopathogenic strains in controlling *Diaphorina citri* Kuwayama was investigated by selecting *B. bassiana* HIB-24, and *C. fumosorosea* HIB-19 isolates with mortality rates of 60.66% and 62.02%, respectively (Gandarilla-Pacheco et al. 2013). In addition, *in vitro* bioassays have been used to evaluate the pathogenicity of 32 and 18 strains of *B. bassiana* and *M. anisopliae*, respectively, on the weed borer *Hedypathes betulinus* (Klug), and the isolates that caused the highest mortality selected to form the basis for virulence, bioassays, and conidial production on different substrates. Strains of *B. bassiana* (UNIOESTE 4, UNIOESTE 52, and UNIOESTE 64) and *M. anisopliae* (IBCB 352) caused confirmed mortality rates equal to or greater than 90% (Fanti and Alves 2013). *C. fumosorosea* CG1228 was selected against nymphs and adults of the whitefly *Bemisia tabaci* (Genn.), which achieved high conidia production in the cadavers and other artificial substrates (Mascarin et al. 2013).

One way to determine the entomopathogenic potential of a fungus is by inoculating suspensions containing the minimum dose or concentration of conidia in the insect, and then assessing the time taken for the development of infection and death. The ability of these fungi to produce large amounts of conidia on the corpses of their hosts and on artificial substrates is a good indicator. In order to select strains of *Cordyceps* against the arboreal termite *Coptotermes gestroi* (Wasmann), *C. javanica* URM 4993 was found to be more virulent to the workers causing 100% mortality after the sixth day of inoculation, and showed high production of conidia on the dead insects (Lopes et al. 2011). Experiments carried out on the production of conidia of

M. anisopliae (Metsch.) Sorok. and *B. bassiana* (Bals.) Vuill on *Coptotermes formosanus* Shiraki showed a significant increase in sporulation after the 11th day of the insect's death (Sun et al. 2002).

Although the selection of virulent strains is based primarily on results of pathogenicity tests, other types of assays can be used for this purpose, such as those used to measure virulence indirectly via the activities of enzymes related to infection pathways. Examples of enzymes that can be useful in this regard are enzymes similar to subtilisin (Pr1), trypsin (Pr2) and chitinases (Jackson et al. 2010; Svedese et al. 2013).

In addition to assessing virulence as a measure of the potential efficacy of fungal entomopathogens, the selection process must include the evaluation of the stability of propagules produced by the fungal isolates, as well as their potential to be economically mass-produced. Moreover, determining acceptable environmental and toxicological profiles is also important, as is the persistence and effectiveness in typical environmental and ecological conditions for the target insect (Jaronski and Mascarin 2017).

2.6.2 *Types of Production for Entomopathogenic Fungi*

The application of fungi in research or on a commercial scale in the control of arthropods requires large amounts of propagules for the infection and colonization of the host and subsequent dispersion of these fungi in the arthropod's habitat. Therefore, artificial media production methods are tested continuously to optimize production and control efficacy (Alves and Pereira 1998; Mascarin and Quintela 2013). Entomopathogenic fungi can be applied in the form of suspensions of conidia, mycelium or blastospores, or using fungal mycotoxins, although the latter option is not considered biological control (Srivastava et al. 2009). Production conditions, moisture content, and temperature for the conservation and storage of conidia can influence these pathogens' viability and virulence (Faria et al. 2009; Blanford et al. 2012).

The production of entomopathogens can be done in host insects (*in vivo*) or in artificial culture media (*in vitro*) through simple or complex fermentation processes. The main advantage of artificial production is its ability to deliver a large amount of fungal inoculum in a short period. However, finding an ideal and inexpensive medium for large-scale production can be a challenge, especially since this medium needs to preserve the pathogens' characteristics, including pathogenicity and virulence. As most entomopathogenic fungi form a mycelium, blastospores, or conidia, there are three methods for their production in artificial media: (1) liquid media or submerged fermentation, solid or semi-solid media; (2) biphasic fermentation, in which liquid or semi-liquid media are used to generate mycelial growth; and (3) solid media used for the production of conidia (Alves and Pereira 1998; Mascarin and Quintela 2013).

The production of infectious and stable inoculants using less expensive artificial substrates has become an essential requirement for production on an industrial scale. This can be done using fermentation methods involving solid or submerged liquid media (Jackson et al. 2010). In addition to the substrate's composition, the storage temperature can influence the viability of the conidia, and consequently, the virulence of the entomopathogenic fungi. Hence, in tropical countries, fungal biopesticides' usefulness is low, due to high ambient temperatures (Faria et al. 2010; Mascarin et al. 2019). There are also reports that substrates with ideal proportions of carbohydrates, nitrogen, and calcium, such as corn and whey are efficient in producing viable conidia for fungal infection (Rangel et al. 2008; Kim et al. 2010). However, previous analyzes of the type of technique used in the production of propagules for large-scale use in pest control are relevant, aiming at the non-waste of material in the production of inoculants and the success of the action of fungi on the target pest. The most common techniques for conidial production of entomopathogenic fungi are described below.

2.6.2.1 In Vitro Production

Continuous fermentation in semi-solid or liquid media produces mainly mycelial mass and, in some specific conditions, blastospores. The choice of the method and substrate is decisive to obtain a high amount of biomass in small fixed spaces and in a short period. However, there is difficulty in obtaining adequate culture media that do not affect virulence, the definition of development conditions, sporulation, and absence of secondary contamination (Alves and Pereira 1998; Ferreira 2004).

The solid substrate fermentation is the method most used for obtaining inoculating propagules, especially aerial conidia because they are more resistant to abiotic factors found in the field. As a consequence most small- to medium-sized enterprises and large multinational companies use this form of production (Jaronski and Mascarin 2017; Muñiz-Paredes et al. 2017). The fungus is inoculated directly into the solid medium, and after growth and sporulation it is separated from the substrate by filtration or, in some cases, the residue from the medium can be used together with the fungal formulation. The production of fungi in solid media is most commonly carried out in polypropylene bags, or in glass bottles or in trays (Alves and Pereira 1998; Alves and Faria 2010; Freitas et al. 2014).

Production using polypropylene bags is a practical and simple method of producing fungi on solid substrates. Generally, parboiled rice is used. The production steps, according to Aquino (1974), are as follows: (1) produce the fungus in solid medium or bottles of serum or Gatorade containing 100 g of autoclaved rice at 120 °C for 20 min, in order to obtain fungal suspensions; (2) add 300 g of rice to polypropylene bags (35 cm × 22 cm) resistant to sterilization, close the bags with twofolds, staple them and autoclave at 120 °C for 30 min; (3) in a vertical flow hood, use the bags with the substrate already chilled for fungal inoculation. Then, stir the bags well to homogenize the inoculum on the rice and keep them in a germination and growth room at 26 °C, exposed to16-h photophase; (4) after 5 days, select the cases with uniform growth, eliminate the contaminated ones and transport the selected bags to the sporulation room at 27 °C and 16-h photophase; (5) after 12 to 16 days of production, discard the contaminated bags and transfer the contents of the selected bags to plastic buckets that will be transferred to vibrating screens to separate the conidia.

In this process, the production cycle lasts from 12 to 16 days, depending on the fungus and the production conditions. To avoid contamination of the bags aseptic conditions and the inoculum's purity should be observed. Studies report that this method of rice sporulation is efficient for most entomopathogenic fungi. The research evaluated the viability of conidia and the insecticidal action of *B. bassiana* ERL836 at different storage temperatures and growing substrates. It showed that the strain, grown on rice stored at low (4 °C) and moderate (25 and 30 °C) temperatures, conserved conidial viability (germination percentage $> 85\%$), and virulence on the host (Kim et al. 2019). It was also demonstrated that *B. bassiana* UNIOESTE produced the largest amount of spores, presenting 1.6×10^8 conidia/g in rice, and on the corpses of *Spodoptera frugiperda* (Smith & Abbott), with an average production of 5.7×10^7 conidia/cadaver. This proved superior to the other strains tested (Thomazoni et al. 2014).

An alternative method of production uses Roux bottles containing 100 g of autoclaved rice at 120 °C for 20 min to obtain a fungal formulation. Conidia production follows the steps described by Alves and Pereira (1998); Aquino (1974); and Aquino et al. (1975): (1) gathering of fungal material for inoculation is carried out according to the protocol used for fungal production in polypropylene bags; (2) add 100 g of raw rice with 30 mL of distilled water to the bottles, close with aluminum foil and string, and then sterilize the material in an autoclave at 120 °C for 30 min; (3) in a vertical flow hood, with the aid of a syringe, inoculate 10 ml of the fungus suspension (6×10^7 conidia/mL), taking care to close the orifice used for inoculation; (4) keep the bottles in a germination room at 27 °C, exposed to 16-h photophase. After 3 days, select the bottles with uniform growth, dispose of the contaminated bottles. After a period of 12 to 15 days, discard the contaminated bottles and transfer the contents of the selected bottles to vibrating screens for the separation of the conidia.

Rice provides nutrients necessary for the growth and reproduction of fungi, and the conidia can be easily separated from the substrate and, in most cases, present good virulence to pests (Alves et al. 2008). Observations on the influence of the substrate on the production of conidia and on the virulence of *B. bassiana* have shown that the conidia of *B. bassiana* (CG 71 and CG 152) produced in rice and synthetic medium were more virulent than those produced in insects. This suggests that there is a difference in virulence between conidia produced in different media. Therefore it is recommended that selection tests are carried out on the same substrate that will be used in fungal production on a large scale (Santoro et al. 2007).

Another alternative for fungus production in a solid medium is tray production. This is carried out in two stages, one aiming at optimizing mycelial growth and the other, sporulation. The culture medium is maintained in different environments in accordance with the different requirements of the fungal development stages. The growth phase is carried out in polypropylene bags, and then the material is deposited in trays inside aseptic rooms. Production takes place in the following four stages (Alves and Pereira 1998): (1) obtaining the fungal material for inoculation is carried out according to the protocol described earlier to produce fungi in polypropylene bags; (2) in a vertical flow hood use polypropylene bags with 300 to 1000 g of sterilized and cooled rice for fungal inoculation. Then keep the bags in a germination room at 21 to 26 °C, depending on the fungus in question, and at 16-h photophase; (3) after the mycelial formation (3–5 days) select the bags without contamination and transfer the material to trays, maintaining a temperature of 21 to 28 °C and a photophase of 16-h; (4) after the sporulation process (12–15 days), the material is dried and processed in a vibrating sieve to separate the conidia.

The techniques used for fungal inoculum production must be of low cost and produce a high concentration of viable and virulent fungal propagules (Loureiro et al. 2005). Research on the production and viability of different strains of *M. anisopliae* in rice has shown that *M. anisopliae* IBCB 425 isolate produced a greater amount of conidia in pre-cooked rice by the tray method and also showed a greater germination capacity of the conidia, reaching 94.84% (Freitas et al. 2014).

For biphasic production, a mycelial mass is obtained in liquid medium and then inoculated on a solid medium to obtain the conidia. The nutritional source used in the preparation of the liquid medium influences the production of conidia in the biphasic process. Media with nitrogen sources provide greater production (between 100 to 1000 times) of conidia when the fungi are inoculated on rice (carbohydrate source) and speed up production time (Santoro et al. 2005). Research on the production of *C. fumosorosea* and *C. farinosa* in biphasic fermentation has been carried out successfully using agro-industrial products and waste as substrate (Mascarin et al. 2010).

These observations indicate that the variation in the production of viable propagules in liquid fermentation and the production of conidia in fermentation in the solid state depends on the species of fungus and the nutritional composition of the culture medium used.

2.6.2.2 In Vivo Production

Some fungi fail to develop in artificial environments; for example, *Nomuraea* Moublanc are produced on their hosts. *In vivo* production can cause the formation of structures that are not easily obtained in artificial media, such as resistance spores. This technique is often used to disperse fungi in the field. You can use the primary host (the one that must be controlled) or an alternative host (used in mass production) due to the ease of monitoring its development in the laboratory (Alves and Pereira 1998).

Contamination of the breeding of the hosts must be avoided, so that part of the breeding should be used to reproduce the hosts and another to carry out the multiplication of the entomopathogens. An example is the *in vivo* production of high quality and virulent conidia obtained from *Nomuraea rileyi* (Farlow) Samson using caterpillars from *Anticarsia gemmatalis* Hübner (Alves and Pereira 1998). Ten isolates of different fungal pathogens were tested on the larval stages of *Musca domestica* L. *M. anisopliae.* CG46 and CG30 were the most virulent, causing up to 60% mortality. *M. anisopliae* CG46 was selected for assay in the field and for producing more conidia on the larval cadavers (Fernandes et al. 2013). *M. anisopliae* IBCB was efficient on *H. betulinus* Klug, causing higher confirmed mortality and conidiogenesis in insect corpses, with high inoculum production (Fanti and Alves 2013).

2.6.2.3 Quality Control and Care with Fungi Produced

After selecting the most virulent and easy to produce entomopathogenic fungal isolates, the productivity of the strains in solid substrates, resistance to climatic factors, the electrophoretic pattern, and the viability of conidia are determined, in addition to the action of fungi on the natural pests enemies. The steps required to verify the quality of the fungal production are described below, according to Alves and Pereira (1998) and Ferreira (2004).

Once the strain selection is complete, the behavior of the strains in the production process must be tested. This is of paramount importance to determining the final cost of the product, since the selected strain must be highly pathogenic to the target pests and have suitable qualities for production. It is also important to maintain the characteristics of the strain during the production stages, and the growth and sporulation of fungi must be compared with pre-established standards. The maintenance of healthy colonies is relevant to mass production success since the elimination of unviable inoculants avoids the loss of material and time in the production process.

During production the quality of the materials used such as culture medium, distilled water, polypropylene bags must be monitored, as well as variables such as temperature of sterilization and growth of fungi, efficiency of the filtration systems, refrigeration. If an error occurs, it must be remedied before compromising the production of the fungus. Different lots must be analyzed for the quality of the fungus, and tests of viability and potency of the inoculum must be carried out to accompany the storage of the product so that its quality does not decrease before being used in the field.

2.6.3 Viability of Stored Conidia

Conidia of entomopathogenic fungi are essential in the infection of hosts. The success in their mass production depends on the use of entomopathogens with preserved characteristics in storage, which favors rapid germination and versatility in developing substrates in solid-state (Muñiz-Paredes et al. 2017). The viability of entomopathogenic conidia is an indicator used by manufacturers to analyze the quality of formulated and unformulated biopesticides. The fungus reactivation technique on the target insects is used to activate these stored pathogens' infection potential on artificial substrates in the Culture Collections before carrying out the pest control bioassays. Concomitantly, fungi are isolated from the insect's cadavers, and the viability of these conidia is tested according to the germination percentage of the fungus. Studies have shown that *Cordyceps* strains show germination greater than 95% after re-isolation in workers of *C. gestroi*, demonstrating the viability and high germination capacity of pathogens (Passos et al. 2014). Studies followed one another, so that *Cordyceps* strains showed germination percentages greater than 90% on the workers of *Nasutitermes corniger* (Motschulsky), after 16 h of inoculation, consuming the ability to germinate conidia and viability after re-isolation (Lopes et al. 2017).

The analysis of fungal viability is done by directly counting germinated and non-germinated conidia under a light microscope. After inoculation of conidia suspension in Petri dishes containing solid medium and incubated for 16 to 24 h in BOD, with pre-set temperature established. The following formula is then used to determine the germination percentage: $G = n \times 100/500$ (G = germination percentage, and n: number of germinated spores) (Goettel and Inglis 1997; Alves and Pereira 1998). Germination capacity is an essential factor in microbial control since it is one of the first and most decisive stages to establish infection (Alves 1998; Svedese et al. 2013; Santos et al. 2016). Therefore, assessing the conidial germination rate before producing and applying entomopathogenic fungi in insect management is particularly important.

2.7 Current Status of Entomopathogenic Fungi as Biopesticides

The use of entomopathogenic fungi to control pests began in the late 1800s. Since then, a significant number of biopesticides have been developed worldwide for controlling insects and mites in agricultural, urban, forest, livestock, and aquatic environments (de Faria and Wraight 2007). Although the market for biopesticides is considered small (approximately 5 to 6% of the global pesticide market) compared to the conventional synthetic pesticides market, there is increasing demand and commercial applications for biopesticides. On average, the production of these products grows around 10% to 20% per year (Marrone 2019). According to Olson (2015), based on projections for the future of both biopesticides and synthetic pesticides it is estimated that at some point the synthetic crop protection market will peak while the biopesticide market continues to grow, equalizing with synthetics between the late 2040s and the early 2050s.

Several factors have contributed to the increased demand for biopesticides, including the need for sustainable and environmentally friendly control alternatives and the consumer demand for health and wellness. Other advantages of biopesticides include the possibility of combining them with synthetic products and obtaining a synergistic control action, reducing the need for synthetic pesticides, delaying the evolution of resistant insects, and reducing the re-entry intervals following applications. In addition, the risk of biopesticides for non-target organisms is lower; and there are applications for biocontrol that are not possible with synthetic crop protection (Olson 2015; Marrone 2019).

To serve this demand for biopesticides, the number of microbial biopesticides, especially mycoinsecticides has dramatically increased (Jaronski and Mascarin 2017; Mascarin et al. 2019). These products are generally developed by smaller and less financially robust companies. However, realizing the potential of this market, large agribusiness and major chemical companies have started to invest in the discovery and licensing of biopesticides (Marrone 2014). Some of these companies have invested hundreds of millions of dollars in the biopesticide area, bringing confidence in the sector and introducing well-established technologies from other industries (Olson 2015). Another attraction of this market is the safety and low-risk of biopesticides, which enable faster development times and more favorable regulatory processes in some countries (Marrone 2019).

Europe, Latin America and North America account for the highest percentage of global biopesticide sales. Most mycoinsecticides are produced in America, Europe, and Asia, with South American and North American companies and institutions developing most commercial products. Africa and Oceania are the smallest producers of biopesticides (de Faria and Wraight 2007; Maina et al. 2018; Marrone 2019). However, several products are available for use in many African countries, which in the future may achieve levels of mycoinsecticides use similar to those observed in countries of Europe and America (Maina et al. 2018).

Although it is estimated that between 20,000 and 50,000 species of insect-associated fungi exist, they are for the most part poorly known (Blackwell 2011). The fraction of species exploited for biological control is relatively small, and most commercial mycoinsecticides available worldwide are based on a modest number of species. Most of these species belong to the phylum Ascomycota (Goettel et al. 2010). They are more readily cultured *in vitro* than some entomopathogenic species of more basal fungi groups, e.g., Entomophthoralean fungi (Charnley and Collins 2007). The genera *Beauveria*, *Metarhizium*, *Cordyceps*, and *Akanthomyces* (formerly known as *Lecanicillium*) comprise the most commonly employed species as the active ingredients in commercial formulations, such as *M. anisopliae*, *Cordyceps fumosorosea*, *B. bassiana* and *A. lecanii* (Table 2.1). Products based on these species are available in many countries on different continents (de Faria and Wraight 2007; Kabaluk et al. 2010; Lacey et al. 2015; Silva et al. 2015).

Other fungal species have also been developed as commercial products for pest control. For example, *Neoconidiobolus thromboides* (=*Conidiobolus thromboides* and *Entomophthora virulenta*) are employed as the active ingredients in products for whitefly control in Colombia (Goettel et al. 2010) and aphids in China. For controlling Hemiptera insects, *Aschersonia aleyrodis* and *Sporothrix insectorum* have been developed in Russia (Lacey et al. 2015) and Brazil (Kabaluk et al. 2010),

Table 2.1 Commercial formulations of entomopathogenic fungi available in different countries for the control of insects and mites

Entomopathogenic fungi	Biopesticide (commercial name)	Formulation	Country of manufacture	Pest controlled
Akanthomyces muscarius (= *Lecanicillium muscarium*)	Mycotal®	Water dispersible granule	Holland	Thrips and whitefly larvae
Akanthomyces lecanii (= *Lecanicillium lecanii* and *Verticillium lecanii*)	Dr. Bacto's Vertigo	Liquid	India	Whitefly, aphids, thrips and mealy bug and other sucking pest
	LECATECH® WP	Wettable powder	Kenya	Whiteflies and other soft bodied insects such as thrips
	Varunastra™	Liquid	India	Aphids, thrips, mealy bugs, whitefly, jassid, scales and all types of mites
	Shock™	Wettable powder	India	Aphids, thrips, mealy bugs, white fly, scale insects and mites
	Verticon	Liquid	India	Whiteflies, thrips, aphids and mealy bugs
	Mealikil®	Wettable powder	India	Mealy bugs, thrips, jassids, aphids, whiteflies and mites
	Sun Bio Vetri	Liquid	India	Aphids, jassids, whitefly, leaf hoppers and mealy bugs
Beauveria bassiana	Ostrinil®	Microgranule	France	*Paysandisia archon*
	AgroNova® WG	Water-dispersible granule	Colombia	*Hypothenemus hampei*
	Boveril® WP	Wettable powder	Brazil	Two-spotted mite
	Bouveriz Biocontrol®	Wettable powder	Brazil	Whitefly, banana weevil, two-spotted mite and leafhopper
	Beauveria Oligos	Wettable powder	Brazil	Whitefly, thrips, banana weevil, two-spotted mite, leafmining fly and corn leafhopper
	BioCeres® WP	Wettable powder	Canada	Whiteflies, aphids, rice root aphids and thrips
	BotaniGard® ES	Emulsifiable suspension	United States	Eucalyptus weevil, coffee berry borer and whitefly

(continued)

Table 2.1 (continued)

Entomopathogenic fungi	Biopesticide (commercial name)	Formulation	Country of manufacture	Pest controlled
	Aprehend™	Oil	United States	Bed bugs
	Mycotrol® ESO	Emulsifiable suspension	United States	Aphids, thrips, white-flies, psyllids, mealy-bugs, leafhoppers and weevils
	BIO-WONDER™	Wettable powder	India	Hairy insects, aphids, White flies, mealy bugs, grasshoppers, Thrips, stem borer, ter-mites, Bettles, caterpil-lars, etc.
	Myco-Jaal®	Emulsified suspension	India	Diamond-Back, aphids and mites
	Bio power	Liquid and powder	India	Borers, cutworms, root grubs, leafhoppers, whitefly, aphids, thrips and mealybug
	Racer™	Wettable powder	India	Whitefly, thrips, aphids, psyllids, mealybugs, plant bugs, weevils and Lepidoptera
	Naturalis®-L	Oil dispersion	United Kingdom	Whitefly and thrips
Beauveria brongniartii	Bas-Eco	Water soluble powder	India	*Helicoverpa armigera*, berry borer and root grubs
Cordyceps fumosorosea (= Paecilomyces fumosoroseus and Isaria fumosorosea)	PFR 97 20% WDG®	Water dis-persible granules	United States	Psyllids, leafhoppers, and aphids
	PreFeRal® WG	Water dis-persible granule	Belgium	Whiteflies (egg, larva, pupa and adult)
	NoFly ™ WP	Wettable powder	United States	Whiteflies, aphids and thrips
	Ancora®	Water dis-persible granules	United States	Whiteflies, aphids, thrips, spider mites, leafminers, citrus leafminers, mealybugs, psyllids, black vine weevil, other root wee-vils, thrips, pupae, grape phylloxera, rootworms,

(continued)

Table 2.1 (continued)

Entomopathogenic fungi	Biopesticide (commercial name)	Formulation	Country of manufacture	Pest controlled
				wireworms, grubs and *Lepidoptera* caterpillars
	Challenger	Concentrated suspension	Brazil	*Psilideo* and *Helicoverpa*
	OCTANE®	Concentrated suspension	Brazil	Corn leafhopper
	Pae-Sin® WP	Wettable powder	Mexico	Whiteflies' (*Bemisia* spp.) eggs, nymphs, and adults
Hirsutella thompsonii	NO-MITE	Aqueous suspension and wettable powder	India	Various mites in cereals and apple maggot flies
	Almite	Liquid	India	Red, pink, yellow mites
	Premium HT™	Water soluble powder	India	Mites
Metarhizium anisopliae	Meta-Sin®	Liquid suspension	Chile	Weevils
	Metarril® WP	Wettable powder	Brazil	Leafhopper complex and mites
	Vantage®	Wettable powder	Brazil	Spittlebugs
	METIÊ®	Wettable powder	Brazil	Spittlebugs and brown root stink bug
	Gr-inn	Wettable powder	Brazil	Spittlebugs
	Metarhyd FR25®	Emulsifiable gel	Brazil	Spittlebugs
	BioMet	Wettable powder and emulsion in water	Guatemala	Bed bug and migratory locusts
	Met52™ EC	Emulsifiable concentrate	Canada	Whiteflies, thrips, chinch bug and ticks
	Ory-X	Powder	Malaysia	Rhinoceros beetles
	Green Meta	Liquid	India	Root weevils, planthopers, japanese beetle, black vine weevil, spittlebug, termites and white grubs
	Pacer®	Powder	India	Termites, root grubs, locusts, root weevils, ants, beetles and caterpillar pests

(continued)

Table 2.1 (continued)

Entomopathogenic fungi	Biopesticide (commercial name)	Formulation	Country of manufacture	Pest controlled
	ABTEC	Powder	India	Root grubs, beetles, weevils, bugs and termites
Metarhizium anisopliae var. *acridum*	Green Guard® SC	Oil suspension	Australia	Locust nymphs, wing-less grasshoppers and other pest grasshoppers
	Green Guard® ULV	Oil suspension	Australia	Locust nymphs and wingless grasshoppers
	Green Muscle ™	Oil suspension	South Africa	Locust nymphs, wing-less grasshoppers
Metarhizium brunneum	ATTRACAP®	Granule	Germany	Wireworms

respectively. *N. rileyi* has been designed to control Lepidoptera insects in Columbia and India (Lacey et al. 2015).

Until 2007, more than 80 companies worldwide have developed 171 mycoacaricides and mycoinsecticides, based on at least 12 species of fungi. Targets of these products include insects of the orders Siphonaptera, Thysanoptera, Lepidoptera, Orthoptera, Hymenoptera, Coleoptera, Isoptera, Hemiptera and Diptera (de Faria and Wraight 2007). Since then, some products have lost share in the international market, but others have been introduced, mainly in Latin American America, China and the Indian subcontinent (Kabaluk et al. 2010; Lacey et al. 2015; Jaronski and Mascarin 2017). For example, according to Jaronski and Mascarin (2017), more than a dozen new mycopesticides have been developed in Ecuador alone. The data suggest that an increasing number of commercial products are expected to be made available on the market worldwide in the coming decades, with more fungal species being explored.

Over time, biopesticides have improved in terms of performance and cost. Investment in science to improve various stages of developing a product, including obtaining more stable formulations, is making an outstanding contribution to this. The entry of large pesticide manufacturers increased notoriety for the category of biopesticides (Marrone 2019). The institutional support offered by countries can be facilitating the development and adoption of microbial pesticides, as the regulatory system of these products influences the development and the international reach of the mycoinsecticide market. Jurisdictions of the countries have established frameworks for the protection from harm while offering effective products to control pests in a promising market (Kabaluk et al. 2010).

2.8 Prospects and Conclusions

Agriculture is one of the largest economic sectors globally, and much of its growth is due to the use of fertilizers, pest management, better variety of crops and disease control. Management of pests is crucial for obtaining a healthy and high crop yield (Thakur et al. 2020) because crop damage caused by insects is responsible for significant losses in agricultural productivity (Culliney 2014). Synthetic chemical pesticides are the most widely applied for controlling insects. However, their extensive use has increased insects' resistance and led to the collapse of pest control in some countries (Naqqash et al. 2016). It also eliminates natural enemies and leads to contamination of water bodies, groundwater, and soil. This results in negative impacts on environmental and human health, due to recalcitrance, toxicity and carcinogenic potential of many of these compounds (Baron et al. 2019).

These factors have driven the search for potent and environmentally friendly control agents (Olson 2015; Hyde et al. 2019). Entomopathogenic fungi effectively reduce or eradicate insect populations, are safer for the environment and non-target organisms and more specific for target species. They are biodegradable and suitable for the integrated pest management programs (Altschul et al. 1990; Kumar and Singh 2015). Thus, these organisms are attracting global attention due to their potential to manage pest populations as a sustainable and eco-friendly alternative to chemical pesticides (Damalas and Koutroubas 2018; Maina et al. 2018).

Despite the great potential of insect-associated fungi, they are poorly known considering their global diversity (Blackwell 2011). Entomopathogenic fungi have been found in many fungal taxonomic groups (Charnley and Collins 2007; Mora et al. 2017), but most commercially produced fungi used for controlling insect pests belong to a restricted number of species (Vega et al. 2009; Kabaluk et al. 2010). Therefore, there is a need for studies that expand the number of species applicable in pest management, particularly ones that evaluate and improve their performance in the field under different critical factors that can impact their potential and viability. The success of biological control depends on the competence of the fungal isolates to function and persist in the environment of the target pest (Lacey et al. 2015).

Several technological advances can expand the use of fungi in the management of pests (Baron et al. 2019; Hyde et al. 2019). Entomopathogenic fungi can be applied directly or in formulations developed to enhance their shelf life and environmental stability (Jaronski and Mascarin 2017). Research and development efforts have led to an increase in mycoinsecticides (de Faria and Wraight 2007). Advances in knowledge about epizootiology, host defence, mechanisms of pathogenesis, genetic and physiological engineering, formulation, mass production and application of fungi suggest an optimistic future for the mycoinsecticide market (Charnley and Collins 2007). However, despite these advancements leading to an increase in mycoinsecticides' effectiveness, further progress with registration, quality control, and economic issues are needed to maximize their widespread adoption (Kumar and Singh 2015; Mascarin et al. 2019; Hyde et al. 2019).

Therefore, in addition to investing in research to improve these biological agents' performance in the control of insects in the field, it is also important to enhance developing countries' capability to manufacture and use biopesticides. Furthermore, it is essential to develop public policies that ensure crucial regulatory mechanisms to maintain the quality of the products at affordable cost in developing countries (Kumar and Singh 2015). Mainly because the cost, complexity, and inconsistency of registration between the different countries can hamper the development and use of biopesticides (Charnley and Collins 2007). As biopesticides are still largely regulated by the system designed for chemical pesticides, burdensome costs are imposed on the biopesticides industry, creating market entry barriers. Policy measures need to help to create a biopesticide profile and hence reduce barriers to promote the use of sustainable alternatives (Kumar and Singh 2015).

Efforts should be made to enhance the popularity of the entomopathogenic fungi as pest controllers because gaps in understanding their basic biological functioning contribute to this technology being little exploited compared to conventional means of control (Hyde et al. 2019). The lack of knowledge regarding the use, application, and handling of biopesticides can lead to inconsistent results. Hence, the users lose confidence in adopting these eco-friendly alternatives (Mishra et al. 2015). On the other hand, when users understand the modes of action, specific timing, methods, and environmental conditions for application, integration, and monitoring of entomopathogenic fungi, the control effectiveness tends to be enhanced (Marrone 2014; Maina et al. 2018). Thus, it is imperative to raise awareness among the users regarding the use, efficacy, and benefits of biopesticides, with guidance, explanation, and monitoring done regularly. This includes agronomists, farmers, sellers, crop consultants, and researchers supported by government and corporate houses and can be achieved by introducing extension activities, teaching programs, workshops, etc. (Mishra et al. 2015).

Although progress on several aspects is required for fungi to be effectively adopted in pest management and widespread worldwide, mycoinsecticides' future is promising. An increase in efforts worldwide for the development, commercialization, and use of biopesticides has been observed. Large agribusiness companies have been moving into the sector, ringing greater financial investment and well-established technologies from other industries that will increase the production capacity of biocontrol products. As the sector of mycoinsecticides requires the interaction of different areas, e.g. chemistry, biology, agronomy, physiology, it is ripe for partnering (Olson 2015). This characteristic can result in more robust and consolidated progress.

In addition, awareness of the benefits and the need for more sustainable and environmentally friendly production is increasing among consumers, and the demand for health and more conscious consumption. The use of fungi in pest control plays a significant role in this regard. As fungi are incorporated into pest management programs, they are recognized for their qualities, such as efficiency, shorter worker re-entry, residue and resistance management, reduced risk for non-target organisms and the environment (Marrone 2019). Entomopathogenic fungi are promising and strategic for the necessary establishment of sustainable and economically viable agricultural production, and therefore it should receive the internationally deserved consideration.

References

Alexopoulos CJ, Mims CW, Blackwell M (1996) Introductory mycology, 4th edn. Wiley, New York, 870 p

Ali S, Huang Z, Ren S (2010) Production of cuticle degrading enzymes by *Isaria fumosorosea* and their evaluation as a biocontrol agent against diamondback moth. J Pestic Sci 83:361–370

Altinok HH, Altinok MA, Koca AS (2019) Modes of action of Entomopathogenic Fungi. Curr Trends Nat Sci 8:117–124

Altre JA, Vandenberg JD (2001) Penetration of cuticle and proliferation in hemolymph by *Paecilomyces fumosoroseus* isolates that differ in virulence against lepidopteran larvae. J Invertebr Pathol 78:81–86. https://doi.org/10.1006/jipa.2001.5046

Altschul SF, Gish W, Miller W, Myers EW, Lipman DJ (1990) Basic local alignment search tool. J Mol Biol 215:403–410

Alves SB (1998) Fungos entomopatogênicos. In: Alves SB (ed) Controle microbiano dos insetos, 2nd edn. FEALQ, Piracicaba, pp 289–381

Alves RT, Faria M (2010) Pequeno manual sobre fungos entmopatogênicos. Documentos 286. Embrapa Cerrado, Brazil, pp 1–50. https://ainfo.cnptia.embrapa.br/digital/bitstream/item/77782/1/doc-286.pdf

Alves SB, Pereira RM (1998) Produção de fungos entomopatogênicos. In: Alves SB (ed) Controle Microbiano de Insetos, 2nd edn. FEALQ, Piracicaba, pp 845–869

Alves SB, Leite LG, Batista Filho A, Almeida JEM, Marques EJ (2008) Produção massal de fungos entomopatogênicos na América Latina. In: Alves SB, Lopes RB (eds) Controle Microbiano de Pragas na América Latina. FEALQ, Piracicaba, pp 215–237

Ambethgar V (2009) Potential of entomopathogenic fungi in insecticide resistance management (IRM) a review. J Biopest 2:177–193

Aquino MLN (1974) O fungo entomopatógeno *Metarhizium anisopliae* (Metsch.) SoroKin, no Estado de Pernambuco, vol 72. Boletim Técnico do Instituto de Pesquisa Agronômicas, Recife, pp 1–26

Aquino MLN, Cavalcanti VALB, Sena RC, Queiroz GF (1975) Nova tecnologia de multiplicação do fungo *Metarhizium anisopliae*. Boletim Técnico da CODECAP 4:1–31

Araújo JPM, Hughes DP (2016) Chapter one. Diversity of entomopathogenic fungi: which groups conquered the insect body? In: Lovett B, Leger RJS (eds) Genetics and molecular biology of entomopathogenic fungi, Advances in genetics, vol 94, pp 1–39

Baron NC, Rigobelo EC, Zied DC (2019) Filamentous fungi in biological control: current status and future perspectives. Chil J Agric Res 79:307–315

Barr DJS (2001) Chytridiomycota. In: Systematics and evolution. Springer, Berlin, Heidelberg, pp 93–112

Barr DJS, Désaulniers NL, Knox JS (1987) *Catenochytridium hemicysti* N. Sp.: morphology, physiology and zoospore ultrastructure. Mycologia 79(4):587–594. https://doi.org/10.1080/00275514.1987.12025428

Behie SW, Bidochka MJ (2014) Ubiquity of insect-derived nitrogen transfer to plants by endophytic insect-pathogenic fungi: an additional branch of the soil nitrogen cycle. Appl Environ Microbiol 80:1553–1560. https://doi.org/10.1128/AEM.03338-13

Beilharz VC, Beilharz DG, Parbery DG, Swart HJ (1982) Dodine: a selective agent for certain soil fungi. Trans Br Mycol Soc 79:507–511

Benny GL, Humber RA, Voigt K (2014) 8 Zygomycetous Fungi: phylum Entomophthoromycota and subphyla Kickxellomycotina, Mortierellomycotina, Mucoromycotina, and

Zoopagomycotina. In: McLaughlin D, Spatafora J (eds) Systematics and evolution. The Mycota (a comprehensive treatise on Fungi as experimental Systems for Basic and Applied Research), vol 7A. Springer, Berlin, Heidelberg. https://doi.org/10.1007/978-3-642-55318-9_8

Bhattacharyya A, Samal AC, Kar S (2004) Entomophagous fungus in pest management. News Lett 5:1–4

Biswas T, Joop G, Rafaluk-Mohr C (2018) Cross-resistance: a consequence of bi-partite host-parasite coevolution. Insects 9:28

Blackwell M (2011) The Fungi: 1, 2, 3… 5.1 million species? Am J Bot 98(3):426–438. https://doi.org/10.3732/ajb.1000298

Blanford S, Jenkins NE, Christian R et al (2012) Storage and persistence of a candidate fungal biopesticide for use against adult malaria vectors. Malar J 11:354

Boomsma J, Jensen A, Meyling N, Eilenberg J (2014) Evolutionary interaction networks of insect pathogenic fungi. Environ Entomol 59:467–485

Boucias DR, Pendland JC (1998) Entomopathogenic fungi; Fungi imperfecti. In: Boucias DR, Pendland JC (eds) Principles of insect pathology. Kluwer Academic, Dordrecht, pp 321–359

Browne N, Heelan M, Kavanagh K (2013) An analysis of the structural and functional similarities of insect hemocytes and mammalian phagocytes. Virulence 4(7):597–603

Buchon N, Silverman N, Cherry S (2014) Immunity in *Drosophila melanogaster*—from microbial recognition to whole-organism physiology. Nat Rev Immunol 14(12):796–810

Butt TM (2002) Use of entomogenous fungi for the control of insect pests. In: Agricultural applications. Springer, Berlin, pp 111–134

Castro T, Mayerhofer J, Enkerli J, Eilenberg J et al (2016) Persistence of Brazilian isolates of the entomopathogenic fungi *Metarhizium anisopliae* and *M. robertsii* in strawberry crop soil after soil drench application. Agric Ecosyst Environ 233:361–369

Chandler D (2017) Basic and applied research on entomopathogenic fungi. In: Lacey LA (ed) Microbial control of insect and mite pests. Academic, San Diego, CA, pp 69–89

Charnley AK, Collins SA (2007) 10 Entomopathogenic fungi and their role in pest control. In: Kubicek CP, Druzhinina IS (eds) Environmental and microbial relationships, 2nd edn. Springer, Berlin, Heidelberg, pp 159–187

Culliney T (2014) Crop losses to arthropods. In: Pimentel D, Peshin R (eds) Integrated pest management. Springer, Dordrecht, pp 201–225

Da Silva P, Jouvensal L, Lamberty M, Bulet P, Caille A, Vovelle F (2003) Solution structure of termicin, an antimicrobial peptide from the termite *Pseudacanthotermes spiniger*. Protein Sci 12 (3):438–446

Da Silva RA, Quintela ED, Mascarin GM, Pedrini N, Liao LM, Ferri PH (2015) Unveiling chemical defense in the rice stalk stink bug against the entomopathogenic fungus *Metarhizium anisopliae*. J Invertebr Pathol 127:93–100

Damalas CA, Koutroubas SD (2018) Current status and recent developments in biopesticide use. Agriculture 8:13. https://doi.org/10.3390/agriculture8010013

De Faria MR, Wraight SP (2007) Mycoinsecticides and mycoacaricides: a comprehensive list with worldwide coverage and international classification of formulation type. Biol Control 43:237–256

Delgado PAM, Murcia OP (2011) Hongos entomopatógenos: uma alternativa para la obtención de Biopesticidas. Ambi-Agua 6:77–90

Dewel RA, Joines JD, Bond JJ (1985) A new chytridiomycete parasitizing the tardigrade *Milnesium tardigradum*. Can J Bot 63(9):1525–1534. https://doi.org/10.1139/b85-211

Doberski JW, Tribe HT (1980) Isolation of entomogenous fungi from elm bark and soil with reference to ecology of *Beauveria bassiana* and *Metarhizium anisopliae*. Trans Br Mycol Soc 74:95–100. https://doi.org/10.1016/S0007-1536(80)80013-1

Dong Y, Taylor HE, Dimopoulos G (2006) AgDscam, a hypervariable immunoglobulin domain-containing receptor of the *Anopheles gambiae* innate immune system. PLoS Biol 4(7):e229

Donzelli BGG, Krasnoff SB (2016) Molecular genetics of secondary chemistry in *Metarhizium* Fungi. In: Lovett B, Leger RJS (eds) Advances in genetics, vol 94. Elsevier, Amsterdam, pp 365–436

Dromph KM (2001) Dispersal of entomopathogenic fungi by collembolans. Soil Biol Biochem 33 (15):2047–2051. https://doi.org/10.1016/s0038-0717(01)00130-4

Eilenberg J (2000) Entomophthorales on Diptera. In: Papp L, Darvas B (eds) Manual of Palaearctic Diptera. Science Herald, Budapest, pp 521–533

Evans HC (1988) Coevolution of entomogenous fungi and their insect hosts. In: Pirozynski KA, Hawksworth DL (eds) Coevolution of fungi with plants and animals. Academic, Nueva York, EEUU, pp 149–171

Evans CJ, Banerjee U (2003) Transcriptional regulation of hematopoiesis in *Drosophila*. Blood Cells Mol Dis 30(2):223–228

Evans HC, Elliot SL, Barreto RW (2018) Entomopathogenic fungi and their potential for the management of *Aedes aegypti* (Diptera: Culicidae) in the Americas. Mem Inst Oswaldo Cruz 113(3):206–214. https://doi.org/10.1590/0074-02760170369

Falagas ME, Grammatikos AP, Michalopoulos A (2008) Potential of oldgeneration antibiotics to address current need for new antibiotics. Expert Rev Anti Infect Ther 6:593–600

Fan Y, Liu X, Keyhani NO et al (2017) Regulatory cascade and biological activity of *Beauveria bassiana* oosporein that limits bacterial growth after host death. Proc Natl Acad Sci U S A 114: E1578–E1586. https://doi.org/10.1073/pnas.1616543114

Fanti ALP, Alves LFA (2013) Entomopathogenic fungi isolates to control the borer of yerba mate (Hedypathes betulinus) Klug (Coleoptera; Cerambycidae). Semina Cienc Agrar 34 (4):1467–1478. https://doi.org/10.5433/1679-0359.2013v34n4p1467

Faria M, Hajek AE, Wraight SP (2009) Imbibitional damage in conidia of the entomopathogenic fungi *Beauveria bassiana*, *Metarhizium acridum*, and *Metarhizium anisopliae*. Biol Control 51 (3):346–354. https://doi.org/10.1016/j.biocontrol.2009.06.012

Faria M, Hotchkiss JH, Hajek AE, Wraight SP (2010) Debilitation in conidia of the entomopathogenic fungi *Beauveria bassiana* and *Metarhizium anisopliae* and implication with respect to viability determinations and mycopesticide quality assessments. J Invertebr Pathol 105:74–83. https://doi.org/10.1016/j.jip.2010.05.011

Fehlbaum P, Bulet P, Michaut L, Lagueux M, Broekaert WF, Hetru C, Hoffmann JA (1994) Insect immunity. Septic injury of *Drosophila* induces the synthesis of a potent antifungal peptide with sequence homology to plant antifungal peptides. J Biol Chem 269(52):33159–33163

Feng P, Shang Y, Cen K, Wang C (2015) Fungal biosynthesis of the bibenzoquinone oosporein to evade insect immunity. Proc Natl Acad Sci 112(36):11365–11370

Fernandes ÉK, Keyser CA, Rangel DE, Foster RN, Roberts DW (2010) CTC medium: a novel dodine-free selective medium for isolating entomopathogenic fungi, especially *Metarhizium acridum*, from soil. Biol Control 54(3):197–205. https://doi.org/10.1016/j.biocontrol.2010.05.009

Fernandes EG, Valério HM, Borges MAS, Mascarin GM, Silva CE, Van Der Sand ST (2013) Selection of fungi for the control of *Musca domestica* in aviaries. Biocontrol Sci Technol 23 (11):1256–1266. https://doi.org/10.1080/09583157.2013.826343

Ferreira JMS (2004) Protocolo para a produção massal de fungos entomopatogênicos 1: *Beauveria bassiana* (Vuill.). In: Boletim de Pesquisa e Desenvolvimento, vol 3. Embrapa Tabuleiros Costeiros, Aracaju, pp 1–28

Freitas AF, Loureiro ES, Almeida MEB, Pessoa LGA (2014) Yield of conidia and germination of different isolates of *Metarhizium anisopliae* (Metsch) Sorok. (Ascomycota: Clavicipitaceae) grown on rice. Arq Inst Biol 81:75–78. https://doi.org/10.1590/S1808-16572014000100014

Gandarilla-Pacheco FL, Galán-Wong LJ, López-Arroyo JI, Rodríguez-Guerra R, Quintero-Zapata I (2013) Optimization of pathogenicity tests for selection of native isolates of entomopathogenic fungi isolated from *Citrus* growing areas of México on adults of *Diaphorina citri* Kuwayama (Hemiptera: Liviidae). Fla Entomol 96:187–195. https://doi.org/10.1653/024.096.0125

Gleason FH, Lilje O (2009) Structure and function of fungal zoospores: ecological implications. Fungal Ecol 2:53–59. https://doi.org/10.1016/j.funeco.2008.12.002

Gleason FH, Marano AV, Johson P, Martin WW (2010) Blastocladian parasites of invertebrates. Fungal Biol Rev 24:56–67

Goble TA, Dames JF, Hill MP, Moore SD (2010) The effects of farming system, habitat type and bait type on the isolation of entomopathogenic fungi from citrus soils in the eastern Cape Province, South Africa. BioControl 55:399–412. https://doi.org/10.1007/s10526-009-9259-0

Goettel MS, Inglis GD (1997) Fungi: hyphomycetes. In: Lacey LA (ed) Manual of techniques in insect pathology. Academic, New York, NY, pp 213–249. https://doi.org/10.1016/B978-012432555-5/50013-

Goettel MS, Ellenberg J, Glare T (2010) 11 Entomopathogenic fungi and their role in regulation of insect populations. In: Gilbert LI, Gill SS (eds) Insect control: biological and synthetic agents. Elsevier, London, pp 387–437

Greenfield BPJ, Lord AM, Dudley E, ButtTM (2014) Conidia of the insect pathogenic fungus, *Metarhizium anisopliae*, fail to adhere to mosquito larval cuticle. R Soc Open Sci 1:140193. https://doi.org/10.1098/rsos.140193

Gryganskyi AP, Humber RA, Smith ME, Hodge K, Huang B, Voigt K, Vilgalys R (2013) Phylogenetic lineages in Entomophthoromycota. Persoonia 30:94–105. https://doi.org/10.3767/003158513X666330

Gul HT, Saeed S, Khan FA (2014) Entomopathogenic fungi as effective insect pest management tactic: a review. Appl Sci Bus Econ 1:10–18

Haelewaters D, Boer P, Noordijk J (2015) Studies of Laboulbeniales (Fungi, Ascomycota) on Myrmica ants: *Rickia wasmannii* in the Netherlands. J Hymenopt Res 44(June 11):39–47. https://doi.org/10.3897/jhr.44.4951

Hajek AE, Delalibera I (2010) Fungal pathogens as classical biological control agents against arthropods. BioControl 55:147–158

He M, Zhao RL, Hyde KD et al (2019) Notes, outline and divergence times of Basidiomycota. Fungal Divers 99:105–367. https://doi.org/10.1007/s13225-019-00435-4

Hibbett D, Binder M, Bischoff JF, Blackwell M, Cannon PF, Eriksson OE, Huhndorf S, James TY, Kirk PM et al (2007) A higher-level phylogenetic classification of the Fungi. Mycol Res 111:509–547

Hultmark D (2003) *Drosophila immunity*: paths and patterns. Curr Opin Immunol 15(1):12–19

Humber RA (1997) Fungi-Identification. In: Lacey L (ed) Manual of techniques in insect pathology. Academic, London, pp 153–185

Humber RA (2008) Evolution of entomopathogenicity in fungi. J Invertebr Pathol 98:262–266

Humber RA (2012a) Entomophthoromycota: a new phylum and reclassification for entomophthoroid fungi. Mycotaxon 120:477–492

Humber RA (2012b) Identification of entomopathogenic fungi. In: Lacey LA (ed) Manual of techniques in invertebrate pathology. Academic, London, pp 151–187

Hyde KD, Xu J, Rapior S, Jeewon R, Lumyong S, Niego AGT, Chaiyasen A (2019) The amazing potential of fungi: 50 ways we can exploit fungi industrially. Fungal Divers 97:1–136

Inglis GD, Enkerli JUERG, Goettel MS (2012) Laboratory techniques used for entomopathogenic fungi: Hypocreales. In: Manual of techniques in invertebrate pathology, vol 2, pp 18–53

Jackson MA, Dunlap CA, Jaronski ST (2010) Ecological considerations in producing and formulating fungal entomopathogens for use in insect biocontrol. BioControl 55:129–145. https://doi.org/10.1007/978-90-481-3966-8_10

James TY, Porter TM, Martin WW (2014) 7 Blastocladiomycota. In: McLaughlin D, Spatafora J (eds) Systematics and evolution. The Mycota (a comprehensive treatise on Fungi as experimental Systems for Basic and Applied Research), vol 7A. Springer, Berlin, Heidelberg. https://doi.org/10.1007/978-3-642-55318-9_7

Jaronski ST, Mascarin GM (2017) Mass production of fungal entomopathogens. In: Lacey LA (ed) Microbial control of insects and mite pests. Academic, Amsterdam, pp 141–155

Jerônimo GH, Jesus AL, Marano AV, James TY, Souza JI, Rocha SCO, Pires-Zottarelli CLA (2015) Diversidade de Blastocladiomycota e Chytridiomycota do Parque Estadual da Ilha do Cardoso, Cananéia, SP, Brasil. Hoehnea 42:135–163. https://doi.org/10.1590/2236-8906-32/2014

Kabaluk JT, Svircev AM, Goettel MS, Woo SG (2010) The use and regulation of microbial pesticides in representative jurisdictions worldwide. IOBC Global, Rome, pp 1–99

Karling JS (1981) In: Cramer J (ed) Predominantly holocarpic and eucarpic simple biflagellate Phycomycetes, 2nd edn, 252 p

Keller S, Kessler P, Schweizer C (2003) Distribution of insect pathogenic soil fungi in Switzerland with special reference to *Beauveria brongniartii* and *Metharhizium anisopliae*. BioControl 48:307–319

Kendrick B (2017) The fifth kingdom: an introduction to mycology, 4th edn. Hackett Publishing, Indianapolis/Cambridge

Kepler RM, Luangsa-ard JJ, Hywel-Jones NL et al (2017) A phylogenetically-based nomenclature for Cordycipitaceae (Hypocreales). Ima Fungus 8:335–353. https://doi.org/10.5598/imafungus.2017.08.02.08

Kim JS, Skinner M, Parker B (2010) Influence of whey permeate and millet as substrates on thermotolerance of *Beauveria bassiana* and *Metarhizium anisopliae* conidia during storage. Biocontrol Sci Tech 20(8):859–863. https://doi.org/10.1080/09583157.2010.486072

Kim JC, Lee MR, Kim S, Lee SJ, Park SE, Baek S, Gasmi L, Shin TY, Kim JS (2019) Long-term storage stability of *Beauveria bassiana* ERL836 granules as fungal biopesticide. J Asia Pac Entomol 22:537–542. https://doi.org/10.1016/j.aspen.2019.04.001

Kirk PM, Cannon PF, Minter DW, Stalpers JA (2008) Dictionary of fungi, 10th edn. CABI, Wallingford

Kocks C, Cho JH, Nehme N, Ulvila J, Pearson AM, Meister M, Strom C, Conto SL, Hetru C, Stuart LM (2005) Eater, a transmembrane protein mediating phagocytosis of bacterial pathogens in *Drosophila*. Cell 123(2):335–346

Kumar S, Singh A (2015) Biopesticides: present status and the future prospects. J Fertil Pestic 6: e129. https://doi.org/10.4172/jbfbp.1000e129

Lacey L, Grzywacz D, Shapiro-Ilan D, Frutos R, Brownbridge M, Goettel M (2015) Insect pathogens as biological control agents: back to the future. J Invertebr Pathol 132:1–41

Lamberty M, Ades S, Uttenweiler-Joseph S, Brookhart G, Bushey D, Hoffmann JA, Bulet P (1999) Insect immunity isolation from the lepidopteran *Heliothis virescens* of a novel insect defensin with potent antifungal activity. J Biol Chem 274:9320–9326

Lanot R, Zachary D, Holder F, Meister M (2001) Postembryonic hematopoiesis in *Drosophila*. Dev Biol 230:243–257

Lemaitre B, Hoffmann J (2007) The host defense of *Drosophila melanogaster*. Annu Rev Immunol 25:697–743

Liu ZY, Milner RJ, McRae CF, Lutton GG (1993) The use of dodine in selective media for the isolation of *Metarhizium* spp. from soil. J Invertebr Pathol 62:248–251

Lopes RS, Svedese VM, Portela APAS, Albuquerque AC, Luna-Alves Lima EA (2011) Virulence and biological aspects of *Isaria javanica* (Frieder & Bally) Samson & Hywell-Jones in *Coptotermes gestroi* (Wasmann) (Isoptera: Rhinotermitidae). Arq Inst Biol 78:565–572

Lopes RS, Lima G, Correia MTS, Costa AF, Luna-Alves Lima EA, Lima VLM (2017) The potential of *Isaria* spp. as a bioinsecticide for the biological control of *Nasutitermes corniger*. Biocontrol Sci Tech 27(9):1038–1048. https://doi.org/10.1080/09583157.2017.1380163

Lopes RS, Martins MCB, Oliveira LG et al (2020) Termiticidal activity of Libidibia ferrea var. ferrea and of the association with *Isaria* spp. against *Nasutitermes corniger*. J Agric Sci 12:159–170. https://doi.org/10.5539/jas.v12n1p159

Loureiro ES, Batista Filho A, Almeida JEM, Pessoa LGA (2005) Produção de isolados de Metarhizium anisopliae, selecionados para o controle de Mahanarva fimbriolata (Stal, 1854). Arq Inst Biol 72:469–472

Lu H-L, Leger RS (2016) Insect immunity to entomopathogenic fungi. In: Advances in genetics, vol 94. Elsevier, Amsterdam, pp 251–285

Lu H-L, Wang JB, Brown MA, Euerle C, Leger RJS (2015) Identification of *Drosophila mutants* affecting defense to an entomopathogenic fungus. Sci Rep 5:12350

Luz C, Bastos Netto MC, Rocha LFN (2007) In vitro susceptibility to fungicides by invertebrate-pathogenic and saprobic fungi. Mycopathologia 164:39–47

Maina UM, Galadima IB, Gambo FM, Zakaria D (2018) A review on the use of entomopathogenic fungi in the management of insect pests of field crops. J Entomol Zool Stud 6:27–32

Mannino MC, Huarte-Bonnet C, Davyt-Colo B, Pedrini N (2019) Is the insect cuticle the only entry gate for fungal infection? Insights into alternative modes of action of entomopathogenic fungi. J Fungi 5:33

Marrone PG (2014) The market and potential for biopesticides. In: Gross AD, Coats JR, Seiber JN, Duke SO (eds) Biopesticides: state of the art and future opportunities. American Chemical Society, Washington, DC, pp 245–258

Marrone PG (2019) Pesticidal natural products–status and future potential. Pest Manag Sci 75:2325–2340

Martin WW (1978) Two additional species of *Catenaria* (Chytridiomycetes, Blastocladiales) parasitic in midge eggs. Mycologia 70:461–467

Mascarin GM, Jaronski ST (2016) The production and uses of *Beauveria bassiana* as a microbial insecticide. World J Microbiol Biotechnol 32:1–26

Mascarin GM, Quintela ED (2013) Técnica de Produção do Fungo Entomopatogênico *Metarhizium anisopliae* para Uso em Controle Biológico. Documentos, vol 289. Embrapa Arroz e Feijão, pp 1–17

Mascarin GM, Alves SB, Lopes RB (2010) Culture media selection for mass production of *Isaria fumosorosea* and *Isaria farinosa*. Braz Arch Biol Technol 53(4):753–761. https://doi.org/10.1590/S1516-89132010000400002

Mascarin GM, Kobori NN, Quintela ED, Delalibera I Jr (2013) The virulence of entomopathogenic fungi against *Bemisia tabaci* biotype B (Hemiptera: Aleyrodidae) and their conidial production using solid substrate fermentation. Biol Control 66(3):209–218. https://doi.org/10.1016/j.biocontrol.2013.05.001

Mascarin GM, Lopes RB, Delalibera Í, Fernandes ÉKK, Luz C, Faria M (2019) Current status and perspectives of fungal entomopathogens used for microbial control of arthropod pests in Brazil. J Invertebr Pathol 165:46–53. https://doi.org/10.1016/j.jip.2018.01.001

Matsuura K, Tanaka C, Nishida T (2000) Symbiosis of a termite and a sclerotium-forming fungus: Sclerotia mimic termite eggs. Ecol Res 15:405–414

Meyling NV (2007) Methods for isolation of entomopathogenic fungi from the soil environment-laboratory manual. University of Copenhagen, Denmark, p 18. http://orgprints.org/11200

Mishra J, Tewari S, Singh S, Arora NK (2015) Biopesticides: where we stand? In: Arora NK (ed) Plant microbes symbiosis: applied facets. Springer, New Delhi, pp 37–75

Mora MAE, Castilho AMC, Fraga ME (2017) Classification and infection mechanism of entomopathogenic fungi. Arq Inst Biol 84:1–10

Muñiz-Paredes F, Miranda Hernández F, Loera O (2017) Production of conidia by entomopathogenic fungi: from inoculants to final quality tests. World J Microbiol Biotechnol 33(57):1–9. https://doi.org/10.1007/s11274-017-2229-2

Naqqash MN, Gokçe A, Bakhsh A, Salim M (2016) Insecticide resistance and its molecular basis in urban insect pests. Parasitol Res 115:1363–1373

Naranjo-Ortiz MA, Gabaldón T (2019) Fungal evolution: diversity, taxonomy and phylogeny of the fungi. Biol Rev 94:2101–2137. https://doi.org/10.1111/brv.12550

Nehme NT, Liégeois S, Kele B, Giammarinaro P, Pradel E, Hoffmann JA, Ewbank JJ, Ferrandon D (2007) A model of bacterial intestinal infections in *Drosophila melanogaster*. PLoS Pathog 3 (11):e173

Olson S (2015) An analysis of the biopesticide market now and where it is going. Outlooks Pest Manag 26:203–206

Osta MA, Christophides GK, Vlachou D, Kafatos FC (2004) Innate immunity in the malaria vector *Anopheles gambiae*: comparative and functional genomics. J Exp Biol 207(15):2551–2563
Ownley BH, Griffin MR, Klingeman WE, Gwinn KD, Moulton JK, Pereira RM (2008) *Beauveria bassiana*: endophytic colonization and plant disease control. J Invertebr Pathol 98:267–270
Ozdal M, Incekara U, Polat A, Gur O, Tasar EBK, Erhan G (2012) Isolation of filamentous fungi associated with two common edible aquatic insects, *Hydrophilus piceus* and *Dytiscus marginalis*. J Microbiol Biotechnol Food Sci 9(5):95–105
Pal S, Leger RJS, Wu LP (2007) Fungal peptide Destruxin a plays a specific role in suppressing the innate immune response in *Drosophila melanogaster*. J Biol Chem 282(12):8969–8977
Passos EM, Albuquerque AC, Marques EJ, Teixeira VW, Silva CC, Oliveira MAP (2014) Effects of *Isaria* (Persoon) isolates on the subterranean termite *Coptotermes gestroi* (Wasmann) (Isoptera: Rhinotermitidae). Arq Inst Biol 81(3):232–237. https://doi.org/10.1590/1808-1657000642012
Pedrini N, Villaverde M, Fuse C, Dal Bello G, Juárez M (2010) *Beauveria bassiana* infection alters colony development and defensive secretions of the beetles *Tribolium castaneum* and *Ulomoides dermestoides* (Coleoptera: Tenebrionidae). J Econ Entomol 103:1094–1099
Pell JK, Eilenberg J, Hajek AE, Steinkraus DC (2001) Biology, ecology and pest management potential of Entomophthorales. In: Butt MT, Jackson C, Maga N (eds) Fungi as biocontrol agents. CAB International, Wallingford, pp 71–115
Pell JK, Hannam JJ, Steinkraus DC (2010) Conservation biological control using fungal entomopathogens. In: The ecology of fungal entomopathogens. Springer, Dordrecht, pp 187–198
Pestano F, Liverpool E, Gomathinayagam S, Ansari A (2017) Investigation of entomopathogenic fungi from insects at University of Guyana Turkeyen campus. Curr Trends Biomed Eng Biosci 5:001–005. https://doi.org/10.19080/CTBEB.2017.05.555656
Pringle A, Patek SN, Fischer M, Stolze J, Money NP (2005) The captured launch of a ballistospore. Mycologia 97(4):866–871. https://doi.org/10.1080/15572536.2006.11832777
Qu S, Wang S (2018) Interaction of entomopathogenic fungi with the host immune system. Dev Comp Immunol 83:96–103
Rafaluk-Mohr C, Wagner S, Joop G (2018) Cryptic changes in immune response and fitness in *Tribolium castaneum* as a consequence of coevolution with *Beauveria bassiana*. J Invertebr Pathol 152:1–7
Rangel DE, Alston DG, Roberts DW (2008) Effects of physical and nutritional stress conditions during mycelial growth on conidial germination speed, adhesion to host cuticle, and virulence of *Metarhizium anisopliae*, an entomopathogenic fungus. Mycol Res 112(11):1355–1361. https://doi.org/10.1016/j.mycres.2008.04.011
Rocha LFN, Luz C (2009) Utility of six fungicides for selective isolation of *Evlachovaea* spp. and *Tolypocladium cylindrosporum*. Mycopathologia 167:341–350
Rosales C (2005) Molecular mechanisms of phagocytosis. Eurekah.com and Springer Science+ Business Media, New York
Rosales C (2011) Phagocytosis, a cellular immune response in insects. Invertebr Surviv J 8 (1):109–131
Rossman AY (1978) *Podonectria*, a genus in the Pleosporales on scale insects. Mycotaxon 7:163–182
Samson RA, Evans HC, Latgé JP (1988) Atlas of entomopathogenic fungi. Springer-Verlag GmbH. KG, Berlin, Alemanha, 197 p
Sánchez-Pérez Andersen M, Magan N, Mead A, Chandler D (2006) Development of a population-based threshold model of conidial germination for analyzing the effects of physiological manipulation on the stress tolerance and infectivity of insect pathogenic fungi. Environ Microbiol 8:1625–1634
Sánchez-Pérez LC, Barranco-Florido JE, Rodríguez-Navarro S, Cervantes-Mayagoitia JF, Ramos-López MÁ (2014) Enzymes of entomopathogenic fungi, advances and insights. Adv Enzyme Res 2:65–76. https://doi.org/10.4236/aer.2014.22007

Sandhu SS, Shukla H, Aharwal RP, Kumar S, Shukla S (2017) Efficacy of entomopathogenic fungi as green pesticides: current and future prospects. In: Panpatte DG et al (eds) Microorganisms for green revolution, microorganisms for sustainability, vol 6. Springer, Singapore, pp 327–349

Santoro PH, Neves PMOJ, Silva RZ, Akimi S, Janaína Zorzetti J (2005) *Beauveria bassiana* (Bals.) Vuill. Spores production in biphasic process utilizing different liquid media. Semina Ciênc Agrár 26(3):313–320. https://doi.org/10.5433/1679-0359.2005v26n3p313

Santoro PH, Neves PMOJ, Alexandre TM, Alves LFA (2007) Interferência da metodologia nos resultados de bioensaios de seleção de fungos entomopatogênicos para o controle de insetos. Pesq Agropec Bras 42(4):483–489. https://doi.org/10.1590/S0100-204X2007000400005

Santos ACS, Oliveira RLS, Costa AF, Tiago PV, Oliveira NT (2016) Controlling *Dactylopius opuntiae* with *Fusarium incarnatum–equiseti* species complex and extracts of *Ricinus communis* and *Poincianella pyramidalis*. J Pestic Sci 89:539–547

Schabel S (1976) Oral infection of *Hylobius pales* by *Metarhizium anisopliae*. J Invertebr Pathol 383:377–383

Scholte E-J, Knols BGJ, Samson RA, Takken W (2004) Entomopathogenic fungi for mosquito control: a review. J Insect Sci 4:1–24. https://doi.org/10.1093/jis/4.1.19

Schuhmann B, Seitz V, Vilcinskas A, Podsiadlowski L (2003) Cloning and expression of gallerimycin, an antifungal peptide expressed in immune response of greater wax moth larvae, *galleria mellonella*. Arch Insect Biochem Physiol 53(3):125–133

Shakeel M, Xu X, Xu J, Zhu X, Li S, Zhou X, Yu J, Xu X, Hu Q, Yu X (2017) Identification of immunity-related genes in *Plutella xylostella* in response to fungal peptide destruxin a: RNA-Seq and DGE analysis. Sci Rep 7(1):1–11

Shakeel M, Xu X, Xu J, Li S, Yu J, Zhou X, Xu X, Hu Q, Yu X, Jin F (2018) Genome-wide identification of Destruxin A-responsive immunity-related microRNAs in diamondback moth, *Plutella xylostella*. Front Immunol 9:185

Sharma L, Oliveira I, Torres L, Marques G (2018) Entomopathogenic fungi in Portuguese vineyards soils: suggesting a 'galleria-Tenebrio-bait method'as bait-insects *galleria* and *Tenebrio* significantly underestimate the respective recoveries of *Metarhizium* (*robertsii*) and *Beauveria* (*bassiana*). MycoKeys 38:1–23. https://mycokeys.pensoft.net/issue/1534/

Shearer CA, Descals E, Kohlmeyer B, Kohlmeyer J, Marvanová L, Padgett D, Porter D, Raja HA, Schmit JP, Thorton HA, Voglymayr H (2007) Fungal biodiversity in aquatic habitats. Biodivers Conserv 16:49–67

Shin TY, Choi JB, Bae SM, Cha YR, Oh JM, Koo HN, Woo SD (2009) Isolation and identification of entomopathogenic fungus from the pine wilt disease vector, *Monochamus alternatus* Hope (Coleoptera: Cerambycidae) in Korea. Int J Ind Entomol 18:125–129

Silva APAP, Alves RT, Lima EALA, Lima VLM (2015) Bioformulations in pest control—a review. Annu Res Rev Biol 5(6):535–543. https://doi.org/10.9734/ARRB/2015/12395

Silva APS, Cunha RX, Bezerra JDP, Brito TGS, Fonseca CSM, Santos IP, Lima VLM (2020) Secondary metabolites by endophytic fungi: potential for bio-based products development. In: Chaurasia PK, Bharati SL (eds) Research advances in the fungal world: culture, isolation, identification, classification, characterization, properties and kinetics, 1st edn. Nova Science, New York, pp 115–137. http://novapublishers.com/shop/research-advances-in-the-fungal-world-culture-isolation-identification-classification-characterization-properties-and-kinetics/

Skinner M, Parker BL, Kim JS (2014) Role of entomopathogenic fungi. In: Abrol DP (ed) Integrated pest management. Academic, Cambridge, pp 169–191

Sosa-Gomez DR, Lastra CCL, Humber RA (2010) An overview of arthropod-associated fungi from Argentina and Brazil. Mycopathologia 170:61–76

Sparrow FK (1960) Aquatic Phycomycetes, 2nd rev edn. University of Michigan Press, Ann Arbor, MI

Srivastava CN, Maurya P, Preeti Sharma P, Mohan LM (2009) Prospective role of insecticides of fungal origin: review. Entomol Res 39:341–355. https://doi.org/10.1111/j.1748-5967.2009.00244.x

Strand MR (2008) The insect cellular immune response. Insect Sci 15(1):1–14

Sun J, Fuxa JR, Henderson G (2002) Sporulation of *Metarhizium anisopliae* and *Beauveria bassiana* on *Coptotermes formosanus* and in vitro. J Invertebr Pathol 81:78–85. https://doi.org/10.1016/S0022-2011(02)00152-0

Sung G-H, Hywel-Jones NL, Sung J-M, Luangsa-Ard JJ, Srestha B, Spatafora JW (2007) Phylogenetic classification of *Cordyceps* and the clavicipitaceous fungi. Stud Mycol 7:55–59

Svedese VM, Tiago PV, Bezerra JDP, Paiva LM, Luna-Alves Lima EA, Porto ALF (2013) Pathogenicity of *Beauveria bassiana* and production of cuticle-degrading enzymes in the presence of *Diatraea saccharalis* cuticle. Afr J Biotechnol 12(46):6491–6497. https://doi.org/10.5897/AJB2013.11972

Tanzini M, Alves S, Setten A, Augusto N (2001) Compatibilidad de agent estensoactivos com *Beauveria bassiana* y *Metarhizium anisopliae*. Manejo Integrado de Plagas 59:15–18

Tedersoo L, Sánchez-Ramírez S, Kõljalg U et al (2018) High-level classification of the fungi and a tool for evolutionary ecological analyses. Fungal Divers 90:135–159

Thakur N, Kaur S, Tomar P, Thakur S, Yadav AN (2020) Microbial biopesticides: current status and advancement for sustainable agriculture and environment. In: Rastegari AA, Yadav AN, Yadav N (eds) New and future developments in microbial biotechnology and bioengineering. Elsevier, Amsterdam, pp 243–282

Thomazoni D, Formentini MA, Alves LFA (2014) Patogenicity of entomopathogenic fungi to *Spodoptera frugiperda* (Smith) (Lepidoptera: Noctuidae). Arq Inst Biol 81:126–133. https://doi.org/10.1590/1808-1657001162012

Tretter ED, Johnson EM, Benny GL, Lichtwardt RW, Wang Y, Kandel P, Novak SJ, Smith JF, White MM (2014) An eight-gene molecular phylogeny of the Kickxellomycotina, including the first phylogenetic placement of Asellariales. Mycologia 106(5):912–935

Tzou P, Reichhart J-M, Lemaitre B (2002) Constitutive expression of a single antimicrobial peptide can restore wild-type resistance to infection in immunodeficient *Drosophila mutants*. Proc Natl Acad Sci 99(4):2152–2157

Vega FE, Goettel MS, Blackwell M, Chandler D, Jackson MA, Keller S, Pell JK (2009) Fungal entomopathogens: new insights on their ecology. Fungal Ecol 2(4):149–159. https://doi.org/10.1016/j.funeco.2009.05.001

Vega FE, Meyling NV, Luangsa-Ard JJ, Blackwell M (2012) Fungal entomopathogens. In: Vega F, Kaya HK (eds) Insect pathology, 2nd edn. Academic, San Diego, CA, pp 171–220

Vey A, Matha V, Dumas C (2002) Effects of the peptide mycotoxin destruxin E on insect haemocytes and on dynamics and efficiency of the multicellular immune reaction. J Invertebr Pathol 80(3):177–187

Wang C, Leger RJS (2006) A collagenous protective coat enables *Metarhizium anisopliae* to evade insect immune responses. Proc Natl Acad Sci 103(17):6647–6652

Wang C, Leger RJS (2007) The MAD1 adhesin of *Metarhizium anisopliae* links adhesion with blastospore production and virulence to insects, and the MAD2 adhesin enables attachment to plants. Eukaryot Cell 6:808–816. https://doi.org/10.1128/EC.00409-06

Warcup JH (1950) The soil-plate method for isolation of fungi from soil. Nature 166:117e118

Webster S, Mitchell PJ, Hampson NA, Dyson JI (1984) A short review of electrocrystallization and its applications to the lead-acid battery. Surf Technol 23:105–116. https://doi.org/10.1016/0376-4583(84)90117-1

Wei G, Lai Y, Wang G, Chen H, Li F, Wang S (2017) Insect pathogenic fungus interacts with the gut microbiota to accelerate mosquito mortality. Proc Natl Acad Sci U S A 114:5994–5999

Weiser J, Batko A (1966) A new parasite of *Culex pipiens* L., Entomophthora destruens sp. nov. (Phycomycetes, Entomophthoraceae). Folia Parasitol 13:144–149

Wijayawardene NN, Pawłowska J, Letcher PM et al (2018) Notes for genera: basal clades of Fungi (including Aphelidiomycota, Basidiobolomycota, Blastocladiomycota, Calcarisporiellomycota, Caulochytriomycota, Chytridiomycota, Entomophthoromycota, Glomeromycota, Kickxellomycota, Monoblepharomycota, Mortierellomycota, Mucoromycota, Neocallimastigomycota, Olpidiomycota, Rozellomycota and Zoopagomycota). Fungal Divers 92:43–129

Xu J, Xu X, Shakeel M, Li S, Wang S, Zhou X, Yu J, Xu X, Yu X, Jin F (2017) The entomopathogenic fungi *Isaria fumosorosea* plays a vital role in suppressing the immune system of *Plutella xylostella*: RNA-Seq and DGE analysis of immunity-related genes. Front Microbiol 8:1421

Zhao RL, Li GJ, Sánchez-Ramírez S, Stata M, Yang ZL, Wu G et al (2017) A six-gene phylogenetic overview of Basidiomycota and allied phyla with estimated divergence times of higher taxa and a phyloproteomics perspective. Fungal Divers 84:43–74. https://doi.org/10.1007/s13225-017-0381-5

Zimmermann G (1986) The galleria bait method for detection of entomopathogenic fungi in soil. J Appl Entomol 102:213–215

Zimmermann G (2007) Review on safety of the entomopathogenic fungi *Beauveria bassiana* and *Beauveria brongniartii*. Biocontrol Sci Tech 17(6):553–596

Chapter 3
Utilization of Entomopathogenic Bacteria for Modern Insect Pest Management

Sajjad Ali, Muhammad Anjum Aqueel, Muhammad Farhan Saeed, Qaiser Shakeel, Muhammad Raheel, and Muhammad Irfan Ullah

Abstract Biopesticides, using living microbial bodies and their bio-active composites against insects, are potential replacements for synthetic insecticides for safer and modern food production systems. Entomopathogenic bacteria (EPB) are important biological control agents of insect pests since the last century. Though bacterial species have been documented to be used against insects for developing symbiotic relationships, only a few of them are identified as entomopathogens. Most of these are members of the family Bacillaceae, Enterobacteriaceae, Pseudomonadaceae, Clostridiaceae, and Neisseriaceae. More than 100 bacterial species have been reported to infect various arthropods. *Bacillus thuringiensis* (*Bt*), *B. sphaericus*, *B. cereus,* and *B. popilliae* are the most appreciated microbial pest control agents. However, new bacterial species also need to be explored for their entomopathogenic role and materialized as new biopesticide products. The commercial biopesticides based on novel EPBs with improved genetic materials must be a part of future research for effective integrated pest management programs. This present chapter highlights the classification, infection, replication, transmission mechanisms, and important EPB in integrated pest management.

Keywords Entomopathogenic bacteria · EPB · *Bacillus* spp. · IPM · Biopesticides

S. Ali · M. A. Aqueel
Department of Entomology, Faculty of Agriculture and Environment, The Islamia University of Bahawalpur, Bahawalpur, Pakistan

M. F. Saeed
Department of Environmental Sciences, COMSATS University, Vehari, Pakistan

Q. Shakeel (✉) · M. Raheel
Department of Plant Pathology, Faculty of Agriculture and Environment, The Islamia University of Bahawalpur, Bahawalpur, Pakistan
e-mail: qaiser.shakeel@iub.edu.pk

M. I. Ullah
Department of Entomology, College of Agriculture, University of Sargodha, Sargodha, Pakistan

S. De Mandal et al. (eds.), *New and Future Development in Biopesticide Research: Biotechnological Exploration*, https://doi.org/10.1007/978-981-16-3989-0_3

3.1 Introduction

Biological insecticides or bioinsecticides are bio-based drugs that act with different action mechanisms to kill various insect pests. These are categorized into three major groups based on technical descriptions given by the US Environmental Protection Agency (EPA): (1) Natural biochemicals that operate under non-toxic process; (2) Entomopathogens of the microbial origin; and (3) Proteins originating from genetically modified plants introduced in plants (Kachhawa 2017).To date, over 3000 types of microorganisms have been reported to cause insect disorders leading to their mortality, and more than 100 bacteria were identified as insect pathogens, among which *Bacillus thuringiensis* Berliner (*Bt*) is regarded as the best microbial control agent (Koul 2011). Two species of bacteria, such as spore-forming *Bacillus thuringiensis* (Bizzarri et al. 2008; Porcar et al. 2008) and the non-spore-forming *Serratia entomophila* (Inglis and Lawrence 2001; O'Callaghan and Gerard 2005) have gained more attention as pest control agents. Other important EPB like *B. popilliae*, *Pseudomonas alcaligenes, P. aureofaciens, Clostridium bifermentans, Streptomyces avermitilis,* and *Saccharopolyspora spinosa* were also exploited as potential biocontrol agents.

Bt is most commonly used for the control of lepidopterans (*Spodoptera exempta, Cydia pomonella, Helicoverpa armigera,* etc.), Dipterans (*Anopheles albimanus, Culex obscures, Aedes aegypti,* etc.), Coleopterans (*Popillia japonica, Leptinotarsa decemlineatam, Tribolium confusum,* etc.), and Hymenopterans (*Megachile frontalis, Megastigmus spermotrophus, Xylocopa aruana*, etc.) (Bravo et al. 2007). *Bt* var. *israelensis* and *B. sphaericus* strains produce specific endotoxin, which is used worldwide to eradicate the mosquito larva, particularly in malaria and human lymphatic filariasis endemics. Moreover, black fly (*Simulium spp*) larva, which serves as a vector for the river blindness of man (onchocerciasis) in Africa's tropical regions, is also controlled by israelensis. More than 40 *Bt* products, like Larvect 50, Mosquito Dunks, Monterey *Bt*, are used to control caterpillars, beetles, and mosquitoes, which constitutes 1% of the overall insecticide business (Bizzarri et al. 2008). Beta-proteobacteria supports another group, including species with significant implications as biocontrol agents (BCA) against various insect pests. *Burkholderia rinojensis* strain has been used for insect control that works through ingestion and interaction with various insects and mites (Cordova et al. 2013). These insecticides' action is based on different metabolites, and the commercial product mostly focuses on heat-killed cells and fermentation media. Another commercially active beta-proteobacterium is a strain of *Chromobacterium subtsugae* whose metabolites show wide-spectrum insecticide activity against Lepidoptera, Hemiptera, Coleoptera, and Diptera insect species (Martin et al. 2007). Several *Streptomyces* species possess various insecticidal toxins, particularly macrocyclic lactone derivatives, which act mostly on insects' peripheral nerves in almost the same phylum. Similarly, *Saccharopolyspora spinosa* releases potent and wide-spectrum insecticidal toxins known as spinosin, which are natural and semi-synthetic compounds of economic importance (Kirst 2011).

3.2 Classification of Entomopathogenic Bacteria (EPB)

The details of EPB groups are shared below.

3.2.1 Family: Bacillaceae

3.2.1.1 Genus: *Bacillus*

This genus contains the catalase-positive bacteria that induce sporulation that protects them under unfavorable conditions. They can form oval-shaped endospores, which may be followed by parasporal bodies in some organisms. *Bacillus* contains the essential entomopathogenic species for insect control. Important species include *B. thuringiensis, B. weihenstephanensis, B. pseudomycoides,* and *B. mycoides*. The plasmids of *Bacillus* sp. contain genes that play a vital role in their pathogenicity. This leads to the development of several bio-insecticides using *Bacillus* sp. (Raymond et al. 2010a). Conversely, increased saprophytic development is preferred when insecticidal toxin-containing plasmids are lacking in the bacterium (Gohar et al. 2005).

3.2.1.1.1 *Bacillus thuringiensis* (*Bt*)

Several authors have reviewed *Bacillus thuringiensis* (*Bt*) and its toxins (Burges 2001; Sanchis 2011). The pathogen was previously derived from diseased silkworm *(Bombyx mori)* larvae by Shigetane Ishiwata in 1901 (Ishiwata 1901). Furthermore, studies of Ishiwata have shown that its toxicity is correlated with proteins in sporozoite cultures but not in plant cells (Aoki and Chigasaki 1916). Ernst Berliner extracted it from the Mediterranean flour moth's diseased larvae, formally described and named the bacterium in Thuringia, Germany (Berliner 1915). Berliner reported the existence of a spindle cell's body or crystalline participation that was later identified as protein and solubilized in an alkali solution (Hannay and Fitz-James 1955). The only phenotypic characteristic unique to *Bt* appears to be forming these structural proteins in spindle cells. This feature is used to differentiate this bacterium from many other *Bacillus* (Vilas-Boas et al. 2007). To identify and classify new *Bt* segregates, several biochemical, morphological, and antigenic strategies were used (Heimpel and Angus 1958). The existence of the Pasteur Institute serotyping infrastructure has contributed significantly to the widespread implementation of the H-serotyping method to identify *Bt* isolates (Dulmage et al. 1981; Lecadet et al. 1999). Identification of mutations based on the fundamental susceptibility to the different phages of bacterial strains and Ribosome RNA gene typing was also used to identify strains (Akhurst et al. 1997; Ackermann et al. 1995). Conservation of certain extragenic palindromic elements, DNA interbreeding analysis, and improved spontaneous polymorphism analysis with appropriate systems have been suggested to endorse H-serotyping, which remains the most widely accepted method for the identification of *Bt* isolates (Reyes-Ramırez and Ibarra 2005; Hansen et al. 1998). A maximum of 82 *Bt* serotypes or even more commonly utilized subspecies have been described in its most current version of an H-antigen category (Lecadet et al. 1999). This category contained 69 classes of receptors and 13 subsets. *Serotype mogi* extracted through leaf litter and displaying virulence to mosquito larvae (Roh et al. 2009). All known *Bt* isolates initially showed pathogenicity towards *Lepidoptera embryos* (pathotype A) (Goldberg and Margalit 1977). *Bt* strains have also been pathogenic to the Hemiptera, Hymenoptera, Isoptera, Orthoptera, Coleoptera, and nematodes (Quesada-Moraga et al. 2004; Lima et al. 2008; Garcia-Robles et al. 2001; de Castilhos-Fortes et al. 2002; Quesada-Moraga et al. 2004; Bottjer et al. 1985). The potential for distinguishing between pathotype-based *Bt* varieties has been reported to be highly threatened by the development of numerous poisons with different isolation characteristics and the growing number of recorded pathotypes. Consequently, many *Bt* serotypes produce strains that are pathogenic to species of various taxonomic groups. *Bt* toxins are highly selective, and even a specific *Bt* gene or crystal toxin, even within the genus, can only be effective against such a limited number of insect species. For example, Tolworthi (Bt) is exceptionally infectious to military worm embryos (*Spodoptera frugiperda*) and East Asian leafworm (*Spodoptera litura*) (Hernandez 1988; Amonkar et al. 1985).

3.2.1.1.2 *Bacillus cereus*

Most of the strains under *Bacillus cereus* associated with insects are saprophytic or symbiotic bacteria inhabiting the insect digestive system. They showed high genetic similarity to *Bt*; however, *B. cereus* does not produce a crystalline parasporal poison that restricts its toxicity to arthropod hosts. *B. cereus* isolates can induce natural or artificial diseases in different pest species such as scarab beetle larvae (Selvakumar et al. 2007); flour beetle *Tribolium castaneum* (Kumari and Neelgund 1985); spruce budworms *C. Fumiferana* (Strongman et al. 1997); *Anopheles* mosquito (Chatterjee et al. 2010); *Trichoplusiani* (Wai Nam et al. 1975); and *Glossina morsitans* (Kaaya and Darji 1989). The vegetative insecticide proteins (Vip) were initially known to be present in the *B. cereus* strain that causes pathogenesis to maize rootworms (*Diabrotica* spp.) (Warren et al. 1996). Vip in *B. cereus* poisons are similar to Clostridium's secondary toxins and consist of a subunit of a toxin bound to the target polymer toxin cell (Vip1 toxin) and a second toxin displaying patterns of actin-ADP-ribosylating activity (Vip2 toxin) to avoid actin polymerization and kill the host insect tissues (Han et al. 1999). The Vip1 and Vip 2 toxins are crucial for their toxicity. Recently, the Vip2 toxin receptor expression in plants as a non-active proenzyme obtained after ingestion by the rootworm helped develop transgenic crops to control pests (Jucovic et al. 2008). Heat tolerant *B. cereus* variants have been identified for cotton boll weevil (*Anthonomus grandis*). However, less effect has been observed for cotton leafworm (*Spodoptera littoralis*) or black bean aphid,

Aphis fabaee (Perchat et al. 2005; Yuan et al. 2007; Luxananil et al. 2001; Jucovic et al. 2008). Transgenic plants that transmit toxin Vip from *B. cereus* were created; however, such productions' protection would require careful consideration before commercialization.

3.2.1.1.3 *Lysinibacillus*

Lysinibacillus are strictly aerobic and include both a saprophytic and pathogenic bacteria group (Hu et al. 2008). *L. sphaericus* is heterogenous species identical to *B. sphaericus* (Ahmed et al. 2007). The distinctive feature of this bacterium is the development of the spherical spore swollen sporangium in a terminal location. Strains of *B. sphaericus* have been well established for dipteran infections. Out of 49 serotypes, 9 are mosquitocidal strains (H1, H2, H3, H5, H6, H9, H25, H26, and H48), are recognized by *sphaericus*. However, serotype is not a good indicator of mosquitocidal behavior or the production of virulence factors (Priest and Dewar 2000; Hu et al. 2008). Strain 2362 and C-41 are the most active ingredients in commercial products of *B. Sphaericus* (Park et al. 2010). In mosquito species, the resistance of *B. sphaericus* appears to be highly variable depending on the strain (Wraight et al. 1987). This is attributable to toxin development (Davidson et al. 1975). *Anopheles, Culex, Mansonia,* and *Aedes* show the regular declining order of sensitivity against *B. sphaericus*. The large difference in sensitivity to *Culex* and *Aedes* mosquitoes may be due to variations in the binding poison to the intestinal epithelium (Davidson 1989). Histological research has shown that bacteria are contained in the peritrophic matrix when consumed by the host, which might be associated with the toxicity development (Davidson et al. 1975). The toxin appears to attach carbohydrate residue to intestinal receptors entrance and growth of mosquito larvae within midgut cells (Davidson 1989; Oliveira et al. 2009). Pathogens colonize to enter the body after host death, the foliage cells multiply, and the cycle ends with magnifying spore formation around the area. Mortality is delayed at low bacterial levels; however, long-term impacts on population growth are detected (Davidson 1989).

3.2.2 Family: **Paenibacillaceae**

3.2.2.1 Genus: *Paenibacillus*

Milk disease was identified as a mere infection of *P. japonica*, a Japanese beetle that entered the US in the 1930s via hemolymph of scarab larvae (Dutky 1940). There are two types of infectious bacteria: *B. popilliae*, a distinctive parasporal body inside the sporangium, and *B. lentimorbus*, without the parasporal body. Many other types of bacteria-causing milk disease concentrate on a particular dwelling phase with even a spore and parasporal structure embedded in a thick sporangium, giving a footprint-like appearance to these cells until seen by an optical microscope. Strains can be isolated from the infected morphotype (Milner 1981) based on the nature and shape of a spindle cell structure, including these specific entities. Spores are ingested by *Scarabaeidae* members that feed upon plant roots as well as organic material. Spores grow during higher pH in enzyme-rich intestine scarab conditions (Jackson et al. 2004). Inside the hemolymph, vegetative cells grow, causing little or no toxaemia, allowing the larva to remain active. Infection with milk disease typically consists of foliage rods and spores. When infected, a larva never molds, there is almost no evidence of the melanization response to an infection, and larval mortality is mostly due to the lack of nutrients and fat stores in the body. Hemocoel is filled with nearly 20 billion spores at the end of the infection duration (Sharpe and Detroy 1979). These are introduced to the soil after the death of a contaminated larva. While in the ground, spores live for even a more extended period and live till new susceptible larvae accessed at the same site (Franken et al. 1996). This is also proof that strain-host compatibility is characterized by a bacteria's ability to cross an intestinal wall boundary before optimum haemolymph conditions are multiplied. Discrepancies of response to an infection can be due to problems in achieving apparent spore germination. Heat operation to excess nutrients (Stahly and Klein 1992) and pressure use with suspension in intestinal fluids (Krieger et al. 1996) help in germination. The resistance of vancomycin seems to be a common function for *B. popilliae* (Pettersson et al. 1999). Even so, the resistance to vancomycin to *P. popilliae* is shown to be contradictory as repeated studies have also shown that the resistance gene was not found throughout the American *P. popilliae* strains (Harrison et al. 2000). Therefore, neither the presence of paraspore nor vancomycin tolerance has been shown to become a decisive differentiating factor.

3.2.2.2 Genus: *Brevibacillus*

Brevibacillus laterosporus, a spore-producing pathogen distinguished by a unique lamellar parasporal structure were reported. Based on the 16S rRNA, such species were primarily classified under Brevibacillus (Shida et al. 1996). Its insecticidal activity was first used to target the *A. aegypti* and *Anopheles stephensi*. However, insecticidal activities have also been reported against Coleopteran and other insect species (de Oliveira et al. 2004). Several *B. laterosporus* variants have crystalline additions, which are released through the lyses of sporangium (Smirnova et al. 1996). Crystals containing variants 921 and 615 are toxic to insects. Toxic effects among these strains were associated with parasporal crystals (special 130 kDa protein) poisonous to *Aedes* larvae (Zubasheva et al. 2010). Genetic findings indicate a high similarity between isolates in this population, suggesting a small type of genetic polymorphism (de Oliveira et al. 2004; Ruiu et al. 2007).

3.2.3 *Family:* Enterobacteriaceae

3.2.3.1 Genus: *Serratia* spp.

The genus *Serratia* consists of ten species widespread in soil and water (Grimont and Grimont 2006). *S. marcescens* is known to colonize an extensive range of insects' digestive tract. It can produce a potent toxin that can kill insects with a lower lethal dose (Tan et al. 2006). Several insects are susceptible to this toxin, which enters via the oral pathway and acts against *Glossina* spp. (Flies tsetse) (Poinar et al. 1979); *Lucilia sericata* (blowfly) (O'Callaghan et al. 1996); and *Melolontha* spp. (May beetles) (Jackson et al. 2004). Another *S. Sntomophilia* or *Serratia proteamaculans* has been proven to be responsible for causing amber infection in New Zealand Grass grub larvae. *S. proteamaculans* and *S. entomophilia*. The LD_{50} for *S. entomophila* strain 154 was estimated to be 2–4 × 10^4 cells per larvae (Jackson et al. 2004). The inhalation of microbes has a significant impact mostly on the appearance of a diseased larva. After the intake of amber disease-causing bacteria, within 1–3 days, larvae of *C. zealandica* avoid eating; furthermore, the rates of stomach acid, trypsin, and chymotrypsin drop drastically throughout the intestinal tract (Jackson et al. 2001, 2004). During intake, *S. entomophilia* colonizes certain insects and attach to a cuticle of the foreskin (Jackson et al. 2001). The reduction in enzyme dilution in the intestine allowed transmission of serine protease enzymes responsible for ingestion into *C. zealandica* larvae (Marshall et al. 2008). Also, protein content, including protein synthesis concentrations, has been shown to rise throughout the intestinal tract of infected insects (Gatehouse et al. 2008).

3.2.3.2 Genus: *Yersinia* spp.

The species named, *Yersinia pestis* is best known for its invasion on the food track of rat flea (*Xenopsylla cheopis*) (Jarrett et al. 2004). Genetic polymorphisms have proved that the Tc gene's genome homologs are normal among Yersinia strains and express a great diversity for insecticide actions (Fuchs et al. 2008). The *Yersinia* strain containing TC pathogenicity was obtained from the sixth larvae (Bresolin et al. 2006). Cooler temperatures are critical for both the activation of TC gene expression and infectivity of *Y. enterocolitica* (Champion et al. 2009). It has shown to have a diverse range of target insects (Hurst et al. 2011). About 10^6 colony-forming units (CFU) of *Y. pseudotuberculosis* IP32953 are thought to cause *G. mellonella* larvae mortality. A new entomopathogenic microbe, *Y. entomophaga*, has also been extracted from New Zealand Grass Grub (*C. zealandica*). *Yersinia* spp. is pathogenic to a wide range of Coleopteran, Lepidopteran, and Orthopteran insect species (Hurst et al. 2011). The targeted insects stop eating after consumption of this EPB. Following absorption, *Y. entomophaga* induces rapid epithelial membrane degradation in the digestive tract by hemocoel invasion, causing septicemia and destruction.

Y. entomophaga can be distinguished from other forms of *Yersinia* spp. based on the 16S rRNA gene and hybridization of DNA (Hurst et al. 2011).

3.3 Mechanism of Infection, Replication, and Transmission of Entomopathogenic Bacteria (EPB)

Most types of EPB typically have specific pathogenicity characterized by certain virulence factors consisting of insecticidal proteins (IPs), which play a vital role in the mortality of infected insects. The IPs attaches receptor throughout the midgut that interprets to kill gut cells. Disruption of the intestinal endothelial lining allows bacteria to grow a nutrient-rich lymph vessel. The IPs largely determine their accuracy by binding to specific midgut epithelial membrane receptors of the host insects (Djukic et al. 2011). Many *Bt* strains also synthesize cytolytic (Cyt) proteins that bind to lipid sites in the midgut membrane to generate detergent-like defects contributing towards cells cytolysis (Adang et al. 2014; Lee et al. 2003). A range of receptors was identified as Cry toxin receptors, including alkaline phosphatase (ALP), ATP binding cassette (ABC), aminopeptidase N (APN), and cadherin carriers and glycolipids (Adang et al. 2014). Prominent IP families produced by the EPB are the Cry, Cyt, Vip, and Bin protein toxins (Adang et al. 2014).

3.3.1 Cry Toxins

Many microbes can produce an insecticidal toxin that kills the insects and are specifically toxic to the insect orders Lepidoptera Coleoptera, Hymenoptera, and Diptera, and also to nematodes (Bravo et al. 2007). Such microbes may not always be infectious in certain instances but do not always inhabit the recipient insect cadaver. Insecticidal Cry toxins are well researched and characterized toxins reported from a vast number of EPB. Cry toxins have been recorded from many bacteria as a secretory protein (Varani et al. 2013; Barloy et al. 1996; Crickmore et al. 1994). Identifying the amino acid composition is presently used as a base for Cry toxin classification (Crickmore et al. 2015). On consumption of susceptible individuals, parasporal *Bt* crystals become dissolved in intestine physicochemical conditions but converted through an effective poison center of *Bt* proteins or proteases within the host gastrointestinal tract (Waterfield et al. 2004) the immediate action of cry toxins to lyse midgut epithelial cells in the target insect by forming pores in the apical microvilli membrane of the cells (Bravo et al. 2002). Cry proteins pass from crystal inclusion protoxins into membrane-inserted oligomers that cause ion leakage and cell lysis. After cell lysis and the midgut disrupt, epithelium releases the cell contents providing spores a germinating medium leading to severe septicemia and insect death by osmotic shock (Bravo et al. 2005).

3.3.2 Vegetative Insecticidal Proteins (Vip)

Vip protein is produced during Bt's vegetative growth stage and includes Vip1, Vip2, and Vip3. The pathogenicity of Vip protein is based on the insect gut paralysis and lysis of gut epithelium. That available Vip toxicity data shows the Vip1, and Vip2 function as just dual poisons While Vip3 toxin was dependent upon pore formation of cytotoxic effects (Lee et al. 2003; Leuber et al. 2006; Liu et al. 2011; Singh et al. 2010; Barth et al. 2004). Vip3 have high larvicidal activity *against Agrotis ipsilon* and *S. frugiperda* whereas, *S. litura* and *P. xylostella* dramatically depend on pore formation for cytotoxicity (Lee et al. 2003). Unlike cry proteins, Vip3 toxins do not seem to have a protease-resistant toxin core, yet the mechanism leading to enterocyte death remains mostly unknown (Liu et al. 2011).

3.3.3 Bin Toxins

Binary or bin toxins, produced by *L. sphaericus*, are shown as one needle-shaped crystal comprising equimolar quantities of two protein component, 42-kDa (BinA and P42) and 51-kDa (BinB as well as P51) (Broadwell et al. 1990) and has been widely used for control of mosquitoes. These two subunits (Bin A and Bin B) work together to exert maximal toxicity against mosquito larva through pore formation and induction of apoptosis (Boonserm et al. 2006). Both Bin-A and bin-B toxins get a comparatively low sequence similarity, yet they shared several domains that are essential for the activity of the toxin.). Gastrointestinal liquids converted both Bin-A and Bin-B via 40 kDa (Bin-A) and 43 kDa (Bin-B) proteins (Broadwell et al. 1990). The Bin-B protein includes a lectin-like N-terminal domain and a C terminal domain (Srisucharitpanit et al. 2014). It is associated with the initial receptor binding followed by Bin A's interaction before internalization of the toxin complex (Boonserm et al. 2006).

3.3.4 Mtx Toxins

Several *L. Sphaericus* strains develop mosquitocidal toxin (Mtx) and are highly toxic to mosquito larvae, causes morphological changes that lead to loss of the typical cell shape and cluster formation (Carpusca et al. 2006). Mtx matured proteins contain an N-terminal 27 kDa segment of ADP ribosyl transferase action and a C-terminal 70 kDa segment with even a sequence identical to a lectin-like ricin bind component (Thanabalu et al. 1993; Hazes and Read 1995). It is also assumed that after attaching via associations between both the 70 kDa segment and unidentified receptor, the toxin becomes internalized through endosomes whose lower pH emission help the diffusion of a 27 kDa segment into ADP ribosylate proteins cytosol (Schirmer et al.

2002). The mixture of Mtx and Cry toxin through *Bt* spp. Israelensis has shown slight synergy towards *C. quinquefasciatus* (Wirth et al. 2014).

3.3.5 Toxin Complex

High molecular weight insecticidal toxins secreted by bacteria comprised of multiple protein subunits, termed the Toxin Complexes or Tc's. They usually consist of three protein subunits (TcA, TcB and TcC) that effectively harm the insects (Ffrench-Constant et al. 2007). Information of a Tc toxicity mechanism of action has recently been clarified. "TcC" subunit contains, in addition to cytotoxicity, an ADP ribosyl transferase activity which is placed inside such a groove developed through TcB and TcC subsets (Busby et al. 2013). Such (TcB and TcC) heterodimer is bound to a p-start formation through TcA subsets, which bind the unknown heretofore receptors mostly to the central cell surface. Tc toxins are composed of three subunits that perforate the host membrane, similar to a syringe, and translocate toxic enzymes into the host cell. The reactions of the toxic enzymes lead to deterioration and, ultimately, the cell (Meusch et al. 2014; Gatsogiannis et al. 2013).

3.4 EPB Based Commercial Biopesticides

Biopesticides have great potential to control pests due to their target specificity. They are non-toxic to human health and environment-friendly, easily degradable with the right mortality level of insect pests (Kumar and Singh 2015). In plants protection strategy, the biopesticides cover just 2% of pesticides used for insect pest management, but their use is increasing every year. The global biopesticides market in 2021 is predicted to reach approximately 7.7 billion USD (Ruiu 2018). Among biopesticides, 90% are produced from EPB (Hubbard et al. 2014). Biopesticides are attracting the world due to their better pest control and management by pretending fewer effects on the environment and human health. These are classified into microbial, biochemical, and plant-incorporated protectants. The efficacy level of biopesticides is better than traditional pesticides (Kumar and Singh 2015). Bacterial pesticides reduce plant damage and maintain the pest population below the ETL. Bacterial entomopathogenic first enters the host body, avoids host defensive reaction, and produces virulence factors that produce diseases in the host and eventually kills the host (waterfield et al., 2004). EPB commercially has been developed for the control of insect pests of field crops. Bacterial species such as *Lysinbacillus sphaericus, Paenibaccilus* spp. *Serrata entomophila* and *Bacillus thuringiensis* subspecies kurstaki are mostly used to control the pests of field crops and forests (Lacey et al. 2015). Biopesticides, derived from *B. thuringiensis,* are very useful in the control of particular insect pests. *Bt* commercially used for forests and crops insect pests. *Bt* subspecies due to high host range (Lepidoptera, Diptera, Coleoptera, and other insects) are widely used to produce bacterial pesticides (van Frankenhuyze 2009). It is fast-reacting, readily available at a low price, long-lasting shelf life, and easy to formulate. It is also highly selective in target pests and has fewer negative impacts on the environment. *Bt* is used until harvesting begins, and it does not kill the beneficial insects. It is degraded in sunlight; therefore, it can be applied frequently (Glare et al. 2012). *Bt* is used to control pests in vegetables and lepidopteran pests in crops such as cotton, cucurbits, corn, and legumes (Lacey et al. 2015). Control of pests within Coleoptera by *Bt* is limited in the *Chrysomelidae* family (Wraight and Hajek 2009). Two main toxins are produced by *Bt,* which are Cry and Cyt. Vegetative insecticidal proteins (Vip) are also produced and secreted by *Bt* cells (Crickmore et al. 2015; Raymond et al. 2010b). Entomopathogenic *Paenibacillus* are classified into *Paenibacillus larvae, P. popilliae,* and *P. lentimorbus. P. Popilliae and P. lentimorbus* are very distinct at the molecular level. Parasporal body contains toxins that disrupt the gut epithelial barrier and facilitate the attack on the hemocoel (Zhang et al. 1997). The affected larvae cannot molt, become retorted, and eventually died. *Paenibaccilus larvae* are the etiological mediator of core bacteriological honey bee pathology (Gende et al. 2011). *L. sphaericus* produces spherical spores present in the terminal within a sporangium (Nakamura 2000). Their attack depends on the production of toxins that invade midgut cells in host larvae. This *Bacillus* species are pathogenic to insect orders Coleoptera, Diptera, and Lepidoptera (de Oliveira et al. 2004). *Serratia entomophila* and *S. proteamaculans* are non-spore-forming bacteria with limited stable life stages. These are used to control damaging pests of pasture and grass grub. These bacteria cause amber disease in *C. zealandica* larvae with chronic pathology. When larvae feed on this bacterium, their feeding stops, and the larval midgut is cleared, which results in the coloration of amber disease (Jackson et al. 2004). Table 3.1 described the *Bt* subspecies topical insecticidal products based on various transconjugant and recombinant strains (Sanahuja et al. 2011).

3.5 Potential of EPB as a Biological Control Agent

Biopesticides and their by-products are used for plant protection with less injurious effects. The biological control system of crop pests is changed after discovering the EPB (Glare and O'Callaghan 2000). They can control the target pest population as a natural enemy (Mampallil et al. 2017). Once enter into the insect cell, they affect the midgut epithelial cells and cause the host's death by producing different toxins (Mampallil et al. 2017). Most of the bacterial pathogens of insect pests are present under families such as Bacillaceae, Pseudomonadaceae, Enterobacteriaceae, Streptococcaceae, and Micrococcaceae. The species belonging to the family Bacillaceae are highly effective against arthropods. Among EPB, mainly the spore developing species such as *Bt, B. sphaericus,* and *B. popilliae* is mostly used for biological control of pests (Lacey et al. 2015). *Bt* is the most successful biological control agent used to control insects. It comprises species that are naturally existing

Table 3.1 Entomopathogenic bacteria (EPB) based commercial insecticidal products

Trade name	Entomopathogenic bacteria	Target pests	Reference
Xen Tari DF, Agree WP, NB200 FC®	*Bt*. Subspecies *aizawai*	Armyworm, Diamondback moth	Shelton et al. (1993)
Mosquito Beater WSP	*Bt*. Subspecies *israelensis*	Diptera	Tikar and Prakash (2017)
Costar, HIL®Monterey *Bt*., Thuricide, Dipel ES, Biobit, Halt®	*Bt*. Subspecies *kurstaki*	Coleoptera Lepidoptera	Mohan and Gujar (2001) and Sanahuja et al. (2011)
Novodor FC Trident	*Bt*. Subspecies *tenebrionis*	Lepidoptera Coleoptera	Wraight and Ramos (2005)
Milky Spore Powder	*Paenibacillus popilliae*	Japanese beetle	Sanahuja et al. (2011)
VectoMax, VectoLex	*Bt*. Subspecies *sphaericus*	Mosquito	Sanahuja et al. (2011)
M-Trak, Match, MVP	*Bt*. Subspecies *pseudomonas*	Coleoptera Lepidoptera	Sanahuja et al. (2011)
Majestene, Venerate	*Burkholderia spp.*	Chewing and sucking insects and mites	Ruiu (2018)
Tracer™ 120, Conserve	*Saccharopolyspora spinosa*	Many insects	Ruiu (2018)
Grandevo	*Chromobacterium subtsugae*	Chewing and sucking insects and mites	Ruiu (2018)

and are supplementary in the ecosystem for the control of insect pests. *Bt* produces toxins that contain an insecticidal protein called endotoxin. It attacks the host in larval condition, inserts in the host's body, invades the midgut tissue, which causes death (Bravo et al. 2007). *B. popilliae,* which forms spores, causes milky disease in phytophagous coleopteran larvae. Spores are ingested by the host, which germinate into the midgut. It causes milky spore disease in the Scarabaeidae family (Evans 2008). Gram-negative bacteria such as *Serratia* and *Enterobacter* within the family *Enterobacteriaceae* were reported to possess entomopathogenic activity. *Serratia* is a facultative, anaerobic bacteria and proliferates in insects' midgut and causes septicemia, leading to insect death. This bacterium is also secluded from diseased insects. Many bacteria interconnected with plants in the soil, which utilize beneficial effects like development, encouraged conflict to pathogens and pest control ability. Bacteria that subordinate with plants have different names like rhizosphere bacteria, endophytic bacteria present in their natural environmental conditions (Bostock et al. 2001).

Due to environmental and socioeconomic advantages, globally acreage cultivation with herbicide-and pest-resistant genetically modified crops has drastically increased since 1996. Until 2018, 191.7 million hectares and 825 GM varieties were confirmed to be released and cultivated. Specifically, *Bt* was used in 103 million hectares. Out of 103 million hectares, 80 million hectares were cultivated by crops containing stacked *Bt* herbicide tolerance genes, while 23 million hectares were cultivated with those crops containing only the *Bt* gene for resistance to coleopterans and lepidopterans insects-pests (ISAAA 2018). Since 1996, 304 *Bt*-GM varieties and lines of 10 plant species, including 1 tomato (*Lycopersicon esculentum*), 30 potato (*Solanum tuberosum*), 208 maize (*Zea mays*), 49 cotton (*Gossypium hirsutum*), 3 rice (*Oryza sativa*), 6 soybean (*Glycine* max), 3 sugarcane (*Saccharum sp.*), 2 poplar (*Populus sp.*), 1 cowpea (*Vigna unguiculata*) and 1 eggplant (*Solanum melongena*) variety, have been authorized for commercial release in 27 different countries around the world (ISAAA's GM Approval Database 2020). Among them, 243 crop varieties are resistant to lepidopteran pests and contain anti-lepidopteran *cry* and *vip* genes, including *cry1Ab, cry1Ac, cry1C, cry1F, cry1Fa2, cry2Ab2, cry2Ae, cry9C*, and *vip3A*.

3.6 Genetic Improvements of EPB

Genetic technology is a potent way to improve good characters in these living entities' desire species for their better utilization. It comprised the selection of growing populations, revealing the chosen qualities, artificial assortment, cross-hybridization, and further genetic management like gene mutations and genetic engineering (Karabörklü et al. 2018). Genetically improved bioinsecticides comprised of genetically modified entomopathogens are among the supreme essentials with fast and durable pest control (Azizglu et al. 2020). They are less toxic and easily degradable and decompose very fast than conventional pesticides (Arora et al. 2016). Scientists are continually developing and discovering new, environmentally friendly biopesticides that can be used alone or in association with chemical pesticides (Ruiu et al. 2013). Recombinant DNA technology has been widely used to make novel fusion proteins and bind toxin genes to transporter proteins that can show lethal effects after entering the pest cell (Fitches et al. 2004). EPB haves been utilized to manage insect pests of crops and control the mosquitoes that cause serious disease (Fitches et al. 2004). However, the main problem associated with the application of EPB is their poor stability and are commonly affected by adverse ecological aspects that affect their pathogenicity. EPB can be genetically modified to resist adverse environmental conditions (Karabörklü et al. 2018). Moreover, these genetically modified bacteria have fewer adverse effects on the beneficial organism (Azizglu et al. 2020). Genetically modified strains of EPBs have shown higher efficiency against target pests (Arora et al. 2016). Several EPB such as *Photorhabdus, Bacillus, Serratia, Pseudomonas, Lysinibacillus* has already been genetically modified.

The limitation of field stability is one of the biggest problems of EPB applications. Furthermore, pathogenicity is also affected by adverse environmental factors. In this case, genetic modification can increase the resistance of EPB to adverse environmental conditions (Karabörklü et al. 2018). Genetic modification of wild-type EPB is very important to increase effectiveness against the target insect and develop broad-spectrum insecticides (Wang et al. 2008; Patel et al. 2015). Lower insect resistance, higher pathogenicity, low spraying requirements, and longer-term efficacy are the advantages of genetically modified EPB over wild type EPBs (Sharma 2009; Castagnola and Jurat-Fuentes 2012; Karabörklü et al. 2018). There are several concerns and risks regarding the use of genetically modified EPBs. The major concern is the effects of modified EPBs on human health and the environment. Others include the development of resistance in target insect-pest, possible gene flow to wild species, and effects on non-target beneficial species, last but not the least impact on the rhizospheric microbial population (Castagnola and Jurat-Fuentes 2012; Karabörklü et al. 2018; Amarger 2002).

3.7 Conclusion

Future research in agriculture is mainly looking for eco-friendly pest and disease management practices. Such tools are being established and assessed globally to lessen the environmental and health-risks because the higher use of synthetic chemicals in agriculture is alarming the natural biodiversity. Under this scenario, EPBs have a massive scope as bio-control agents and an excellent source for exploring insecticidal toxin genes. Many *Bt* strains have already been approved, whereas many other strains are being described but not commercially developed. There is a broader scope for identifying new bacterial toxin genes from the under-research strains for their development into successful EPB based biopesticides.

References

Ackermann HW, Azizbekyan RR, Bernier RL, de Barjac H, Saindoux S, Valero JR, Yu MX (1995) Phage typing of *Bacillus suBtilis* and *B. thuringiensis*. Res Microbiol 146:643–657

Adang M, Crickmore N, Jurat-Fuentes JL (2014) Diversity of *Bacillus thuringiensis* crystal toxins and mechanism of action. In: Dhadialla TS, Gill S (eds) Advances in insect physiology. Insect midgut and insecticidal proteins, vol 47. Academic, San Diego, CA, pp 39–87

Ahmed I, Yokota A, Yamazoe A, Fujiwara T (2007) Proposal of *Lysinibacillus boronitolerans* gen. Nov. sp. nov., and transfer of *Bacillus fusiformis* to *Lysinibacillus fusiformis* comb. nov. and *Bacillus sphaericus* to *Lysinibacillus sphaericus* comb. nov. Int J Syst Evol Microbiol 57:1117–1125

Akhurst RJ, Lyness EW, Zhang QY, Pinnock DE (1997) A 16S rRNA gene oligonucleotide probe for identification of *Bacillus thuringiensis* isolates from sheep fleece. J Invertebr Pathol 69:24–30

Amarger N (2002) Genetically modified bacteria in agriculture. Biochimie 84:1061–1072. https://doi.org/10.1016/S0300-9084(02)00035-4

Amonkar SV, Kulkarni U, Anand A (1985) Comparative toxicity of *Bacillus thuringiensis* subspecies to *Spodoptera litura* (F.). Curr Sci 54:475–478

Aoki K, Chigasaki Y (1916) Uber die pathogenitat der sog. Sottobacillen (Ishiwata) bei seidenraupen. Bull Imperial Sericult Exp Sta 1:97–139

Arora N, Verma M, Prakash J, Mishra J (2016) Regulation of biopesticides: global concerns and policies. In: Bioformulations: for sustainable agriculture. Springer, New Delhi, pp 283–299
Azizglu U, Jouzani SG, Yilmaz N, Baz E, Ozkok D (2020) Genetically modified bacteria, recent developments, benefits and impact. Sci Total Environ 734:139–169
Barloy F, Delecluse A, Nicolas L, Lecadet MM (1996) Cloning and expression of the first anaerobic toxin gene from *Clostridium bifermentans* subsp. *Malaysia*, encoding a new mosquitocidal protein with homologies to *Bacillus thuringiensis* delta-endotoxins. J Bacteriol 178:3099–3105
Barth H, Aktories K, Popoff MR, Stiles BG (2004) Binary bacterial toxins: biochemistry, biology, and applications of common *Clostridium* and *Bacillus* proteins. Microbiol Mol Biol Rev 68:373–402
Berliner E (1915) Uber die schlaffsucht der mehlmottenraupe (*Ephestia kuhniella*, Zell.) und ihren erreger *B. thuringiensis* n. sp. Z. Angew Entom 2:29–56
Bizzarri MF, Prabhakar A, Bishop AH (2008) Multiple-locus sequence typing analysis of *Bacillus thuringiensis* recovered from the phylloplane of clover (*Trifolium hybridum*) in vegetative form. Microb Ecol 55:619–625
Boonserm P, Moonsom S, Boonchoy C, Promdonkoy B, Parthasarathy K, Torres J (2006) Association of the components of the binary toxin from *Bacillus sphaericus* in solution and with model lipid bilayers. Biochem Biophys Res Commun 342:1273–1278
Bostock RM, Karban R, Thaler JF, Weyman PD, Gilchrist D (2001) Signal interactions in induced resistance to pathogens and insect herbivores. Eur J Plant Pathol 107:103–111
Bottjer KP, Bone LW, Gill SS (1985) Nematoda: susceptibility of the egg to *Bacillus thuringiensis* toxins. Exp Parasitol 60:239–244
Bravo A, Gill SS, Soberón M (2005) Comprehensive molecular insect science. In: Bacillus thuringiensis mechanisms and use. Elsevier BV, pp 175–206
Bravo A, Sanchez J, Kouskoura T, Crickmore N (2002) N-terminal activation is an essential early step in the mechanism of action of the *Bacillus thuringiensis* Cry1Ac insecticidal toxin. J Biol Chem 277:23985–23987
Bravo A, Gill S, Soberon M (2007) Mode of action of *Bacillus thuringiensis* cry and Cyt toxins and their potential for insect control. Toxicology 49:423–435
Bresolin G, Neuhaus K, Scherer S, Fuchs TM (2006) Transcriptional activity of long-term adaptation of *Yersinia enterocolitica* to low-temperature growth. J Bacteriol 188:2945–2958
Broadwell AH, Baumann L, Baumann P (1990) The 42-kilodalton and 51-kilodalton mosquitocidal proteins of *Bacillus sphaericus* 2362-construction of recombinants with enhanced expression and *in vivo* studies of processing and toxicity. J Bacteriol 172:2217–2223
Burges HD (2001) *Bacillus thuringiensis* in pest control. Pestic Outlook 12:90–98
Busby PE, Zimmerman N, Weston DJ, Jawdy SS, Houbraken J, Newcombe G (2013) Leaf endophytes and Populus genotype affect severity of damage from the necrotrophic leaf pathogen, Drepanopeziza populi. Ecosphere 4:125
Carpusca I, Jank T, Aktories K (2006) *Bacillus sphaericus* mosquitocidal toxin (MTX) and pierisin: the enigmatic offspring from the family of ADP-ribosyltransferases. Mol Microbiol 62:621–630
Castagnola AS, Jurat-Fuentes JL (2012) Bt crops: past and future. In: Sansinenea E (ed) *Bacillus thuringiensis* biotechnology. Springer, Dordrecht, Heidelberg, London, New York, pp 283–304. https://doi.org/10.1007/978-94-007-3021-2_15
Champion OL, Cooper IAM, James SL, Ford D, Karlyshev A, Wren BW, Duffield M, Oyston PCF, Titball RW (2009) *Galleria mellonella* as an alternative infection model for *Yersinia pseudotuberculosis*. Microbiology 155:1516–1522
Chatterjee S, Ghosh TS, Das S (2010) Virulence of *Bacillus cereus* as natural facultative pathogen of *Anopheles subpictus* Grassi (Diptera: Culicidae) larvae in submerged rice-fields and shallow ponds. Afr J Biotechnol 9:6983–6987
Cordova AL, Fernandez LE, Koivunen M, Yang A, Flor-Weiler L, Marrone PG (2013) Isolation and characterization of *Burkholderia rinojensis* sp. nov., a non-*Burkholderia cepacia* complex soil bacterium with insecticidal and miticidal activities. Appl Environ Microbiol 79:7669–7678

Crickmore N, Wheeler VC, Ellar DJ (1994) Use of an operon fusion to induce expression and crystallisation of a *Bacillus thuringiensis* delta-endotoxin encoded by a cryptic gene. Mol Gen Genet 242:365–368

Crickmore N, Zeigler DR, Schnepf E, Baum J, Bravo A, Dean DH (2014) Bacillus thuringiensis toxins nomenclature. www.btnomencluture.infolifesci.sussex.ac.uk/Home/Neil_Crickmore/Bt. Accessed 7 Nov 2014

Crickmore N, Baum J, Bravo A, Lereclus D, Narva K, Sampson K, Schnepf E, Sun M, Zeigler DR (2015) *Bacillus thuringiensis* toxin nomenclature

Davidson EW (1989) Variation in binding of *Bacillus sphaericus* toxin and wheat germ agglutinin to larval midgut cells of six species of mosquitoes. J Invertebr Pathol 53:251–259

Davidson EW, Singer S, Briggs JD (1975) Pathogenesis of *Bacillus sphaericus* strain SSII-1 infections in *Culex pipiens* quinquefasciatus (¼ C. *pipiens fatigans*) larvae. J Invertebr Pathol 25:179–184

de Castilhos-Fortes R, Matsumura ATS, Diehl E, Fiuza LM (2002) Susceptibility of *Nasutitermes ehrhardti* (Isoptera: Termitidae) to *Bacillus thuringiensis* subspecies. Braz J Microbiol 33:219–222

de Oliveira EJ, Rabinovitch L, Monnerat RG, Passos LKJ, Zahner V (2004) Molecular characterization of *Brevibacillus laterosporus* and its potential use in biological control. Appl Environ Microbiol 70:6657–6664

Djukic M, Poehlein A, Thurmer A, Daniel R (2011) Genome sequence of *Brevibacillus laterosporus* LMG 15441, a pathogen of invertebrates. J Bacteriol 193:5535–5536

Dulmage HT, de Barjac H, Krywienczyk J et al (1981) Insecticidal activity of isolates of Bacillus thuringiensis and their potential for pest control. In: Burges HD (ed) Microbial control of pests and plant diseases, 1970–1980. Academic, New York, pp 193–222

Dutky SR (1940) Two new spore-forming bacteria causing milky diseases of Japanese beetle larvae. J Agric Res 61:57–68

Evans J (2008) Biopesticides: from cult to mainstream. Agrow 2008:11–14

Ffrench-Constant RH, Dowling A, Waterfield NR (2007) Insecticidal toxins from *Photorhabdus* bacteria and their potential use in agriculture. Toxicon 49:436–451

Fitches E, Edwards MG, Mee C, Grishin E, Gatehouse AMR (2004) Fusion proteins containing insect-specific toxins as pest control agents: snowdrop lectin delivers fused insecticidal spider venom toxin to insect haemolymph following oral ingestion. J Insect Physiol 50:61–71

Franken E, Krieger L, Schnetter W (1996) *Bacilluspopilliae*: a difficult pathogen. Bull OILB/SROP 19:40–45

Fuchs TM, Bresolin G, Marcinowski L, Schachtner J, Scherer S (2008) Insecticidal genes of Yersinia spp.: taxonomical distribution, contribution to toxicity towards *Manduca sexta* and *galleria mellonella*, and evolution. BMC Microbiol 8:214

Garcia-Robles I, Sanchez J, Gruppe A, Martınez-Ramırez AC, Rausell C, Real MD, Bravo A (2001) Mode of action of *Bacillus thuringiensis* PS86Q3 strain in hymenopteran forest pests. Insect Biochem Mol Biol 31:849–856

Gatehouse HS, Marshall SD, Simpson RM, Gatehouse LN, Jackson TA, Christeller JT (2008) *Serratia entomophila* inoculation causes a defect in exocytosis in *Costelytra zealandica* larvae. Insect Mol Biol 17:375–385

Gatsogiannis C, Lang AE, Meusch D, Pfaumann V, Hofnagel O, Benz R, Aktories K, Raunser S (2013) A syringe-like injection mechanism in *Photorhabdus luminescens* toxins. Nature 495:520–523

Gende L, Satta A, Ligios V, Ruiu L, Buffa F, Fernandez N, Churio S, Eguaras M, Fiori M, Floris I (2011) Searching for an American foulbrood early detection threshold by the determination of Paenibacillus larvae spore load in worker honey bees. Bull Insectol 64:229–233

Glare TR, O'Callaghan M (2000) *Bacillus thuringiensis*: biology, ecology and safety. Wiley, Chichester

Glare T, Caradus J, Gelernter W, Jackson T, Keyhani N, Kihl J, Marrone P, Morin L, Stewart A (2012) Have biopesticides come of age? Trends Biotechnol 30:250–258

Gohar M, Gilois N, Graveline R, Garreau C, Sanchis V, Lereclus D (2005) A comparative study of *Bacillus cereus, Bacillus thuringiensis* and *Bacillus anthracis* extracellular proteomes. Proteomics 5:3696–3371

Goldberg LJ, Margalit J (1977) A bacterial spore demonstrating rapid larvicidal activity against *Anopheles sergentii, Uranotaenia unguicalata, Culex univitattus, Aedes aegyptii* and *Culex pipiens*. Mosq News 37:355–358

Grimont F, Grimont PAD (2006) The genus Serratia. In: Dworkin M, Falkow S, Rosenberg E, Schleifer KH, Stackebrandt E (eds) The prokaryotes, vol 6, 3rd edn. Springer, New York, pp 219–244

Han S, Craig JA, Putnam CD, Carozzi NB, Tainer JA (1999) Evolution and mechanism from structures of an ADPribosylating toxin and NAD complex. Nat Struct Biol 6:932–936

Hannay CL, Fitz-James P (1955) The protein crystals of *Bacillus thuringiensis* Berliner. Can J Microbiol 1:694–710

Hansen BM, Damgaard PH, Eilenberg J, Pedersen JC (1998) Molecular and phenotypic characterization of *Bacillus thuringiensis* isolated from leaves and insects. J Invertebr Pathol 71:106–114

Harrison H, Patel R, Yousten AA (2000) Paenibacillus associated with milky disease in central and south American scarabs. J Invertebr Pathol 76:169–175

Hazes B, Read R (1995) A mosquitocidal toxin with a ricin-like cell-binding domain. Nat Struct Mol Biol 2:358–359. https://doi.org/10.1038/nsb0595-358

Heimpel AM, Angus TA (1958) The taxonomy of insect pathogens related to *Bacillus cereus* Frankland and Frankland. Can J Microbiol 4:531–541

Hernandez JLL (1988) Evaluation de la toxicite de *Bacillus thuringiensis* sur *Spodoptera frugiperda*. Entomophaga 33:163–171

Hu X, Fan W, Han B, Liu H, Zheng D, Li Q, Dong W, Yan J, Gao M, Berry C, Yuan Z (2008) Complete genome sequence of the mosquitocidal bacterium *Bacillus sphaericus* C3-41 and comparison with those of closely related Bacillus species. J Bacteriol 190:2892–2902

Hubbard M, Hynes RK, Erlandson M, Bailey KL (2014) The biochemistry behind the biopesticide efficacy. Sustain Chem Process 2:18

Hurst MR, Becher SA, Young SD, Nelson TL, Glare TR (2011) *Yersinia entomophaga* sp. nov., isolated from the New Zealand grass grub *Costelytra zealandica*. Int J Syst Evol Microbiol 61:844–849

Inglis GD, Lawrence AM (2001) Effects of *Serratia marcescens* on the F1 generation of laboratory-reared *Heliothis virescens* (Lepidoptera: Noctuidae). J Econ Entomol 94:362–366

ISAAA (2018) Global status of commercialized biotech/GM crops in 2018: biotech crops continue to help meet the challenges of increased population and climate change. ISAAA brief no. 54. ISAAA, Ithaca, NY, https://www.isaaa.org/resources/publications/briefs/54/executivesummary/pdf/B54-ExecSum-English.pdf

ISAAA's GM Approval Database (2020). http://www.isaaa.org/gmapprovaldatabase/. Accessed 18 Feb 2020

Ishiwata S (1901) On a kind of flacherie (sotto disease). Dainihon Sanshi Keiho 114:1–5

Jackson TA, Boucias DG, Thaler JO (2001) Pathobiology of amber disease, caused by Serratia spp., in the New Zealand grass grub, *Costelytra zealandica*. J Invertebr Pathol 78:232–243

Jackson TA, Christeller JT, McHenry JZ, Laing WA (2004) Quantification and kinetics of the decline in grass grub endopeptidase activity during initiation of amber disease. J Invertebr Pathol 86:72–76

Jarrett CO, Deak E, Isherwood KE, Oyston PC, Fischer ER, Whitney AR, Kobayashi SD, DeLeo FR, Hinnebusch BJ (2004) Transmission of *Yersinia pestis* from an infectious biofilm in the flea vector. J Infect Dis 190:783–792u7j7

Jucovic M, Walters FS, Warren GW, Palekar NV, Chen JS (2008) From enzyme to zymogen: engineering Vip2, an ADP-ribosyltransferase from *Bacillus cereus*, for conditional toxicity. Protein Eng Des Sel 21:631–638

Kaaya GP, Darji N (1989) Mortality in adult tsetse, *Glossina morsitans* morsitans, caused by entomopathogenic bacteria. J Invertebr Pathol 54:32–38

Kachhawa D (2017) Microorganisms as a biopesticides. J Entomol Zool Stud 5(3):468–473

Karabörklü S, Azizoglu U, Azizoglu ZB (2018) Recombinant entomopathogenic agents: a review of biotechnological approaches to pest insect control. World J Microbiol Biotechnol 34:1–12. https://doi.org/10.1007/s11274-017-2397-0

Kirst HA (2011) The spinosyn family of insecticides: realizing the potential of natural products research. J Antibiot 63:101–111

Koul O (2011) Microbial biopesticides: opportunities and challenges. In: CAB reviews: perspectives in agriculture, veterinary science, nutrition and natural resources, vol 6, pp 1–2

Krieger L, Franken E, Schnetter W (1996) *Bacillus popilliae* var. melolontha H1, a pathogen for the may beetles, Melolontha spp. In: Jackson TA, Glare TR (eds) Proceedings of the 3rd international workshop on microbial control of soil dwelling pests. AgResearch, Lincoln, pp 79–87

Kumar S, Singh A (2015) Biopesticides: present status and the future prospects. J Fertil Pestic 6 (2):1–7

Kumari SM, Neelgund YF (1985) Preliminary infectivity tests using six bacterial formulations against the red flour beetle, Tribolium castaneum. J Invertebr Pathol 46:198–199

Lacey LA, Grzywacz D, Shapiro-llan DI, Frutos R, Brownbridge M, Goettel MS (2015) Insect pathogens as a biological control agent: back to the future. J Invertebr Pathol 132:1–41

Lecadet MM, Frachon E, Dumanoir VC, Ripouteau H, Hamon S, Laurent P, Thiery I (1999) Updating the H-antigen classification of *Bacillus thuringiensis*. J Appl Microbiol 86:660–672

Lee MK, Walters FS, Hart H, Palekar N, Chen JS (2003) The mode of action of the *Bacillus thuringiensis* vegetative insecticidal protein Vip3A differs from that of Cry1Ab delta-endotoxin. Appl Environ Microbiol 69:4648–4657

Leuber M, Orlik F, Schiffler B, Sickmann A, Benz R (2006) Vegetative insecticidal protein (Vip1Ac) of *Bacillus thuringiensis* HD201: evidence for oligomer and channel formation. Biochemistry 45:283–288

Lima GMS, Aguiar RWS, Correa RFT, Martins ES, Gomes ACM, Nagata T, De-Souza MT, Monnerat RG, Ribeiro BM (2008) Cry2A toxins from *Bacillus thuringiensis* expressed in insect cells are toxic to two lepidopteran insects. World J Microbiol Biotechnol 24:2941–2948

Liu JG, Yang AZ, Shen XH, Hua BG, Shi GL (2011) Specific binding of activated Vip3Aa10 to *Helicoverpa armigera* brush border membrane vesicles results in pore formation. J Invertebr Pathol 108:92–97

Luxananil P, Atomi H, Panyim S, Imanaka T (2001) Isolation of bacterial strains colonizable in mosquito larval guts as novel host cells for mosquito control. J Biosci Bioeng 92:342–345

Mampallil JL, Faizal MH, Anith KN (2017) Bacterial bioagents for insect pest management. J Entomol Zool Stud 5(6):2237–2244

Marshall SD, Gatehouse LN, Becher SA, Christeller JT, Gatehouse H, Hurst MR, Boucias DG, Jackson TA (2008) Serine proteases identified from a *Costelytra zealandica* (white) (Coleoptera: Scarabaeidae) midgut EST library and their expression through insect development. Insect Mol Biol 17:247–259

Martin PAW, Gundersen-Rindal D, Blackburn M, Buyer J (2007) *Chromobacterium subtsugae* sp. nov., a betaproteobacterium toxic to Colorado potato beetle and other insect pests. Int J Syst Evol Microbiol 57:993–999

Meusch D, Gatsogiannis C, Efremov RG, Lang AE, Hofnagel O, Vetter IR, Aktories K, Raunser S (2014) Mechanism of Tc toxin action revealed in molecular detail. Nature 508:61–65

Milner RJ (1981) Identification of the *Bacillus popilliae* group of insect pathogens. In: Burges HD (ed) Microbial control of pests and plant diseases 1970–1980. Academic, London, pp 45–59

Mohan M, Gujar GT (2001) Toxicity of *Bacillus thuringiensis* strains and commercial formulations to the diamondback moth, *Plutella xylostella* (L.). Crop Prot 20(4):311–316

Nakamura LK (2000) Phylogeny of *Bacillus* sphaericus-like organisms. Int J Syst Evol Microbiol 50(5):1715–1722

O'Callaghan M, Gerard EM (2005) Establishment of *Serratia entomophila* in soil from a granular formulation. NZ Plant Prot 58:122–125

O'Callaghan M, Garnham ML, Nelson TL, Baird D, Jackson TA (1996) The pathogenicity of Serratia strains to *Lucilia sericata* (Diptera: Calliphoridae). J Invertebr Pathol 68:22–27

Oliveira CD, Tadei WP, Abdalla FC (2009) Occurrence of apocrine secretion in the larval gut epithelial cells of *Aedes aegypti* L., *Anopheles albitarsis lynch-Arribalzaga* and *Culex quinquefasciatus* say (Diptera: Culicidae): a defense strategy against infection by *Bacillus sphaericus* Neide? Neotrop Entomol 38:624–631

Park HW, Bideshi DK, Federici BA (2010) Properties and applied use of the mosquitocidal bacterium, *Bacillus sphaericus*. J Asia Pacif Entomol 13:159–168

Patel A, Pathak L, Parvez N, Panpatte D, Khatri K, Jani J (2015) Molecular approaches for the improvement of *Bacillus thuringiensis* against pests. In: Chakravarthy AK (ed) New horizons in insect science: towards sustainable pest management, pp 179–185. https://doi.org/10.1007/978-81-322-2089-3_17

Perchat S, Buisson C, Chaufaux J, Sanchis V, Lereclus D, Gohar M (2005) *Bacillus cereus* produces several nonproteinaceous insecticidal exotoxins. J Invertebr Pathol 90:131–133

Pettersson B, Rippere KE, Yousten AA, Priest FG (1999) Transfer of *Bacillus lentimorbus* and *Bacillus popilliae* to the genus Paenibacillus with emended descriptions of *Paenibacillus lentimorbus* comb. nov. and *Paenibacillus popilliae* comb. nov. Int J Syst Bacteriol 49:531–540

Poinar GO, Wassink HJM, Leegwater-van der Linden ME, van der Geest LPS (1979) *Serratia marcescens* as a pathogen of tsetse flies. Acta Trop 36:223–227

Porcar M, Gomez F, Gruppe A, Gomez-Pajuelo A, Segura I, Schroder R (2008) Hymenopteran specificity of *Bacillus thuringiensis* strain PS86Q. Biol Control 45:427–432

Priest FG, Dewar SJ (2000) Bacteria and insects. In: Priest FG, Goodfellow M (eds) Applied microbial systematics. Kluwer Academic, Dordrecht, pp 165–202

Quesada-Moraga E, Garcıa-Tovar E, Valverde-Garcıa P, Santiago A, Lvarez C (2004) Isolation, geographical diversity and insecticidal activity of *Bacillus thuringiensis* from soils in Spain. Microbiol Res 159:59–71

Raymond B, Wyres KL, Sheppard SK, Ellis RJ, Bonsall MB (2010a) Environmental factors determining the epidemiology and population genetic structure of the *Bacillus cereus* group in the field. PLoS Pathog 6:e1000905

Raymond B, Johnston PR, Nielsen-LeRoux C, Lereclus D, Crickmore N (2010b) *Bacillus thuringiensis*: an impotent pathogen? Trends Microbiol 18:189–194

Reyes-Ramırez A, Ibarra JE (2005) Fingerprinting of *Bacillus thuringiensis* type strains and isolates by using *Bacillus cereus* group-specific repetitive extragenic palindromic sequence-based PCR analysis. Appl Environ Microbiol 71:1346–1355

Roh JY, Liu Q, Lee DW, Tao X, Wang Y, Shim HJ, Choi JY, Seo JB, Ohba M, Mizuki E, Je YH (2009) *Bacillus thuringiensis* serovar mogi (flagellar serotype 3a3b3d), a novel serogroup with a mosquitocidal activity. J Invertebr Pathol 102:266–268

Ruiu L (2018) Microbial biopesticides in agroecosystems. Agronomy 8:235

Ruiu L, Satta A, Floris I (2007) Susceptibility of the house fly pupal parasitoid *Muscidifurax raptor* (Hymenoptera: Pteromalidae) to the entomopathogenic bacteria *Bacillus thuringiensis* and *Brevibacillus laterosporus*. Biol Control 43:188–194

Ruiu L, Satta A, Floris I (2013) Emerging entomopathogenic bacteria for insect pest management. Bull Insectol 66:181–186

Sanahuja G, Banakar R, Twyman RM, Capell T, Christou P (2011) *Bacillus thuringiensis*: a century of research, development and commercial applications. Plant Biotechnol J 9:283–300

Sanchis V (2011) From microbial sprays to insect-resistant transgenic plants: history of the biospesticide *Bacillusthuringiensis*. A review. Agron Sustain Dev 31:217–231

Schirmer J, Just I, Aktories K (2002) The ADP-ribosylating mosquitocidal toxin from *Bacillus sphaericus*: proteolytic activation, enzyme activity, and cytotoxic effects. J Biol Chem 277:11941–11948

Selvakumar G, Mohan M, Sushil SN, Kundu S, Bhatt JC, Gupta HS (2007) Characterization and phylogenetic analysis of an entomopathogenic *Bacillus cereus* strain WGPSB-2 (MTCC 7182) isolated from white grub, *Anomala dimidiata* (Coleoptera: Scarabaeidae). Biocontrol Sci Tech 17:525–534

Sharma HC (2009) Genetic engineering of entomopathogenic microbes for pest management. In: Sharma HC (ed) Biotechnological approaches for pest management and ecological sustainability. CRC, Taylor and Francis Group, Boca Raton, pp 255–292. https://doi.org/10.1201/9781420088489

Sharpe ES, Detroy RW (1979) Fat body depletion, a debilitating result of milky disease in Japanese beetle larvae. J Invertebr Pathol 34:92–94

Shelton AM, Robertson JL, Tang JD, Perez C, Eigenbrode SD, Preisler HK, Wilsey WT, Cooley RJ (1993) Resistance of diamondback moth (Lepidoptera: Plutellidae) to *Bacillus thuringiensis* subspecies in the field. J Econ Entomol 86:697–705

Shida O, Takagi H, Kadowaki K, Komagata K (1996) Proposal for two new genera, Brevibacillus gen. Nov. and Aneurinibacillus gen. Nov. Int J Syst Bacteriol 46:939–946

Singh G, Sachdev B, Sharma N, Seth R, Bhatnagar RK (2010) Interaction of *Bacillus thuringiensis* vegetative insecticidal protein with ribosomal S2 protein triggers larvicidal activity in *Spodoptera frugiperda*. Appl Environ Microbiol 76:7202–7209

Smirnova TA, Minenkova IB, Orlova MV, Lecadet MM, Azizbekyan RR (1996) The crystal-forming strains of *Bacillus laterosporus*. Res Microbiol 147:343–350

Srisucharitpanit K, Yao M, Promdonkoy B, Chimnaronk S, Tanaka I, Boonserm P (2014) Crystal structure of BinB: a receptor binding component of the binary toxin from *Lysinibacillus sphaericus*. Proteins 82:2703–2712

Stahly DP, Klein MG (1992) Problems with in vitro production of spores of *Bacillus popilliae* for use in biological control of the Japanese beetle. J Invertebr Pathol 60:283–291

Strongman DB, Eveleigh ES, van Frankenhuyzen K, Royama T (1997) The occurrence of two types of entomopathogenic bacilli in natural populations of the spruce budworm, *Choristoneura fumiferana*. Can J For Res 27:1922–1927

Tan B, Jackson TA, Hurst MR (2006) Virulence of Serratia strains against *Costelytra zealandica*. Appl Environ Microbiol 72:6417–6418

Thanabalu T, Hindley J, Brenner S, Oei C, Berry C (1993) Expression of the mosquitocidal toxins of Bacillus sphaeicus and Bacillus thuringiensis subsp. israelensis by recombinant Caulobacter crescentus, a vehicle for biological control of aquatic insect larvae. Appl Environ Microbiol 58:905–910

Tikar S, Prakash S (2017) Fly ash-based *Bacillus thuringiensis israelensis* formulation: an ecofriendly approach. Indian J Med Res 146(6):680–682

van Frankenhuyze K (2009) Insecticidal activity of *Bacillus thuringiensis* crystal proteins. J Invertebr Pathol 101:1–16

Varani AM, Lemos MV, Fernandes CC, Lemos EG, Alves EC, Desiderio JA (2013) Draft genome sequence of *Bacillus thuringiensis* var. *thuringiensis* strain T01–328, a Brazilian isolate that produces a soluble pesticide protein, Cry1Ia. Genome Announc 1:5

Vilas-Boas GT, Peruca APS, Arantes OMN (2007) Biology and taxonomy of *Bacillus cereus*, *Bacillus anthracis*, and *Bacillus thuringiensis*. Can J Microbiol 53:673–687

Wai Nam T, Gudauskas RT, Harper JD (1975) Pathogenicity of *Bacillus cereus* isolated from *Trichoplusia ni* larvae. J Invertebr Pathol 26:135–136

Wang G, Zhang J, Song F, Gu A, Uwais A, Shao T, Huang D (2008) Recombinant *Bacillus thuringiensis* strain shows high insecticidal activity against *Plutella xylostella* and *Leptinotarsa decemlineata* without affecting nontarget species in the field. J Appl Microbiol 105:1536–1543. https://doi.org/10.1111/j.1365-2672.2008.03866.x

Warren GW, Koziel MG, Mullins MA, Nye GJ, Carr B, Desai NM, Kostichka K, Duck NB, Estruch JJ (1996) Novel pesticidal proteins and strains. WIP organization. WO 96/10083

Waterfield NR, Daborn PJ, Ffrench-Constant RH (2004) Insect pathogenicity islands in the insect pathogenic bacterium Photorhabdus. Physiol Entomol 29:240–250

Wirth MC, Berry C, Walton WE, Federici BA (2014) Mtx toxins from *Lysinibacillus sphaericus* enhance mosquitocidal cry-toxin activity and suppress cry-resistance in *Culex quinquefasciatus*. J Invertebr Pathol 115:62–67

Wraight SP, Hajek AN (2009) Manipulation of arthropod pathogen for IPM. In: Radcliff EB, Hutchison WD, Cancelado RE (eds) Integrated pest management: concept, tactics, strategies, and case studies. Cambridge University Press, Cambridge, pp 131–150

Wraight SP, Ramos ME (2005) Synergistic interaction between *Beauveria bassiana*- and *Bacillus thuringiensis tenebrionis*-based biopesticides applied against field populations of Colorado potato beetle larvae. J Invertebr Pathol 90(3):139–150

Wraight SP, Molloy DP, Singer S (1987) Studies on the culicine mosquito host range of *Bacillus sphaericus* and *Bacillus thuringiensis* var. israelensis with notes on the effects of temperature and instar on bacterial efficacy. J Invertebr Pathol 49:291–302

Yuan YM, Hu XM, Liu HZ, Hansen BM, Yan JP, Yuan ZM (2007) Kinetics of plasmid transfer among *Bacillus cereus* group strains within lepidopteran larvae. Arch Microbiol 187:425–431

Zhang J, Hodgman TC, Krieger L, Schnetter W, Schairer HU (1997) Cloning and analysis of the first cry gene from *Bacillus popilliae*. J Bacteriol 179:4336–4341

Zubasheva MV, Ganushkina LA, Smirnova TA, Azizbekyan RR (2010) Larvicidal activity of crystal-forming strains of *Brevibacillus laterosporus*. Appl Biochem Microbiol 46:755–762

Chapter 4
Entomopathogenic Nematodes (EPNs): A Green Strategy for Management of Insect-Pests of Crops

Qaiser Shakeel, Muhammad Shakeel, Muhammad Raheel, Sajjad Ali, Waqas Ashraf, Yasir Iftikhar, and Rabia Tahir Bajwa

Abstract The successful control of many insect-pests makes entomopathogenic nematodes (EPNs) among one of the best biocontrol agents for insect pests. Moreover, the ability of EPNs to seek out their hosts and kill them in those habitats where chemicals fail makes them even more attractive. The EPNs-bacterial mutualistic association helps them kill their hosts in a relatively shorter period than other necromenic or parasitic nematode associations. In addition to this end-user safety, hotspot application which allows minimizing treated area, natural enemies' safety, withholding period absence, and environmental protection are a few of many advantages over chemical pesticides. Two important genera of EPNs, i.e., Heterorhabditid and Steinernematids, are associated with symbiotic bacteria Photorhabdus and Xenorhabdus, respectively, while bacterial symbiont of neosteinernamatids is yet to be described. About 21 species of Heterorhabditis and 100 species of Steinernema have been isolated and identified worldwide. With the increasing environmental concerns and low efficacy of synthetic pesticides, agriculturists and researchers have a growing interest in finding alternatives to synthetic pesticides. Several EPNs can be widely used in place of synthetic pesticides in agro-ecosystem. There is still a need to improve several aspects of EPNs, such as efficacy and efficiency, reduced costs, mass production, and formulation technology.

Q. Shakeel (✉) · M. Raheel · W. Ashraf · R. T. Bajwa
Department of Plant Pathology, Faculty of Agriculture and Environment, The Islamia University of Bahawalpur, Bahawalpur, Pakistan
e-mail: qaiser.shakeel@iub.edu.pk

M. Shakeel
Key Laboratory of Bio-Pesticide Innovation and Application of Guangdong Province, College of Plant Protection, South China Agricultural University, Guangzhou, China

S. Ali
Department of Entomology, Faculty of Agriculture and Environment, The Islamia University of Bahawalpur, Bahawalpur, Pakistan

Y. Iftikhar
Department of Plant Pathology, College of Agriculture, University of Sargodha, Sargodha, Pakistan

S. De Mandal et al. (eds.), *New and Future Development in Biopesticide Research: Biotechnological Exploration*, https://doi.org/10.1007/978-981-16-3989-0_4

Furthermore, their potential for recycling in the host population beckons them to be further exploited for sustainable pest control. This chapter will emphasize the use and potential of EPNs as an integral part of integrated pest management. To aid with understanding the potential of EPNs, this chapter will also provide an overview of ecology and biology, mass production, application strategies, and integration with other management tools.

Keywords Biocontrol · Safety · Entomopathogenic nematodes · Heterorhabditis · Steinernema

4.1 Entomopathogenic Nematodes

Nematodes have a worm-like, round-shaped, colorless, and unsegmented body with no appendages belonging to the kingdom "Animalia." These may be saprophytic, predacious, or parasitic in nature. Parasitic nematodes may cause different diseases in humans, animals, and plants. Entomopathogenic nematodes (EPNs) are obligated parasites of the insects. These nematodes may cause severe, long-lasting effects on their hosts like sterility, minimized fecundity, reduced longevity and flight activity, rapid mortality, and delayed development of other physiological or morphological abnormalities. In 1923, Gotthold Steiner described the EPNs for the first time. The term EPNs was given because they cause disease in insects since they have a symbiotic association with bacteria. Almost 23 nematode families are involved in the association with insects, out of which seven families have the potential for their usage as biological control agents against insect pests. Almost 35 species of EPNs have been described and identified almost from everywhere (Koppenhofer et al. 2000). A minimal number of nematodes are responsible for killing the insect-pests. These are difficult to use, expensive to produce in mass, and have a limited host range of pests with very low economic importance. EPNs are the only ones to fulfill biological control attributes against insects, and these mainly belong to Genera *Steinernema* and *Heterorhabditis*, as they possess the qualities to kill the insect quickly.

Management of insect-pests has mostly been dependent on synthetic insecticides/ pesticides. Considering the massive impact of these synthetic chemicals on society's safety, environmental pollution, insecticidal resistance, reduction in natural enemies' population has alarmed scientists to explore alternative eco-friendly management methods. So, scientists have a propensity towards the use of biological control methods. Currently, the application of EPNs, a green strategy (i.e., eco-friendly or positively impacting the environment), plays a crucial role in Integrated Pest Management (IPM) system. EPNs possess a broad range of other features, i.e., they are highly lethal for the targeted insects and safe for non-target; they might directly or indirectly influence the population of plant pathogens. Steinernema and Heterorhabditis nematodes are the only insect antagonists that maintain a proper balance of biocontrol aspects. Some species can be recycled and persist in the environment; they can also be involved indirectly in improving the quality of the soil and are congruent with a broad range of pesticides, including both chemical and biological pesticides applied in IPM due to which EPNs are considered as useful or practical applicants against a variety of agricultural insect pest (Lacey and Georgis 2012; Lortkipanidze et al. 2019).

4.1.1 Advantages

- United States Environmental Protection Agency (U.S EPA) has exempted the EPNs from pesticide registration.
- EPNs have a broad insect-pest host range and are very harmful to the targeted host but are safer for the non-targeted host (Georgis et al. 1991).
- Compared to synthetic chemical, EPNs pose more safety as the people are not bound to wear masks or other safety equipment. There are no chances of residual toxic effects on the environment, groundwater contamination, pollination, and chemical trespass.
- Insect-pest will not produce resistance against EPNs because of their constant recycling in the host, unlike synthetic pesticides.
- EPN-bacteria symbiosis helps them kill their host more rapidly than other biological control agents (Akhurst and Smith 2002).
- The *in-vitro and in-vivo* suitability of EPNs for the mass production.
- EPNs do not require specialized equipment for their spread; instead, conventional application equipment can be used, including different sprays like mist, fan, aerial, backpack, etc., and irrigation systems.

4.1.2 Disadvantages

- Mass production of EPNs is not cost-effective.
- Lack of skilled personal with essential Nematological knowledge.
- A requirement of refrigerated storage and limited shelf life.
- Difficulties in formulations and quality control.
- EPNs infectivity and survival have been adversely affected by inadequate temperature and moisture.
- The application of pesticides also has an adverse effect on the infectivity and survival of EPNs.
- Soil properties and ultraviolet radiation had a lethal impact on EPNs (Shapiro-Ilan et al. 2012a, b).

4.2 Biology and Lifecycle

The EPNs belonging to the family Steinernematidae and Heterorhabditidae have synonymous life history. The non-feeding, free-living third stage of the nematode is responsible for initiating the parasitic cycle and is termed as the 'dauer' or 'infective juvenile' or 'IJ' stage. The IJ have their unique role in insect predation. The chemoreceptors produce the required biological signal, which helps them in movement and host-seeking. Once they get into their host, highly virulent EPNs can kill their host within 24–48 h. These nematodes can easily be cultured on artificial media and have a high reproductive rate. The life cycle or reproduction mechanism of both nematodes' genera, i.e., Steinernematid and Heterorhabditids, is similar but varies only in the first generation of their life cycle. Steinernematid infective juveniles of all generations are developed into males and females (amphimictic). In contrast, Heterorhabditids develop into self-fertilizing hermaphrodites and then in the next generation mixture of adult males, adult females, and hermaphrodites (Grewal et al. 2005). Whenever the EPNs find their host, it quickly penetrates the host's body cavity through natural openings or by cuticle (if genus *Heterorhabditis* is involved). *S. glaseri* can penetrate its hosts through the gut while *H. bacteriophora* penetrates through intersegmental membranes (Wang and Gaugler 1999). When a symbiotic bacterium (that is gram-negative, facultatively anaerobic rods belongs to the family Enterobacteriaceae) is released by EPNs from their intestine into the gut of the insect, it easily multiplies there and quickly causes septicemia, resulting in the killing of the insect pest. The symbiotic bacteria serve as food for EPNs, which decomposes host tissues and become mature or adult. Within the host carcass, about two to three EPN's generations are completed. Once the host gets infected, EPNs take 1–3 weeks to develop and reproduce within the host carcass (Stock 1995). After depletion of food reserves, the hundreds and thousands of nematode's offspring are released, developing into resistant IJs that are able to persist in the environment and hunt for a new host, and the cycle repeats.

The pathogenicity and reproduction of EPNs are facilitated by the symbiotic association with bacteria (serving as food). The nematodes without symbiotic bacteria may rarely be able to kill the insect host but do not usually reproduce. Moreover, bacteria alone cannot penetrate the insect body cavity or are unable to enter the insect's hemocoel. Therefore, in the symbiotic association of nematodes with bacteria, nematodes play a role as vectors to carry the bacteria into the body of the host for their proliferation; meanwhile, the bacteria create a suitable environment within the host carcass for the survival and reproduction of EPNs by the rapid killing of an insect. In infective juveniles of Steinernematid, intestinal vesicle (modified ventricular portion of the intestine) is the place reserved for the storage of symbiotic bacteria while the location of symbiotic bacteria in infective stage of Heterorhabditid nematodes is esophagus as well as a ventral portion of the intestine (Poinar and Leutenegger 1968; Poinar et al. 1977). *Xenorhabdus* is the genus of bacteria associated with all species of Genus *Steinernema*, while Genus *Photorhabdus* of bacteria is associated with all *Heterorhabditis* species of nematodes (Boemare et al.

1993). Each EPN species has a symbiotic relationship with one bacterium, while each bacterial strain can have a symbiotic association with several nematode species (Akhurst and Boemare 1990; Grewal et al. 1997a). On the basis of insect carcass color, the genera of involved nematodes can be predicted, i.e., tan or brown color of insect carcass indicates that *Heterorhabditids* are involved in killing this insect while the red color of host carcass shows the involvement of Steinernematids as a cause of death. The insect carcass color is symptomatic of metabolites or pigments produced by symbiotic bacteria developing in the host insect (Kaya and Gaugler 1993).

There are three levels according to which the symbiotic relationship between bacteria and nematodes is operated.

1. Provision of factors increasing the recovery of the infective juvenile from dauer stage, i.e., non-feeding.
2. Provision of essential nutrients by symbiotic bacterium for the consumption of nematodes.
3. Retention of symbiotic bacterium in the dauer (non-feeding) infective juvenile's intestine (Grewal et al. 2005).

4.3 Host Locating Strategy

EPNs naturally exist in the soil and trace their host in the reaction to vibration, CO_2, chemical stimulus, or by sensing the insect's body structure (Kaya and Gaugler 1993). To locate a host, one of the two strategies, i.e., ambushers or cruisers, is used by EPNs (Grewal et al. 1994a). Ambushers having energy-conserving tactic, lie down with blinking or nictitating eyes in the upper soil layer and wait for mobile insects to attack them. This type of host searching strategy is used by *Steinernema carpocapsae*. *S. glaseri* and *Heterorhabditis bacteriophora* show cruisers strategy to locate their host. They are usually found under the earth's surface. They are too active that by using volatile signals, they can move significant distances in search of their subterranean host. They can also attack less mobile insects like scarab beetles, commonly known as white grubs, and are efficient EPNs in pest control. An intermediate host locating strategy (i.e., a combination of cruiser and ambrush) is used by some other species, including *S. riobrave* and *S. feltiae*.

4.4 Habitat

Steinernematids and Heterorhabditids are found in soils (deserts, cultivated land, forests, grasslands etc.) all around the world. EPNs have been isolated even from beaches and oceans. When the survey was conducted, 02%–45% EPNs were found from the soils of sampled sites (Khatri-Chhetri et al. 2010; Ramliana and Yadav 2010; Hominick 2002). However, some EPNs have also been isolated from naturally infected insects in the field. For the isolation of these EPNs, susceptible insects are used as 'baits' in the soil, and the larvae of Greater Wax moth is the most common insect used as bait (Bedding and Akhurst 1974).

4.5 Effects of Abiotic Factors

Several investigators have recognized the effects of abiotic factors like soil type, soil structure, temperature, aeration, and moisture (Kaya and Gaugler 1993; Georgis et al. 2006; Grewal et al. 2005; Shapiro-Ilan et al. 2012a, b; Gaugler and Kaya 1990). EPNs survival is highly influenced by soil texture. Their survivability is the lowest in the clayey soil. They prefer sandy soil as the clayey soil has very limited oxygen content with smaller pores. Oxygen is mostly limited in these soils having water-saturated soil texture with some of the organic matter. The pH of the soil has no potential effect on the infective juvenile's survival. The IJs can survive the range from 4–8 pH, and the survival rate is declined at a pH of 10. Similarly, EPN can survive and infect its host in soils with high salinity levels. However, it has also been reported that Heterorhabditis has restricted infectivity in saline soils while their high-temperature tolerance has improved (Finnegan et al. 1999; Thurston et al. 1994). Moreover, Heterorhabditis survival in seawater has no negative effects (Griffin et al. 1994). EPNs are sensitive to temperature extremes, i.e., very high and very low temperatures, UV radiations, and desiccation. The origin place, species, and habitat are the key characteristics for varying optimum temperature range for the survival and infection potential of EPNs (Kaya 1990). For instance, *Steinernema carpocapsae* can be found almost inactive at 10 °C, and *S. feltiae* has a wide temperature range for infectivity, i.e., between 2–30 °C while at a temperature range of 7–35 °C Heterorhabditids nematodes are able to infect insects (Lacey et al. 2006; Georgis et al. 2006; Kaya 1990). Species that are heat tolerant include *S. glaseri*, *S. ribrave* and *H. indica*, on the other hand, cooler temperature is preferred by some other species including *S. feltiae*, *H. marelatus,* and *H. megidis* (Grewal et al. 1994b).

4.6 Host Range

Different sorts of tests were conducted to evaluate the efficiency of the EPNs, which resulted in minimum to maximum efficacy. Moreover, EPNs attack many hosts in the laboratory conditions, as there are optimal temperature and environmental conditions with no developmental barriers to the existing infection. Although, in field conditions, it is challenging to maintain the optimal conditions, and concentration of the EPNs have to be maintained in fields. Table 4.1 elaborates the host range of the EPNs against their targeted insects (Kaya and Gaugler 1993; Finney and Walker 1979).

Table 4.1 Commercial use of EPNs as bio-pesticides against major targeted insects

Nematodes species	Major targeted insects as prescribed by different commercial companies
Steinernema glaceri	White grubs including Japanese beetle, scarabs, *Popillia* spp. and banana root borers
Steinernema feltiae	*Bradysia* spp. (Fungus gnats), western flower thrips, leafminers, shore flies
Steinernema scapterisci	*Scapterisci* spp. including mole crickets
Steinernema carpocapsae	Pests of vegetables and ornamentals (codling moths, banana moths, dogwood borer, cranberry girdlers, peachtree borers, black vine weevils, shore flies) and pests of turfgrass (armyworms, cutworms, sod webworms, chinch bugs, billbugs, crane flies)
Steinernema riobrave	Mole crickets and citrus root weevils (*Diaprepes* spp.)
Steinernema kraussei	Otiorhynchus sulcatus (Black vine weevil)
Heterorhabditis indica	Root mealy bugs, fungus gnats, and grubs
Heterorhabditis bacteriophora	Black vine weevils, scarabs, and cutworms
Heterorhabditis marelatus	Scarabs, black vine weevils, cutworms
Heterorhabditis megidis	Weevils
Heterorhabditis zealandica	Scarab or white grubs

4.7 The Pathogenicity and Reproduction of EPNs

Entomopathogenic nematodes (*Heterorhabditis* and *Steinernema*) has mutualistic symbiotic relationship with *Photorhabdus* and *Xenorhabdus* bacteria respectively (Akhurst 1983). These entomopathogenic bacteria (EPB) are being carried and released by EPNs in insect hemocoel. After killing insects, EPB convert insect cadaver into food sources which is suitable for EPNs growth and development (Shapiro-Ilan et al., 2012b). Before emergence from insect cadavers, EPNs again re-associates with their respective symbiotic bacteria and start search for new host (Goodrich-Blair and Clarke 2007). There are complex chemical communications in multilateral interactions; Nematode-insect for host recognition, Nematode-bacteria for symbiotic relationship and Insect-bacteria for pathogenicity. Among these insect-bacteria pathogenic interaction is well studied. The pathogenic interaction involves the production of metabolites derived from polyketide synthase (PKS) or non-ribosomal peptide synthetase (NRPS). However Other secondary compounds are also produced by bacterial synthetic machineries (Tobias et al. 2017; Mollah et al. 2020). Depending upon the species and strains, EPB virulence is variable. To suppress the insect immunity, EPB secrete reveral virulence factors which ultimately cause fatal septicemia (Sajjadian and Kim 2020; Park et al. 2007; Sergeant et al. 2006; Shrestha and Kim 2007). The activity of Phospholipase A_2 (PLA_2) is usually inhibited by both bacteria to cause septicemia. The arachidonic acid from phospholipids are catslyzed by PLA_2 which is necessary step for production of eicosanoids (Kim et al. 2005; Mouchlis and Dennis 2019). This are more evident by the fact that *X. nematophila* secrete at least eight secondary metabolites to suppress PLA_2. Intra- and inter-specific variation among Xenorhabdus virulence have been explained by their differential inhibitory activity against PLA_2 (Kim et al. 2018; Seo et al. 2012; Ahmed and Kim 2018; Hasan et al. 2019). It was explained that specific outer membrane protein plays very important role in bacterial virulence. Furthermore, these findings were extended that expression level of leucine responsive proteins (*Lrp*) temper pathogenicity of bacteria (Park et al. 2017; Casanova-Torres et al. 2017). The production of secondary metabolites is being modulated by the expression of transcriptional factors including *Lrp* suggesting positive correlation with bacterial pathogenicity (Engel et al. 2017).

Secondary metabolites of Photorhabdus and Xenorhabdus produced by NRPS and PKS are of different composition. These secondary metabolites with different composition enable EPB to induce immunosuppression in insects with diverse physiological molecules. For example, in *Galleria mellonella* phenoloxidase activity is inhibited by rhabducin, an isocyanide-containing compound produced from biosynthetic gene cluster. Xenortide peptides/rhabdopeptide with more than 70 kinds derived from NRPS are structurally similar to protease inhibitors. They might degrade various proteins associated with immunity. Phurealipids produced from NRPS/PKS can prevent the expression of antimicrobial peptide genes. Thus, diverse secondary metabolites produced by entomopathogenic bacteria might effectively suppress insect immune responses to induce septicemia (Shi and Bode 2018; Crawford et al. 2012; Cai et al. 2016; Sussmuth and Mainz 2017; Nollmann et al. 2015).

4.8 Mass Production

Mass production of EPNs as biopesticide takes place by using various *in-vivo* and *in-vitro* methods. *Galleria mellonella* L. (Wax moth) larva is the most common host for EPNs rearing both *in-vitro* and *in-vivo,* even for business processes. The moth larva is the most susceptible host to EPNs. Moreover, wide distribution, easy culturing, and the development of a higher number of EPNs make wax moth the most preferable host. Various protocols for nematodes infection, inoculation, and isolation have been demonstrated by many researchers (Dutky et al. 1964; Howell 1979; Woodring and Kaya 1988; Flanders et al. 1996; Finnegan et al. 1999).

4.9 Mass Production

4.9.1 In-Vivo

Multiple authors have reported various methods for culturing EPNs (Dutky et al. 1964; Kaya and Stock 1997; Poinar 1979). The *in-vivo* process of EPN-mass production is very simple as it does not require any modern or latest technology, just a live infected insect host, a surrogate host (to which IJs from the infected host are to be transmitted), trays and shelves are required. The researchers have described systems supporting the White entice (White 1927), which is supported by the migration of IJs outside the host carcass, which is natural. Under *in vivo* process, per larvae yield of IJs depends on the selection of nematode and host species (ranging from 0.5×105 to 4×105). Host size is directly proportional to the yield of specific EPN species (Flanders et al. 1996; Blinova and Ivanova 1987), but the same is not true for infection susceptibility (Blinova and Ivanova 1987; Dutky et al. 1964). *In vivo* production is not effective for a large scale because it involves high costs due to extensive labor, material (insects), and equipment. Moreover, it needs some technical proficiency to conduct the whole process. This method may only be useful for laboratory studies or small-scale production (Friedman 1990; Shapiro-Ilan and Gaugler 2002; Gaugler and Han 2002).

4.9.2 In Vitro

Rudolf Glaser, in 1931, documented the importance of developing methods of nutritive culture media for rearing EPNs and first used this method for *S. glaseri* (Glaser 1932). Although, he had no awareness about the importance of symbiotic bacteria in the culture media and nematodes pathogenicity that was documented after a long time (Poinar and Thomas 1966). Comprehensive knowledge of EPNs biology and behavior is required for their production. At the commercial level, monoxenically EPNs are produced on a solid medium prepared through liquid fermentation and bedding. Solid media are used for successfully producing pathogenic Steinernematids and Heterorhabditids, but requires a high cost of labor (Bedding 1990). The very efficient liquid-fermentation process for producing many Steinernematids while Heterorhabditids are not produced efficiently (Gaugler and Georgis 1991). In this process, EPNs are to be introduced in a nutritive medium having a pure culture of symbiotic bacteria. The nematode stage that is used commercially is dauer juvenile (DJ), a morphologically different form of juvenile, that is made due to the harsh environmental conditions and depleting food source. For commercial use, large fermenters are applied in culture media that significantly encourages the yield of nematode mass production.

4.10 Storage Technology

Production of EPNs into stable products is very useful in commercializing these nematodes and can be used in various biological terms. Active nematodes should be immobilized to protect the nematode against losing depletion of lipid and glycogen reserves. Different methods have been used for their formulation, storage, and application by using activated charcoal, polysaccharide gel, baits, clay, paste, peat, polyurethane sponge, and water dispersal granules. In contrast to other biocontrol agents (i.e., bacteria, fungi, and viruses), EPNs have no dormant resting phase, so limited energy is utilized during storage. Nematode species can be stored successfully for 1–7 months by using refrigeration. Low temperature increases the shelf life of EPNs by decreasing metabolic activity, but some warm adapted species like *H. indica*, *S. riobrave* are not stored well at a temperature below 10 °C (Strauch et al. 2000; Grewal 2002; Grewal et al. 2005).

4.11 Relative Efficiency and Application Parameters

EPNs are certainly beneficial for insect pest management as they have no post-application effects, unlike chemical insecticides having negative impacts on the environment and humans (Ehlers and Peters 1995). In many cases, EPNs seem very beneficial or efficient against insect pests, but sometimes, these nematodes are unsuccessful in the competition (Georgis et al. 2006). The EPN and chemical success gap has narrowed by advanced technologies in EPN formulation, mass production, application order, optimum habitats, and target pests. The use of *S. feltiae* in the floriculture industry in Germany, the Netherlands, and England has proved to be a very suitable replacement for chemical insecticides (Jagdale et al. 2004). EPNs are versatile in their usage and can be used for almost all insects, especially for soil-dwelling or cryptic insects. An increase of temperature and desiccation is the most important abiotic factors that influence them and inactivate these nematodes compared to chemical insecticides (Glazer 2002). However, nematodes require varying temperature ranges depending on their species and suboptimal soil, depth and irrigation frequency for effective insect management compared to chemical insecticides (Shapiro-Ilan et al. 2012a, b; Georgis and Gaugler 1991).

Application of EPNs at dawn and dusk is preferred as at that time exposure of sunlight is low, and relative humidity is high on the surface of the leaf, which increases their efficacy to infect the host. The soil moisture must be maintained for 2 weeks after applying EPNs (Klein 1993). In the past, the application of EPNs for aboveground target was not suitable as the nematode is sensitive to desiccation, but later on, progress in the studies have produced formulation (i.e., mixing of EPNs with specific surfactants and water-dispersible polymers), improving their application quality for aboveground target pest (Shapiro-Ilan et al. 2012a, b). The most convenient method for applying EPNs is spray i.e. ground or hand spray, mist blower, etc. (Georgis et al. 1991). The sprayer nozzle of 100 μm opening and 168 kPa pressure is most suitable for EPN juveniles. These nematodes can also be applied using irrigation systems like a micro jet, drip sprinkler, etc. (Cabanillas and Raulston 1996; Georgis et al. 1991). The EPNs should be applied in a sufficient amount for killing the insects, i.e., 25 IJs are used per cm^2 of the treated area (Shapiro-Ilan et al. 2002). The IJs must be selected according to the targeted pest. In case if nematodes are not used instantly, aeration can be provided by using an aquarium. By many researchers, EPNs are compatible with many chemicals, including fungicides, herbicides, and insecticides (Shapiro-Ilan et al. 2012a, b). Application of EPNs regarding soil surface habitats to control insect pest of crops is described below:

4.11.1 Foliar Application

Data recorded from various fields treated with EPNs against insects residing above the ground surface showed the lowest efficiency of EPNs because, in the foliar application, EPNs get restricted to infect due to rapid desiccation of IJs on the leaf surface. However, their efficiency can be increased by avoiding the desiccation through anti-desiccants against some insect species. For aboveground or foliar insect control, *S. carpocapsae* species is generally used, as, due to their ambusher host locating behavior, they are very effective agents having the ability to rapidly infect the insect on the surface of the soil when they fall from the foliage (Arthurs et al. 2004).

4.11.2 Application for Insects Residing Above the Soil Surface

For control of insects with epigeal habitat, the surface of the soil is treated with EPNs when they are moving over or passing through the soil surface. This is the most effective application of EPNs for such insect pests. In the citrus plants, Diaprepes root weevil (*Diaprepes abbreviates*), a major pest of citrus in Florida, is controlled successively through EPNs by this method. For about 20 years, for the successful control of citrus weevil, *S. riobrave* and *H. indica* nematodes have been marketed in Florida. The most significant results can be obtained at 27 ± 28 °C temperature in sandy soil (Shapiro-Ilan et al. 2002; McCoy et al. 2002, 2007).

4.11.3 Application for Cryptic Habitat Insects

Leaf litter, fruit and vegetable bins, bark, pruning wounds, and nutshells are included in cryptic habitats. Many researchers have documented the effective control of insects residing in cryptic habitats (Lacey and Shapiro-Ilan 2008; Cross et al. 1999; Lacey et al. 2007; Shapiro-Ilan et al. 2005). EPNs successfully controlled the worldwide distributed insect pest of apple and pome fruit, i.e., codling moth (*Cydia pomonella*) in cryptic habitat. *S. feltiae*, *S. carpocapsae*, *H. zealandica* and *H. bacteriophora* are commonly used for the successful control of codling moth (Lacey and Georgis 2012).

4.11.4 The Efficiency of EPNS in Nursery and Greenhouse

In 1988, over 6.2 billion US$ was the estimated annual crop sales for greenhouse and nursery (van Tol RWHM and Raupp 2005). So, the best option to control disease in greenhouses and nurseries is the use of biological control. Different EPNs are used in greenhouses and nurseries.

4.11.5 The Efficiency of EPNs Against Tomato Leaf Miner

Tomato leaf miner (*Tomato absoluta*) is the most destructive pest of tomato. It was first observed in 2009 in the Urla district of Izmir, Turkey, and has the potential to infect and cause 100% damage at all growing stages of tomato plants. Since then, chemical control was the only control method for Tomato leaf miner as it was the most devastating pest for the tomato field, so the application of chemical insecticide was excessive. Tomato producers used insecticides twice a week or sometimes every 6–7 days/season with 8–25 sprays. Although using an excessive amount of chemicals has caused various problems for the environment, human beings, or the plant itself. So, the best alternative for these chemicals was biological control. Therefore, for this purpose, different species of EPNs were tested against tomato leaf miner and concluded that the best EPNs for the control of tomato leaf miners were *Steinernema affine*, *S. carpocapsae*, *S. feltiae* and *Heterorhabditis bacteriophra* (Gozel and Gozel 2016).

4.12 Appearance

EPNs are physically, psychologically, and morphologically active components of biopesticides and can be applied directly. The most important factor is the location of the IJs from the soil profile (Lewis 2002). Ambushing nematode species are sometimes related to extremely mobile surfaces, i.e., the residence of living host insects. After application, the movement of EPNs is straight and precipitous. Upward movement of *S. carpocapsae* IJs while the downward movement of *H. bacteriophora* and *S. glaseri* in soil have been recorded (Georgis and Poinar 1983). Limited oxygen levels suppress the movement of the active cruiser species. Basically, lack of movement does not mean that the nematode is dead; it means that they want some stimulation before they move or become viable. Some nematodes do not feed but depend on the food for energy. Forty percentage of the nematodes' body constitutes the lipid, which is the major source of energy (Fitters et al. 1999; Selvan et al. 1993); some other energy reserves are obtained from proteins, trehalose, carbohydrates, and glycogen (Qiu and Bedding 2000). High levels of lipoid with a dense look are the characteristic of highly infective IJs, whereas active and clear IJs infection level is very low.

4.13 Conservation

Reports and studies of the natural incidence and ecology of EPNs are comparatively uncommon. After the research of ecological behavior of EPNs, it is ensured that these nematodes are adapted to their environment and has a broad range of host in the laboratory conditions while in field conditions they have a limited host range with a conducive environmental condition. Epizootics diseases due to EPNs are most likely to occur in the soil; however, they are microscopic and remains unrecorded (Kaya 1990). Australia reported two epizootics of undescribed species of Heterorhabditis; one of which shows reduction in larvae of *Graphagnathus leucoloma*. Moreover, in the second case, two undescribed Heterorhabditids were isolated from three contiguous sugarcane which were infecting four Scarabaeus species (Akhurst et al. 1992; Sexton and Williams 1981) have been documented from Australia that contained undescribed Heterorhabditis species. In the first case, severe reduction of white-fringed beetle, *Graphagnathus leucoloma*, larvae have been observed while the adult was observed in Lucerne field (Sexton and Williams 1981). In the second, two undescribed species of Heterorhabditis have been recorded in three contiguous sugarcane, infecting four *Scarabaeus* species, i.e., dung beetles. Under *in-vitro* conditions, the susceptibility of Scarabaeidae larvae is limited; infected larvae can even be recovered in *In-vivo* conditions. New advancement in research gives us information for better conservation of the EPNs.

4.14 Compatibility

The infective juveniles of EPNs cannot tolerate the long-term exposure of different fungicides, chemical insecticides, growth regulators, and fertilizers. Despite this, EPNs can be mixed in a tank and applied together; this is the best alternative method to manage insect pests using the EPNs in integrated pest/disease management systems (Rovesti and Deseo 1990; Ishibashi 1993). Moreover, the actual concentration of chemicals that is required to be mixed with nematodes depends on the volume and the scheme used (Alumai and Grewal 2004). The tests were conducted to examine the compatibility of different EPNs with several chemical pesticides (insecticides, herbicides, and fungicides), surfactants, and fertilizers. They are found to be compatible with all of them (Koppenhofer et al. 2000; Raheel et al. 2017).

In many cases, when EPNs are applied in combination with different chemicals, it results in a synergistic effect on the insects. Additionally, the survival and infection efficacy of EPNs can be reduced by some pesticides (Grewal et al. 1998). Some of the chemicals or even nematicides like diazinon, aldicarb, dodine, carbofuran, and methomyl, etc. should be used with care or avoided. *S. carpocapsae* efficacy was enhanced when applied together with organophosphate oxamyl against *Agrotis segatum*. Still, it will be efficient only in fumigated soil. Failure of synthetic chemical pesticides to manage insect-pests, the combination of chemical and EPNs could be implemented and give successful results against insect pests. Before using this combination as tank mixing, the manufacturer's recommendation should be followed for compatibility and potential. Moreover, two different EPNs should be mixed and utilized against two different pests (Kaya and Gaugler 1993). Similarly, different EPNs are compatible with several insect antagonists like *Bacillus thuringiensis* being used successfully against many lepidopterous pests.

4.15 Commercial Availability

Nearly 100 Steinernematid and 21 Heterorhabditis nematodes have been identified, and 12 species have been commercialized. A list of commercial EPNs is given below in Table 4.2 (Siddiqui et al. 2010). Commercial EPNs can be used in various aspects. Basically, for most soil insects, one billion EPNs should be applied per acre.

4.16 Safety

Warm-blooded vertebrates like birds, reptiles, animals, mammals, human beings, etc., have been observed to be safe from EPNs and their symbiont bacteria. They have been found highly lethal for cold-blooded invertebrates, i.e., insects under *in vitro* experimentation and *in vivo* experimentation did not disagree with these

Table 4.2 Commercial product formulations of EPNs available internationally

EPN species	Formulation	Country
Steinernema carpocapsae	ORTHO Biosafe USA	USA
	Bio Vector USA	USA
	Exhibit USA	USA
	Sanoplant	Switzerland
	Boden Nutzlinge	Germany
	Helix	Canada
	X-GNAT	USA
	Vector TL	USA
Steinernema feltiae	Magent	USA
	Nemasys	UK
	Stealth	UK
	Entonem	USA
Steinernema riobrave	Vector MG	USA
	Bio Vector	USA
Steinernema scapterisci		USA
Heterorhabditis bacteriophora	Otinem	USA
Heterorhabditis megidis	Nemasys	UK
Steinernema carpocapsae	Green commandos	India
	Soil commandos	

results (Bathon 1996; Kermarrec et al. 1991; Boemare et al. 1996; Poinar et al. 1982; Poinar and Thomas 1988; Georgis et al. 1991). All feasible negative effects are restricted by EPNs because of their less mobility, living in the cryptic environment (i.e., soil, tunnels inside the plant residues and other plant growth media), and minimized survival on foliage (Glazer 1992; Downes and Griffin 1996). Plant pathogenic nematodes population has been significantly reduced for the past 5 years due to the commercial application of EPNs. Moreover, after just one application of Heterorhabditis nematodes in turfgrass, the population of plant pathogenic nematode was observed to be minimized significantly while free-living nematodes have been found without any variation or reduction in their population up to 60 days (Grewal et al. 1997b, 2001; Smitley et al. 1992).

4.17 Conclusion

Entomopathogenic nematology has a very short or limited history; the research's main focus was to use these nematodes as a biocontrol against different insect pests. The success or failure of EPNs in the soil surrounding the presence of dominant insect-pest remains unknown. This lack of information underscores the prerequisite to thoroughly exploring EPNs regarding ecology, behavior, biology, and biological mechanisms. Advances regarding the ecology and behavior of EPNs revealed that they are not human pathogens. Instead, their activity is limited to a certain soil spectrum where EPNs could successfully eliminate several host insect-pests.

Further exploration science involved in the biocontrol mechanism of EPNs will enable the researcher to manage the insect-pests within their activity pattern effectively. EPNs are now being used as a successful biocontrol agent against soil-dwelling insect pests and hence distributed in most parts of the world as a commercial interest. Using such nematodes in biocontrol programs has many advantages like; broad host range, host-seeking ability, ease of production and application, and high virulence. These are compatible with several biocontrol agents and chemical pesticides and can be stored for a significant period. With the low population's limitation in the natural soils for causing the epidemic to reduce the host population, their lot can still be achieved. To achieve the required targets or results, it is necessary to analyze the requirements and maintain them to raise the EPNs population. There is no doubt that EPB represent an abundant and valuable source of bioactive and chemically novel compounds with potential for exploitation in agriculture. Indeed, we are yet in an infancy stage, but the promise that EPB secondary metabolites warrant will continue to lead the path for further discoveries and applications into pest management. Finally, these fascinating microorganisms have the potential to become effective biological control agents and can be used exclusively. Furthermore, EPNs are helpful towards an understanding of parasitism and symbiosis through insect resistance to infection.

References

Ahmed S, Kim Y (2018) Differential immunosuppression by inhibiting PLA_2 affects virulence of *Xenorhabdus hominickii* and *Photorhabdus temperata temperata*. J Invertebr Pathol 157:136–146

Akhurst RJ (1983) *Neoaplectana* species: specificity of association with bacteria of the genus *Xenorhabdus*. Exp Parasitol 55:258–263

Akhurst R, Boemare NE (1990) Biology and taxonomy of *Xenorhabdus*. In: Gaugler R, Kaya HK (eds) Entomopathogenic nematodes in biological control. CRC, Boca Raton, pp 75–87

Akhurst R, Smith K (2002) Regulation and safety. In: Gaugler R (ed) Entomopathogenic nematology. CABI, New York, pp 311–332

Akhurst RJ, Bedding RA, Bull RM, Smith DRJ (1992) An epizootic of *Heterorhabditis* spp. (Heterorhabditidae: Nematoda) in sugarcane scarabaeids (Coleoptera). Fund Appl Nematol 15:71–73

Alumai A, Grewal PS (2004) Tankmix-comaptibility of the entomopathogenic nematodes, *Heterorhabditis bacteriophora* and *Steinnernema carpocapsae*, with selected chemical pesticides using in turfgrass. Biocontrol Sci Tech 14:725–730

Arthurs S, Heinz KM, Prasifka JR (2004) An analysis of using entomopathogenic nematodes against aboveground pests. Bull Entomol Res 94:297–306

Bathon H (1996) Impact of entomopathogenic nematodes on non-target hosts. Biocontrol Sci Tech 6:421–434

Bedding RA (1990) Logistics and strategies for introducing entomopathogenic nematodes technology into developing countries. In: Gaugler R, Kaya HK (eds) Entomopathogenic nematodes in biological control. CRC, Boca Raton, FL, pp 233–248

Bedding RA, Akhurst RJ (1974) BiologicalControl of Sirex Wasp. CSIRO Annual Report, p 88
Blinova SL, Ivanova ES (1987) The cultivation of the nematode-bacterial complex of *Neoplectana carpocapsaein* insects. Nauka, Moscow, pp 12–17
Boemare NE, Akhurst RJ, Mourant RG (1993) DNA relatedness between *Xenorhabdus* spp. (Enterobacteriacae) symbiotic bacteria of entomopathogenic nematodes, and a proposal to transfer *Xenorhabdus luminescens* to a new genus, *Photorhabdus* gen. Novo. Int Syst Bacteriol 43:249–255
Boemare NE, Laumond C, Mauleon H (1996) The entomopathogenic nematode-bacterium complex: biology, life cycle and vertebrate safety. Biocontrol Sci Tech 6:333–345
Cabanillas HE, Raulston JR (1996) Effects of furrow irrigation on the distribution and infectivity of *Steinernema riobrave* against corn earworm in corn. Fund Appl Nematol 19:273–281
Cai X, Nowak S, Wesche F, Bischoff I, Kaiser M, Furst R et al (2016) Entomopathogenic bacteria use multiple mechanisms for bioactive peptide library design. Nat Chem 9:379–386
Casanova-Torres AM, Shokal U, Morag N, Eleftherianos I, Goodrich-Blair H (2017) The global transcription factor Lrp is both essential for and inhibitory to *Xenorhabdus nematophila* insecticidal activity. Appl Environ Microbiol 83:e00185–e00117
Crawford JM, Portmann C, Zhang X, Roeffaers MB, Clardy J (2012) Small molecule perimeter defense in entomopathogenic bacteria. Proc Natl Acad Sci U S A 109:10821–10826
Cross JV, Solomon MG, Chandler D, Jarrett P, Richardson PN, Winstanley D, Bathon H, Huber J, Keller B, Langenbruch GA, Zimmerman G (1999) Biocontrol of pests of apples and pears in northern and Central Europe: 1. Microbial agents and nematodes. Biocontrol. Sci Technol 9:125–149
Downes MJ, Griffin CT (1996) Dispersal behaviour and transmission strategies of the entomopathogenic nematodes Heterorhabditis and Steinernema. Biocontrol Sci Tech 6:347–356
Dutky SR, Thompson JV, Cantwell GE (1964) A technique for the mass propagation of the DD-136 nematode. J Insect Pathol 6(4):417–422
Ehlers RU, Peters A (1995) Entomopathogenic nematodes in biological control: feasibility, perspectives and possible risks. In: Hokkanen HMT, Lynch JM (eds) Biological control: benefits and risks. Cambridge University Press, Cambridge, pp 119–136
Engel Y, Windhorst C, Lu X, Goodrich-Blair H, Bode HB (2017) The global regulators Lrp, LeuO, and HexA control secondary metabolism in entomopathogenic bacteria. Front Microbiol 8:209
Finnegan MM, Downes MJ, Regan MO, Griffen CT (1999) Effect of salt and temperature stress on survival and infectivity of *Heterorhabditis* spp. Nematology 1(1):69–78
Finney JR, Walker C (1979) Assessment of a field trial using the DD-136 strain of *Neoaplectana* sp. for the control of *Scolytus scolytus*. J Invertebr Pathol 33:239–241
Fitters PFL, Patel MN, Griffin CT, Wright DJ (1999) Fatty acid composition of *Heterorhabditis* sp. during storage. Comp Biochem Physiol 24:81–88
Flanders KL, Miller JM, Shields EJ (1996) In vivo production of *Heterorhabditis bacteriophora* 'Oswego' (Rhabditida: Heterorhabditidae), a potential biological control agent for soil-inhabiting insects in temperate regions. J Econ Entomol 89(2):373–380
Friedman MJ (1990) Commercial production and development. In: Gaugler R, Kaya HK (eds) Entomopathogenic nematodes in biological control. CRC, Boca Raton, FL, pp 153–172
Gaugler R, Georgis R (1991) Culture method and efficacy of entomopathogenic nematodes (Rhabditida: Steinernematids and Heterorhabditidae). Biol Control 1(4):269–274
Gaugler R, Han R (2002) Production technology. In: Gaugler R (ed) Entomopathogenic nematology. CAB International, Wallingford, pp 289–310
Gaugler R, Kaya HK (1990) Entomopathogenic nematodes in biological control. CRC, Boca Raton
Georgis R, Gaugler R (1991) Predictability in biological control using entomopathogenic nematodes. J Econ Entomol 84:713–720
Georgis R, Poinar GO Jr (1983) Effect of soil texture on the distribution and infectivity of *Neoplectana carpocapsae* (Nematoda: Steinernematidae). J Nematol 15:308–311

Georgis R, Kaya HK, Gaugler R (1991) Effect of Steinernematid and Heterorhabditid nematodes (Rhabditida: Steinernematidae and Heterorhabditidae) on non-target arthropods. Environ Entomol 20:815–822

Georgis R, Koppenhoper AM, Lacey LA, Belair G, Duncan LW, Grewal PS, Samish M, Tan L, Torr P, VanTol RWHM (2006) Successes and failures in the use of parasitic nematodes for pest control. Biol Control 38:103–123

Glaser RW (1932) Studies on *Neoplectana glaseri*, a nematode parasite of the Japanese beetle (*Popillia japonica*). New Jersey Department of Agriculture Circular No 211

Glazer I (1992) Survival and efficacy of *Steinernema carpocapsae* in an exposed environment. Biocontrol Sci Tech 2:101–107

Glazer I (2002) Survival biology. In: Gaugler R (ed) EPN. CAB International, Wallingford, pp 169–187

Goodrich-Blair H, Clarke DJ (2007) Mutualism and pathogenesis in *Xenorhabdus* and *Photorhabdus*: two roads to the same destination. Mol Microbiol 64:260–268

Gozel U, Gozel C (2016) Entomopathogenic nematodes in pest management. In: Integrated Pest management (IPM): environmentally sound pest management, p 55

Grewal PS (2002) Formulation and application technology. In: Gaugler R (ed) EPN. Nematology. CAB International, Wallingford, pp 265–287

Grewal PS, Lewis EE, Gaugler R, Campbell JF (1994a) Host finding behavior as a predictor of foraging strategy in entomopathogenic nematodes. Parasitology 108:207–215

Grewal PS, Selvan S, Gaugler R (1994b) Thermal adaptation of entomopathogenic nematodes-niche breadth for infection, establishment and reproduction. J Thermal Biol 19:245–253

Grewal PS, Matsuura M, Converse KW (1997a) Mechanisms of specificity of association between *Steinenema scopterisci* and its bacteria associated. Parasitology 111:14–17

Grewal PS, Martin WR, Miller RW, Lewis EE (1997b) Suppression of plant parasitic nematode populations in turfgrass by application of entomopathogenic nematodes. Biocontrol Sci Tech 7:393–399

Grewal PS, Webber T, Batterley DA (1998) Compatibility of *Steinernema feltiae* with chemicals used in mushroom production. Mushroom News 46:6–10

Grewal PS, Nardo EAD, Aguillera MM (2001) Entomopathogenic nematodes: potential for exploration and use in South America. Neotrop Entomol 30(2):191–205

Grewal PS, Ehlers RU, Shapiro-Ilan DI (2005) Nematodes as biocontrol agents. CABI, New York, NY

Griffin CP, Finnegon MM, Downes MJ (1994) Environmental tolerances and the dispersal of Heterorhabdites: survival and infectivity of European Heterorhabditis following prolonged immersion in seawater. Fund Appl Nematol 17:415–421

Hasan MA, Ahmed S, Mollah MI, Lee D, Kim Y (2019) Variation in pathogenicity of different strains of *Xenorhabdus nematophila*: differential immunosuppressive activities and secondary metabolite production. J Invertebr Pathol 166:107221

Hominick WM (2002) Biogeography. In: Gaugler R (ed) EPN. CABI, New York, NY, pp 115–143

Howell JF (1979) New storage methods and improved trapping techniques for the parasitic nematode *Steinernema carpocapsae*. J Invertebr Pathol 33:155–158

Ishibashi N (1993) Integrated control of insect pests by *Steinernema carpocapsae*. In: Bedding RA, Akhurst R, Kaya K (eds) Nematodes and biological control of insects. CSIRO, East Melbourne, pp 105–113

Jagdale GB, Casey ML, Grewal PS, Lindquist RK (2004) Eûects of application rate and timing, potting medium and host plant on eûcacy of Steinernema feltiae against fungus gnat, *Bradysia coprophila*, in ûoriculture. Biol Control 29:296–305

Kaya HK (1990) Soil ecology. In: Gaugler R, Kaya HK (eds) Entomopathogenic nematodes in biological control. CRC, Boca Raton, pp 93–115

Kaya HK, Gaugler R (1993) Entomopathogenic nematodes. Annu Rev Entomol 38:181–206

Kaya HK, Stock P (1997) Techniques in insect nematology. In: Lacey LA (ed) Manual of techniques in insect pathology. Academic Press, San Diego, CA, pp 282–324

Kermarrec A, Mauleon H, Sirjusingh C, Bauld L (1991) Experimental studies on the sensitivity of tropical vertebrates (toads, frogs and lizards) to different species of entomoparasitic nematodes of the genera Heterorhabditis and Steinernema. Rencontres Caraibes Lutte Biologique, 1991. In: Proceedings of the Caribbean meetings on biological control, Guadeloupe, pp 193–204

Khatri-Chhetri HB, Waeyenberge L, Manandhar HK, Moens M (2010) Natural occurrence and distribution of entomopathogenic nematodes (Steinernematidae and Heterorhabditidae) in Nepal. J Invertebr Pathol 103:74–78

Kim Y, Cho JD, Park Y (2005) Two groups of entomopathogenic bacteria, *Photorhabdus* and *Xenorhabdus*, share an inhibitory action against phospholipase A_2 to induce host immunodepression. J Invertebr Pathol 89:258–264

Kim Y, Ahmed S, Stanley D, An C (2018) Eicosanoid-mediated immunity in insects. Dev Comp Immunol 83:130–143

Klein MG (1993) Biological control of scarabs with entomopathogenic nematodes. In: Bedding R, Akhurst R, Kaya H (eds) Nematodes and the biological control of insect pests. CSIRO, East Melbourne, pp 49–58

Koppenhofer AM, Grewal PS, Kaya HK (2000) Synergism of imidaclopri., and entomopathogenic nematodes against white grubs: the mechanism. Entomol Exp Appl 94:283–293

Lacey LA, Georgis R (2012) Entomopathogenic nematodes for control of insect pests above and below ground with comments on commercial production. J Nematol 44(2):218

Lacey LA, Shapiro-Ilan DI (2008) Microbial control of insect pests in temperate orchard systems: potential for incorporation into IPM. Annu Rev Entomol 53:121–144

Lacey LA, Arthurs SP, Unruh TR, Headrick H, Fritts R Jr (2006) Entomopathogenic nematodes for control of codling moth (Lepidoptera: Tortricidae) in apple and pear orchards: effect of nematode species and seasonal temperatures, adjuvants, application equipment and post-application irrigation. Biol Control 37:214–223

Lacey LA, Arthurs SP, Knight A, Huber J (2007) Field manual of techniques in invertebrate pathology. In: Lacey LA, Kaya HK (eds) Application and evaluation of pathogens for control of insects and other invertebrate pests, 2nd edn. Springer, Dordrecht, pp 527–546

Lewis E (2002) Behavioural ecology. In: Gaugler R (ed) EPN. CAB International, Wallingford, pp 205–223

Lortkipanidze M, Hwseynov K, Kokhia M, Gorgadze O, Kuchava M (2019) Effect of temperature on the virulence of entomopathogenic nematodes. Adv Ecol Environ Res:32–38

McCoy CW, Stuart RJ, Duncan LW, Nguyen K (2002) Field efficacy of two commercial preparations of entomopathogenic nematodes against larvae of *Diaprepes abbreviatus* (Coleoptera: Curculionidae) in alfisol type soil. Fla Entomol 85:537–544

McCoy CW, Stuart RJ, Duncan LW, Shapiro-Ilan DI (2007) Application and evaluation of entomopathogens for citrus pest control. In: Lacey LA, Kaya HK (eds) Field manual of techniques in invertebrate pathology: application and evaluation of pathogens for control of insects and other invertebrate pests, 2nd edn. Springer, Dordrecht, pp 567–581

Mollah MI, Roy MC, Choi D, Hasan MA, Al Baki MA, Yeom HS et al (2020) Variations of indole metabolites and NRPS-PKS loci in two different virulent strains of *Xenorhabdus hominickii*. Front Microbiol 11:583594

Mouchlis VD, Dennis EA (2019) Phospholipase A_2 catalysis and lipid mediator lipidomics. Biochim Biophys Acta 1864:766–771

Nollmann FI, Heinrich AK, Brachmann AO, Morrisseau C, Mukherjee K, Casanova-Torres AM et al (2015) A *Photorhabdus* natural product inhibits insect juvenile hormone epoxide hydrolase. ChemBioChem 16:766–771

Park Y, Herbert EE, Cowles CE, Cowles KN, Menard ML, Orchard SS et al (2007) Clonal variation in *Xenorhabdus nematophila* virulence and suppression of *Manduca sexta* immunity. Cell Microbiol 9:645–656

Park Y, Kang S, Sadekuzzaman M, Kim H, Jung JK, Kim Y (2017) Identification and bacterial characteristics of *Xenorhabdus hominickii* ANU101 from an entomopathogenic nematode, *Steinernema monticolum*. J Invertebr Pathol 144:74–87

Poinar GO (1979) Nematodes for biological control of insects. CRC, Boca Raton, FL, p 277
Poinar GO Jr, Leutenegger R (1968) Anatomy of the infective and normal third stage juveniles of *Steinernema carpocapsae* Weiser (Steinernematidae: Nematoda). J Parasitol 54:340–350
Poinar GO Jr, Thomas GR, Hess R (1977) Characteristics of the specific bacterium associated with *Heterorhabditis bacteriophora*. Nematologica 23:97–102
Poinar GO Jr, Thomas GR (1966) Significance of *Achromobacter nematophilus* Poinar and Thomas (Achoromobactriaceae: Eubactreiales) in the development of nematode, DD-136 (Neoplectana sp. Steinernematidae). Parasitology 56:385–390
Poinar GO Jr, Thomas GM (1988) Infection of frogs tadpoles (Amphibia) by insect parasitic nematodes (Rhabditidae). Experientia 44:528–531
Poinar GO Jr, Thomas GM, Presser SB, Hardy JL (1982) Inoculation of entomopatogenous nematodes Neoplectana and Heterorhabditis and their associated bacteria, Xenorhabdus spp., into chicks and mice. Environ Entomol 11:137–138
Qiu L, Bedding R (2000) Energy metabolism and its relation to survival and infectivity of infective juveniles of *Steinernema carpocapsae* under aerobic conditions. Nematology 2(5):551–559
Raheel M, Javed N, Khan SA, Aatif HM, Ahmed S (2017) Exploiting the bio control potential of entomopathogenic nematodes in combination with chemicals against greater wax moth (*galleria mellonella* L.). J Anim Plant Sci 27(03):877–881
Ramliana L, Yadav AK (2010) Laboratory evaluation of the pathogenicity of three entomopathogenic nematodes against larvae of cabbage butterfly, *Pieris brassicae* Linnaeus (Lepidoptera: Pieridae). Sci Vision 9(4):166–173
Rovesti L, Deseo KV (1990) Compatibility of chemical pesticides with the entomopathogenic nematodes, *Steinernema carpocapsae* Weiser and *S. feltiae* Filipzev (Nematoda: Steinernematidae). Nematologica 36:237–245
Sajjadian SM, Kim Y (2020) Dual oxidase-derived reactive oxygen species against *bacillus thuringiensis* and its suppression by eicosanoid biosynthesis inhibitors. Front Microbiol 11:528
Selvan S, Gaugler R, Lewis EE (1993) Biochemical energy reserves of entomopathogenic nematodes. J Parasitol 79:167–172
Seo S, Lee S, Hong Y, Kim Y (2012) Phospholipase A_2 inhibitors synthesized by two entomopathogenic bacteria, *Xenorhabdus nematophila* and *Photorhabdus temperata* subsp. *temperata*. Appl Environ Microbiol 78:3816–3823
Sergeant M, Baxter L, Jarrett P, Shaw E, Ousley M, Winstanley C et al (2006) Identification, typing, and insecticidal activity of *Xenorhabdus* isolates from entomopathogenic nematodes in United Kingdom soil and characterization of the *xpt* toxin loci. Appl Environ Microbiol 72:5895–5907
Sexton SB, Williams P (1981) A natural occurrence of parasitism of *Graphagnathus luecoloma* (Boheman) by the nematode *Heterorhabditis* sp. J Austr Entomol Soc 20:253–255
Shapiro-Ilan DI, Gaugler R (2002) Production technology for entomopathogenic nematodes and their bacterial symbionts. J Ind Microbiol Biotechnol 28:137–146
Shapiro-Ilan DI, Gaugler R, Tedders WL, Brown I, Lewis EE (2002) Optimization of inoculation for in vivo production of entomopathogenic nematodes. J Nematol 34:343–350
Shapiro-Ilan DI, Duncan LW, Lacey LA, Han R (2005) Orchard crops. In: Grewal PS, Ehlers R-U, ShapiroIlan DI (eds) Nematodes as biological control agents. CABI, Wallingford, pp 215–229
Shapiro-Ilan DI, Bruck DJ, Lacey LA (2012a) Principles of epizootiology and microbial control. In: Vega FE, Kaya HK (eds) Insect pathology, 2nd edn. Academic, San Diego, pp 29–72
Shapiro-Ilan DI, Han R, Dolinski C (2012b) Entomopathogenic nematode production and application technology. J Nematol 44:206–217
Shi YM, Bode HB (2018) Chemical language and warfare of bacterial natural products in bacteria-nematode-insect interactions. Nat Prod Rep 35:309–335
Shrestha S, Kim Y (2007) An entomopathogenic bacterium, *Xenorhabdus nematophila*, inhibits hemocyte phagocytosis of *Spodoptera exigua* by inhibiting phospholipase A_2. J Invertebr Pathol 96:64–70
Siddiqui AU, Parihar A, Sharma SR (2010) Importance and utilization of entomopathogenic nematodes for the suppression of insect pests. Gupta HC, Siddiqui AV, Parihar A Bio pest

management (entomopathogenic nematodes, microbes and bioagents) Agrotech Publishing Academy, Udaipur, 62-78

Smitley DR, Warner WR, Bird GW (1992) Influence of irrigation Heterorhabditis bacteriphora on plantparasitic nematodes in turf. J Nematol 24:637–641

Stock SP (1995) Natural populations of entomopathogenic nematodes in the Pampean region of Argentina. Nematropica 25:143–148

Strauch O, Niemann I, Newmann A, Schmidt AJ, Peters A, Ehlers RU (2000) Storage and formulation of the entomopathogenic nematodes. BioControl 45:483–500

Sussmuth RD, Mainz A (2017) Nonribosomal peptide synthesis—principles and prospects. Angew Chem Int Ed Engl 56:3770–3821

Thurston GD, Ito K, Hayes CG, Bates DV, Lippmann M (1994) Respiratory hospital admissions and summertime haze air–pollution in Toronto, Ontario: consideration of the role of acid aerosols. Environ Res 65:271–290

Tobias NJ, Wolff H, Djahanschiri B, Grundmann F, Kronenwerth M, Shi YM et al (2017) Natural product diversity associated with the nematode symbionts *Photorhabdus* and *Xenorhabdus*. Nat Microbiol 2:1676–1685

van Tol RWHM, Raupp MJ (2005) Nursery and tree application. In: Grewal PS, Ehlers R-U, Shapiro-Ilan DI (eds) Nematodes as biological control agents. CABI Publishing, Wallingford, pp 167–190

Wang Y, Gaugler R (1999) Host and penetration site location by entomopathogenic nematodes against Japanese beetle larvae. J Invertebr Pathol 72(3):313–318

White GF (1927) A method for obtaining infective nematode larvae from cultures. Science 66 (1709):302–303

Woodring JL, Kaya HK (1988) Steinernematid and Heterorhabditid nematodes: a handbook of biology and techniques. Southern Cooperative Bulletin 331. Arkansas Agricultural Experiment Station, Fayettville, AR, p 30

Chapter 5
Trends in Neem (*Azadirachta indica*)-Based Botanical Pesticides

Patrick Juma, Njeri Njau, Fiona Wacera W., Cyrus M. Micheni, Haris Ahmed Khan, Oscar W. Mitalo, and David Odongo

Abstract Modern agricultural production is dominated by the use of synthetic chemical pesticides, which account for 95% of the global market share of total pesticide use. However, this over-reliance on synthetic pesticides adversely affects and interferes with the functioning of the ecosystem. Neem (*Azadirachta indica*), a botanical biopesticide widely known for its bactericidal, fungicidal, insecticidal, herbicidal, and nematicidal properties, offers an eco-friendly alternative to synthetic pesticides. To date, more than 200 bioactive compounds have been extracted from neem, and several commercial formulations have been developed and registered as broad-spectrum biopesticides. More advanced strategies in the use of neem as a botanical biopesticide have been developed with a focus on developing more innovative and effective approaches. This chapter also covers current advancement on neem bioactive ingredients, their efficacy and extraction methods. In addition, stability of the bioactive compounds and environmental, health and safety issues are discussed.

Keywords Neem safety · Pesticides · Neem bioactive ingredient · Stability · Biopesticides

---

P. Juma (✉) · N. Njau · F. W. W. · D. Odongo
Department of Horticulture and Food Security (HFS), Jomo Kenyatta University of Agriculture and Technology, Nairobi, Kenya
e-mail: patricjumah@gmail.com

C. M. Micheni
Food Crops Research Institute (FCRI), Kenya Agricultural and Livestock Research Organization (KALRO), Nairobi, Kenya

H. A. Khan
Atta-ur-Rahman School of Applied Biosciences (ASAB), National University of Sciences and Technology (NUST), Sector H-12, Islamabad, Pakistan

O. W. Mitalo
Graduate School of Life and Environmental Science, University of Tsukuba, Tsukuba, Japan

S. De Mandal et al. (eds.), *New and Future Development in Biopesticide Research: Biotechnological Exploration*, https://doi.org/10.1007/978-981-16-3989-0_5

5.1 Introduction

The use of synthetic chemical pesticides in modern agriculture has recorded a steady increase over the past decades (Bernhardt et al. 2017). Synthetic chemical pesticides are often considered a quick, easy, and inexpensive solution for controlling plant insect pests, nematodes, and diseases. However, low degradation rates of these chemical pesticides, coupled with their overuse, have led to accumulation of pesticide residues in the environment. Pesticide residues have been found in the soil and air, as well as in surface and groundwater (Arain et al. 2018). It is also evident that some of these chemicals pose potential risks to humans and adverse effects on the environment since a considerable portion of the pesticides are hazardous. Furthermore, the World Health Organization (WHO), together with the Food and Agriculture Organization (FAO), reported pesticides as one of the leading causes of death as a result of self-poisoning and potentially serious health effects, particularly in low- and middle-income countries (WHO and FAO 2019). The above-mentioned health and environmental hazards associated with synthetic pesticides in agricultural production have enhanced the need for safe alternative pest management approaches. Use of botanical biopesticides is one such pest management strategy with potential to replace chemical pesticides.

Neem (*Azadirachta indica*) has emerged as one of the highly potent botanical biopesticides. The neem tree is an evergreen plant in the Meliaceae family that is native to Assam-Burma, and is widely grown in tropical and subtropical regions (National Research Council (US) Panel on Neem 1992). It is currently distributed across 139 countries of the world, with the most recent distribution recorded in tropical Africa, and Central and South America (Tinghui et al. 2001; Invasive Species Compendium 2019). More than 200 bioactive ingredients belonging to a general class of limonoids have been extracted from the seeds, leaves, bark, oil, and flowers of neem (Benelli et al. 2015). Some of the most commonly researched neem compounds include azadirachtin, salannin, meliantriol, and nimbin. These derivatives have a long history of remarkable use as homegrown biodegradable biopesticides with high efficacy against a wide range of targets (Shannag et al. 2014). Neem is classified as a broad spectrum biopesticide and is currently estimated to be effective against more than 600 species of insects, nematodes, bacteria, viruses, and fungi (Nigam et al. 1994; Singh and Raheja 1996; Saxena 2014). In particular, neem has been reported to be effective in the management of important agricultural pests such as whiteflies, aphids, mealybugs, spider mites, fruit flies, leaf miners, and fall armyworms, among others (Giongo et al. 2016; Mamoon-Ur-Rashid et al. 2016; Shah et al. 2017; Kumar et al. 2019; Illakwahhi et al. 2019; Raga et al. 2020; Venzon et al. 2020). Neem oil extracts were also effectively used to control root-knot nematodes in tomato, chili, and brinjal by inhibiting egg hatching and mobility of juveniles (Sivakumar and Gunasekaran 2011; Khalil 2013). Additionally, the efficacy of neem extracts have been demonstrated against important plant pathogens including *Plasmopara halstedii, Pseudomonas syringae*, and *Erwinia carotovora,* among others (Goel and Paul 2015; Ndivo et al. 2018; Doshi et al. 2020).

Fortunately, no pest resistance against neem-based products has been reported to date, as opposed to the conventional synthetic chemical pesticides. This is attributed to the numerous bioactive components found in neem. Furthermore, neem-based products ranging from unprocessed materials, seed oil, aqueous extracts to pure bioactive ingredients such as azadirachtin, have relatively low toxicity with an estimated safe dosage of 0.26–15 mg/kg bw/day (Boeke et al. 2004). Hence, neem remains one of the less toxic and most promising biopesticides for agricultural applications. Despite the several advantages of neem as a biopesticide, its potential remains largely untapped. This chapter offers a critical description of recent strategies and research in the use of neem as a botanical biopesticide.

5.2 Bioactive Ingredients and Mode of Action

Neem is known to contain sufficient amounts of a variety of bioactive ingredients in the leaf, seed, bark, and oil. Over 200 chemically and structurally diverse limonoids have so far been extracted from different parts of the neem tree (Gupta et al. 2017), and they contain, or are derived from a precursor having a 4,4,8-trimethyl-17-furanylsteroid skeleton. The most widely studied limonoids from the neem tree include azadirachtin, meliantriol, salannin, nimbin, nimbidin, and nimbolide, all of which exhibit a wide range of biological activities such as insecticidal, antibacterial, antifungal, and nematicidal (Roy and Saraf 2006).

5.2.1 Azadirachtin

Azadirachtin was first isolated from the neem tree by Butterworth and Morgan in 1968, and is the most important and well-studied bioactive neem compound (Nicoletti et al. 2012). There are several homologues of azadirachtin named alphabetically from A to K (Lokanadhan et al. 2012). Azadirachtin A is the most abundant and biologically active of the homologues (Barceloux 2008), and hence it is the major active ingredient in most commercial insecticides. Azadirachtin causes many physiological effects in the midgut of insects by triggering a reduction in the digestive efficiency post-ingestion. This reduction in digestive efficiency is known as secondary antifeedancy and is caused by disturbances in the hormonal and physiological systems of the insect, such as inhibition of both digestive enzyme production and food movement through the midgut (Schmutterer 1995). Cellular uptake of azadirachtin also inhibits cell division as well as protein synthesis, causing cell necrosis in the midgut and flaccid paralysis of muscles (Nisbet et al. 1994). Antifeedant sensitivity, however, differs greatly among insect species; therefore, the efficacy of using neem insecticides lies in its lethal physiological effects. In a study carried out by Qiao et al. (2014), azadirachtin was shown to interfere with mitosis causing histopathological effects on the insect gut epithelial cells, fatty tissues, and muscles, resulting in constrained movement and reduced flight activity. This study also suggested that azadirachtin affects the pest's central nervous system. Additionally, Shao et al. (2016) showed that azadirachtin A can significantly inhibit cell proliferation via induction of autophagy and apoptotic cell death.

5.2.2 Salannin

Salannin is a C-seco limonoid, ranking as the second most important of the neem bioactive compounds after azadirachtin (Gribble 2012). It has three homologues including salannin, salannol, and 3-*O*-acetyl salannol, all of which exhibit strong insecticidal activity attributed to additional oxidation of some of the carbon molecules (Koul et al. 2004; Schwinger et al. 1984). Salannin was reported to increase larval stage duration, delay molting, reduce pupal weights and subsequently cause larval and pupal mortality in *Pieris brassicae* and *Oxya fuscovittata* (Govindachari et al. 1996; Lin-Er et al. 1995). The negative effect of salannin on insect growth can be attributed to inhibition of ecdysone 20-monooxygenase, an insect cytochrome P4560-dependent hydroxylase which was shown to be responsible for reproductive cycles and postembryonic development in *Aedes aegypti, Manduca sexta* and *Drosophila melanogaster* (Mitchell et al. 1997). Photoproducts of salannin, that is, salanninonide and isosalanninolide, also showed growth inhibition and strong antifeedant activity against *Spodoptera littoralis, S. frugiperda, Helicoverpa armigera, Schistoserca gregaria* and *Locusta migratoria* (Simmonds et al. 2004). Govindachari et al. (1996) also demonstrated that salannin and azadirachtin have comparable growth regulatory capacities.

Salannin also depicted a strong feeding deterrent activity in California red scale, migratory locust, striped cucumber beetle, and the Japanese beetle (National Research Council (US) Panel on Neem 1992). It has also been shown to have antifeedant activity against *Oxya fuscovittata, S. litura* and *P. ricini* although with lower efficacy compared to azadirachtin (Govindachari et al. 1996). Koul et al. (1996) attributed salannin-induced insect antifeedant behavior to an induced response by the insects' chemoreceptors, as opposed to azadirachtin antifeedant activity which is associated with an interference of gut trypsin production. Salannin also exhibited a concentration dependent oviposition deterrence effect against *Helicoverpa armigera* (Board 2004).

5.2.3 Nimbin

Nimbin is one of the bioactive compounds derived from the neem tree specifically from the leaves (Sadeghian and Mortazaienezhad 2007), and the bark (Alzohairy 2016). The compound was isolated in 1945 by Siddiqui and is classified as a C-seco limonoid occurring together with nimbinol, 6-deacetylnimbin and

6-deacetylnimbinal (Sarah et al. 2019). Sadeghian and Mortazaienezhad (2007) reported that aqueous extracts of neem leaves contained 2.6% nimbin compared to 1.3% of azidarachtin A and that it is active in insect control. Moreover, isonimbinolide is a photo-oxidation product of nimbin that was shown to be as potent as azadirachtin, and it worked by inhibiting feeding in *Locusta migratoria* and *Schistocerca gregaria*. Nimbin has also been shown to have antiviral activity against potato virus Y and X (Shafie et al. 2017; Verma 1974). In addition, nimbin displays anti-inflammatory, antipyretic, fungicidal, antihistamine and antiseptic properties (Sarah et al. 2019).

5.2.4 *Nimbolide*

Nimbolide is another bioactive ingredient of neem that mostly demonstrates herbicidal activity. Allelopathic and phytotoxic activities of nimbolide were demonstrated in a study carried out by Kato-Noguchi et al. (2014), where it inhibited the growth of lettuce, crabgrass, alfalfa, jungle rice, and barnyard grass. This allelopathic phenomenon is environmentally friendly and can provide an alternative to pesticides in weed management.

5.2.5 *Propyl Disulphide*

Propyl disulphide is a neem volatile compound which was shown to sufficiently control stem-end rot caused by *Lasiodiplodia theobromae* and *Neofusicoccum parvum* in mango through inhibition of mycelial growth (Khan et al. 2021). The mechanism for mycelial inhibition by propyl disulphide remains unclear but it is thought to be caused by disruption of cytoplasmic membranes, inactivation of DNA replication, or inactivation of cytoplasmic/membrane-bound enzymes, among other mechanisms.

5.2.6 *Meliantriol*

Meliantriol is a simpler triterpenoid found in low concentrations in neem plants. This compound has been shown to effectively control locust chewing on crops (Mordue et al. 2010), through its feeding deterrent activity. In most cases, neem bioactive ingredients are evaluated together against a wide range of pest diseases and parasitic nematodes. Nurmayulis et al. (2019) reported that neem plant extracts containing azadirachtin, salanin, and meliantriol were effective in controlling *Conopomorpha cramerella* cocoa moth pests. A combination of various neem bioactive ingredients can exhibit multiple modes of action against a wide range of pests thus providing an optimal management strategy in the agricultural field.

5.3 Source of Bioactive Ingredients

As mentioned earlier, the neem tree comprises myriads of bioactive ingredients offering immense pesticidal properties (Koul et al. 2004). These bioactive ingredients are present in variable concentrations in different parts of the plant, including seeds, leaves, flowers, bark, and roots (Van der Nat et al. 1991). Extracts from the different plant parts have been or possess the potential to be exploited for the development of efficient pest control strategies.

5.3.1 Seeds/Kernels

Neem seed kernels contain about 40–45% of oil, which is rich in several bioactive compounds such as azadirachtin, propyl disulphide, nimbin, nimbidin, nimbinin and salannin (Melwita and Ju 2010). However, azadirachtin and propyl disulphide are the dominant compounds in neem oil and are thus the major bioactivity contributors (Isman 2006). Neem oil is considered a contact pesticide, exhibiting a broad spectrum of action, particularly against soft-bodied insects and mites (Chaudhary et al. 2017). Different biological actions of neem oil have been reported in different insect groups. For instance, application of neem oil is known to cause severe midgut damage, injury, and cell death in the larvae of neuroptera (Scudeler et al. 2016), antifeeding effect and increased larvae mortality in lepidoptera (Tavares et al. 2010), and early nymph deaths in hemiptera (Formentini et al. 2016). Therefore, many biopesticide formulations that are based on neem oil have been widely used to control several agricultural pests both in the field and during postharvest handling (Campos et al. 2016). Apart from the pesticidal properties, neem oil formulations also exhibit fungicidal activity. Moline and Locke (1993) demonstrated the effectiveness of neem seed oil to reduce gray mold caused by *Botrytis cinerea* and bitter rot caused by *Glomerella cingulata* in apple during postharvest storage. Additionally, the application of neem seed kernel extract to stored plum and 'Yali' pears greatly reduced the growth rates of four fungal pathogens: *Monilinia fructicola*, *Penicillium expansum*, *Trichothecium roseum*, and *Alternaria alternate*, thus preventing the occurrence of diseases (Wang et al. 2010).

5.3.2 Neem Leaves

The major compounds that contribute to the bioactivity of neem leaf are nimbin, nimbinene, 6-desacetyllnimbinene, nimbandiol, nimbolide, nimocinol, quercetin, and beta-sitosterol (Siddiqui et al. 2004). Neem leaf extracts have been used to effectively control pests of multiple crop species. Additionally, Ahmad et al. (2015) observed that the shelf life of mung beans increased upon application of neem leaf powder, which inhibited the growth of pulse beetle (*Callosobruchus chinensis*). Neem leaf powder also exhibited aphid repellant activity when applied to organic fertilizer (Brotodjojo and Arbiwati 2016), and it seemed to facilitate the growth of earthworms when used as vermicompost.

5.3.3 Bark

The neem bark generally exhibits a relatively lower pesticidal activity than the seed and leaves due to low concentrations of bioactive compounds (Sirohi and Tandon 2014). Nevertheless, the neem bark contains considerable amounts of azadirachtin, cyanogenic glucosides, and nimbin that are known to exhibit biological activity. In this sense, neem bark extracts or powder have shown biological action against certain unwanted herbs and weeds during crop production in the field (Ascher 1993). Furthermore, allelopathic properties of the neem bark were demonstrated in different crops, including alfalfa, rice, radish, and beans, among others (Xuan et al. 2004). A study conducted by Ahmad et al. (2015) also reported that fabrics dyed with neem bark extracts could reduce the disastrous effects of lepidopteran larvae in various crops.

5.3.4 Neem Flowers and Roots

There is limited research on the pesticidal activity of neem flowers and roots. However, neem root extracts have been shown to effectively control root-knot nematodes in tomato plants (Kayani et al. 2001). Additionally, neem flower extracts have been shown to exhibit antifertility effects in Sprague-Dawley rats (Gbotolorun et al. 2008); thus, this can be exploited to effectively control both insect and rodent pests in the agricultural field.

5.3.5 Micro-Propagation for Neem Biopesticide Production

Despite the numerous advantages and potential of neem-based biopesticides, conventional propagation through seed sowing has many limitations such as recalcitrant properties and low viability, and it is time-consuming considering the long life cycle of the neem tree. Therefore, efforts have been made to develop propagation techniques that can enable mass and rapid propagation. Plant micro-propagation techniques for the production of secondary metabolites from neem started in 1983, when it was fronted as a sustainable production strategy of neem bioactive compounds (Schluter and Schulz 1983). Thus far, active azadirachtin and nimbin have been isolated from the callus of various neem explants in sufficient amounts (Srivastava and Srivastava 2012). Most recently, Ashokhan et al. (2020) successfully induced neem leaf explant regeneration which achieved a tenfold accumulation of azadirachtin. However, bioactive compounds production from callus was affected by various factors such as position of the explants, the plant age, callus colour and composition of the media used. Hence, although neem callus production is a promising technique, a lot remains to be standardized before it can be used for sustainable production of bioactive compounds. Another possible technique for mass production of neem bioactive compounds includes the use of suspension cultures (Sujanya et al. 2008), and some of the compounds isolated using this technique are azadirachtin I, azadirachtin B, nimbin and salannin.

5.4 Extraction Methods of Functional Ingredients

Extraction of bioactive ingredients from neem can be carried out using several methods including solvent extraction, mechanical pressing, steam pressure extraction, and super critical extraction.

5.4.1 Solvent Extraction

Most of the bioactive compounds present in the above-mentioned neem plant parts have low solubility in water, but they are highly soluble in polar solvents like alcohols, ketones, ethers, benzene, acetone, chloroform, and hydrocarbons. The solvent extraction method involves grating and steeping of neem parts into the solvent.

Water extraction is the simplest neem bioactive ingredient extraction method commonly used by many small-scale farmers, especially in developing countries. In this method, ground leaves or seeds are mixed with water, soaked overnight, and filtered for use as a sprayable emulsion. However, this requires a lot of water because of the low solubility of the bioactive ingredients. The low solubility of the bioactive ingredients in water results in lower efficacy of aqueous extracts than pure ingredients (Coventry and Allan 2001).

On the other hand, methanol and ethanol are commonly used in the extraction of plant polar compounds, although it may be suitable for some non-polar compounds. Methanol extraction of neem oil was reported to contain various bioactive compounds such as azadirachtin, epoxyazadion, gedunin, nimbin, phenolic, and flavonoids (Hallur et al. 2002; Yehia 2016). Use of either ethanol or methanol results in highly concentrated bioactive ingredients, about 50% higher than water extraction (Schumacher et al. 2011).

Hexane is also widely used in the concentration and purification of neem bioactive ingredients due to its very non-polar characteristics. When added to neem oil, a highly non-polar hexane-oil phase is formed, leading to decreased solubility and subsequent precipitation of limonoids such as azadirachtin (Melwita and Ju 2010). Hexane has high efficiency of extraction of neem bioactive ingredients compared to ethanol and methanol. However, a combination of several solvents provides a more effective extraction and purification compared to a single solvent (Ayoola et al. 2014).

Although the type of solvent used greatly affects the bioactive ingredients quality and quantity of extract, other factors such as operating pressure, solvent flow rate, extraction time and temperature have been found to influence extraction efficiency (Dai et al. 2001). Similarly, using pressurized methanol at 50 °C temperature and 50 bar pressure resulted in optimal azadirachtin yield from the seed kernels (Jadeja et al. 2011). At 50 °C, hexane extraction yielded 44.29%, while ethanol recorded 41.11% neem oil (Liauw et al. 2008).

5.4.2 Mechanical Extraction

Mechanical extraction from neem seeds can be performed between 40 °C and 50 °C using hydraulic pressing equipment. This method is especially suitable for neem seeds which contain 33.5% oil (Orhevba et al. 2018). Soetaredjo et al. (2008) reported that pre-treatment conditions had an effect on the yield and quality of neem oil obtained by mechanical pressing, ascertaining that quality and quantity of neem oil obtained by mechanical extraction is a factor of various operating conditions. Furthermore, the yield of mechanical extraction is dependent on the pressure on seed kernel oil; however, 925.84 KN/m^2 was estimated to be the brake-even point (Orhevba et al. 2013).

5.4.3 Supercritical Extraction

Supercritical carbon dioxide or a combination of supercritical carbon dioxide and methanol extraction have been tested on azadirachtin, nimbin, salannin, and oil separately from neem seeds. Johnson and Morgan (1997) demonstrated that azadirachtin was efficiently extracted at a pressure of 34.4 MPa and 20% methanol, while nimbin and salannin were both extracted at 20.6 MPa and 6% methanol. On the other hand, optimum extraction rate for nimbin was observed at 23 MPa and a flow rate of 1.24 mL/min (Tonthubthimthong et al. 2001). Hence, supercritical extraction can play an important role in the separation of the neem bioactive ingredients.

5.5 Commercial Neem-Based Biopesticides

Neem products have a long history of manufacture on an industrial scale, particularly in Asia and the United States. Commercial neem products have played a major role in organic production and have currently been adopted in many countries around the world. Due to the many advantages of neem products and the ease in the extraction of the bioactive ingredients, many commercial products have been developed for use in the management of a wide range of pests (Table 5.1).

5.6 Stability of Neem Bioactive Ingredients

Although neem bioactive ingredients have been shown to be active against a wide range of pests, the stability of many neem formulations is affected by various environmental factors, which poses a challenge to their application in agriculture. Photodegradation remains the major challenge for biopesticide activity after their release into the environment. Azadirachtin, the most abundant limonoid, was found to have a half-life of 48 min under radiation at 254 nm (Johnson and Dureja 2002). Similarly, Zuleta-Castro et al. (2017) evaluated the photosensitivity of the crude neem extracts. This study reported that limonoids present in the crude extract reduced by 83% under 368 nm UV light for a continuous 24 h. These studies indicate that neem bioactive ingredients are subject to degradation by UV radiation spectrum that reaches the Earth's surface. However, addition of stabilizers such as 2,6-di-tert butyl-*p*-cresol and 8-hydroxyquinoline to bioactive neem compounds improved their stability to 24 h under UV-light and up to 30 days under sunlight (Johnson et al. 2003).

Azadirachtin was also reported to be unstable in mildly alkaline and strongly acidic solutions, while it was most stable in mildly acidic solutions of pH 4–6 (Pereira et al. 2019). A study of thermal stabilities of azadirachtin and salannin in neem-based commercial biopesticides reported that the half-lives of limonoidal compounds ranged from 25.6 to 220 days under a temperature of between 30–54 °C (Kim et al. 2015). Oxidation is another problem facing the stability of limonoidal compounds in neem. According to a study by Kim et al. (2014), the half-life of the total neem limonoid was about 43.3–57.7 days in aerated water, while less than 10% degradation was detected after 56 days in deoxygenated water.

Table 5.1 Notable commercial neem-based products around the world

Commercial product	Active ingredient	Manufacturer
Agroneem®	Azadirachtin	Bahar Agrochem and Feeds Pvt. Ltd (India)
Azact CE®	Azadirachtin	EPP Ltd (Brazil)
AZA-Direct®	Azadirachtin	Gowan Company (USA)
Azamax®	Azadirachtin	PARRY AMERICA INC (USA)
Azatrol®	Azadirachtin	PBI-Gordon Corporation USA
AzeraTM®	Azadirachtin	MGK (USA)
Dalneem®	Azadiractin A and Azadiractin B	Dalquim Ltd (Brazil)
Debug ON®	Triterpenoids and neem oil	Agro Logistic Systems Inc. (USA)
End All®	Clarified hydrophobic extract of neem oil	Woodstream Corporation (USA)
Margosom®	Azadirachtin	Agri Life (India)
Molt-X®	Azadirachtin	BioWorks Inc. (USA)
NeemAzal®T/S	Azadirachtin	Trifolio-M (Germany)
Neemazal Technical®	Limonoid and Azadirachtin A	EcoGrape/Sustain-Ability Ltd (India)
Neem Drop®	Azadirachtin	Neem India Products Ltd. (India)
Neemex®	Limonoids	Agro Extracts Limited, India
Neemix®	Azadirachtin	Certis (USA)
Neemix 4.5®	Azadirachtin	Certis (USA)
Neemix®	Azadirachtin	Agro Logistic Systems Inc. (USA)
Organica®	Saponified neem oil	Organica Biotech Pvt. Ltd (India)
Ozoneem Oil®	Azadirachtin	Ozone Biotech (India)
Shubhdeep Neem Oil®	Azadirachtin	King Agro Food (India)
Triact®	Clarified hydrophobic extract of neem oil	Certis (USA)
Trilogy®	Clarified hydrophobic extract of neem oil	Certis (USA)

5.7 Safety of Neem-Derived Pesticides

Neem has been popularized as an effective biopesticide due to the effectiveness of its active compound azadirachtin. It is attractive for inclusion in pest management strategies because it poses less risk to the environment as compared to synthetic pesticides. A lot of research has been carried out to test the effectiveness of neem products as a substitute for synthetic pesticides. The safety of neem products to humans, animals, and beneficial insects is imperative for the successful adoption of neem in crop production. Safety studies are necessary to determine the likely side effects that can be caused by neem-based biopesticides. Factors that are considered in these studies include population dynamics on non-target organisms, pesticide persistence in the environment, lethal and sub-lethal effects, as well as physiological and behavioral effects on non-target organisms. Wide scope and comprehensive research is needed to ascertain objective risk assessment before approval of biopesticides.

Traditionally, raw neem tree plant parts were used in many areas around the world to treat animal and human ailments as well as in crop protection. Neem was widely used as medicine for the treatment of malaria, ulcers, cardiovascular diseases, and dermal problems, among others (Campos et al. 2016). The neem plant parts were either sun-dried and ground (Boeke et al. 2004), or the juice extracted by leaving the plant parts in water overnight. With recent developments, various biopesticides have been formulated through extraction from different neem plant parts. Among the many known biopesticides, neem oil is considered the least toxic to humans and beneficial organisms (Campos et al. 2016). However, the possibility of aflatoxins presence in sun-dried neem poses a health risk (Boeke et al. 2004), especially if used for grain storage. Cold-pressed neem oil is considered to have low toxicity if consumed or inhaled (U.S. Environmental Protection Agency 2007), and slightly toxic to the skin if exposed to the product. Further, Boeke et al. (2004) presented toxicological data from animal and human consumption studies. Non-aqueous neem extracts were shown to be more toxic (estimated safe dose (ESD) of 0.002 and 12.5 μg/kg bw/day) compared to the raw materials (ESD 0.26 mg/kg bw/day), seed oil (0.3 mg/kg bw/day) and aqueous extracts (2 μL/kg bw/day). However, most of the pure compounds exhibited relatively low toxicity (ESD azadirachtin 15 mg/kg bw/day). The low mammalian toxicity highly ranks neem biopesticides as a safe alternative for pest control in organic crop production (Raguraman and Kannan 2014). Research conducted on rats showed that there were no signs of toxicity or death upon feeding on neem extracts (Mossa et al. 2018). Further, there were no changes in body and organ weights in rats exposed to azadirachtin. The reproductive function of rats was also tested over two generations and the results showed no adverse side effects.

The effect of neem-based products has also been tested on beneficial insects, including pollinators, predators, and parasitoids. Shah et al. (2017) reported that there were significantly abundant natural enemies in neem-treated plots compared to those treated with imidacloprid, a synthetic insecticide. The researchers recommended that more research should be carried out with respect to the degradation of biopesticides and the impact on natural enemies. Further, Raguraman and Kannan (2014) reported that a majority of neem-based biopesticides, either formulated or raw, exhibited insignificant to moderate negative effects on parasitoids, predators, and pollinators. In addition, El-Wakeil et al. (2006) evaluated the side effects of neem products on natural enemies of *Helicoverpa armigera* Hüb, *Trichogramma* spp. and *Chrysoperla* spp. The research found that there were no severe negative impacts on parasitism and emergence rates of *Trichogramma* spp. Moreover, the efficiency of *Chrysoperla spp.* was not affected. The research concluded that neem products were compatible with the mass release of *Trichogramma* and *Chrysoperla* for controlling *Helicoverpa* in organic cotton production. Neem seed oil also did not reduce the parasitism rate of *Diaeretiella rapae* Mcintosh on the green peach aphid (*Myzus persicae* Sulzer) (Lowery and Isman 1995). However, the emergence of parasitoid adults from aphid mummies was reduced. The study further recommended that neem biopesticides are suitable for use in integrated pest management. The effects of neem products on a non-target isopod, *Porcellionides pruinosus* were also investigated (Doshi et al. 2018). This research concluded that the neem products did not present any risk to these terrestrial isopods, responsible for decomposition.

Despite the numerous benefits, some researchers have highlighted safety concerns in the use of neem products for agriculture. Research by Scudeler et al. (2016) showed that neem products did not show selectivity for green lacewings (*Chrysoperla claveri*), the natural predators of soft-bodied insect pests including aphids, thrips, whitefly, leafhoppers, spider mites, and mealybugs. All stages of their life cycle were negatively affected through exposure by feeding on poisoned prey during the larval stages. Further, the ingestion of neem oil caused delayed development and the death of *C. claveri* larvae. These effects are likely to endanger the population of this species in an agroecosystem. Neem oil was also found to have lethal effects on the third, fourth, and fifth instar nymphs and adults of the non-target predator *Podisus nigrispinus* (Zanuncio et al. 2016). Mortality rates of third instar nymphs were directly proportional to the neem oil concentrations.

Additionally, exposure of gills of fish to 0.03 g/L neem aqueous extract resulted in epithelial lifting, shortening, and necrosis of the secondary lamellae, cellular hyperplasia of primary filament, and globular haematomas (Alim and Matter 2015). The study recommended precautionary use of aqueous neem extracts. Some studies on mutation have also shown that neem extracts caused increased chromosomal aberration in spermatocytes and bone marrow (Mossa et al. 2018). Collectively, the use of neem products is considered safe, as it has no persistent or long-term residual effects on the environment. However, neem biopesticides should be used with good understanding of the negative effects on the non-target species mentioned above.

5.8 Future Trends

The potential mode of action of the multitude of bioactive ingredients present in neem extracts accounts for the zero pest resistance reported. However, Feng and Isman (1995) highlighted the possibility of neem pesticide resistance under repeated application. Therefore, as a precautionary measure, the current pest management should focus on combining two or more plants or species extracts to boost their bio-efficacy and minimize the risk of resistance development in the future.

In pest management practices, bioactive ingredients found in neem biopesticides have the disadvantage of quick degradation when exposed to varying environmental conditions. This hurdle can be circumvented through the addition of stabilizers to enhance their stability. Therefore, selecting an appropriate stabilizer is important to the improvement of product stability and reduction of the inconsistencies observed in their efficacy. Furthermore, nanotechnology offers great potential in improving the physicochemical stability, degradability, and effectiveness of the biopesticides. This technology has more to offer, including the reduction of toxicity to non-target organisms due to the slow release of the bioactive ingredients and enhancement of product stability (Pasquoto-Stigliani et al. 2017). Nanoparticles, particularly the polymeric ones, have high biodegradability and biocompatibility and are generally inert. Hence, nanotechnology should become an integral component for developing environmentally friendly neem-based pesticide formulations for efficient and effective agriculture applications.

References

Ahmad S, Ansari MS, Muslim M (2015) Toxic effects of neem based insecticides on the fitness of Helicoverpa armigera (Hübner). Crop Prot 68:72–78. https://doi.org/10.1016/j.cropro.2014.11.003

Alim DIM, Matter HMA (2015) Histopathological alteration induced in gills of juvenile Nile Tilapia *Oreochromis niloticus* upon exposure to two bio-pesticides. Int J Fish Aquat Stud 2:80–83. https://www.fisheriesjournal.com/vol2issue5/Pdf/2-5-57.1.pdf

Alzohairy MA (2016) Therapeutics role of *Azadirachta indica* (neem) and their active constituents in diseases prevention and treatment. Evid Based Complement Altern Med 1:1–11. https://doi.org/10.1155/2016/7382506

Arain M, Brohi KM, Channa A, Brohi ROZ, Kumar K, Sameeu A (2018) Analysis of Chlorpyrifos pesticide residues in surface water, ground water and vegetables through gas chromatography. J Int Environ Appl Sci 13:167–173. https://dergipark.org.tr/en/pub/jieas/issue/39948/454015

Ascher KRS (1993) Nonconventional insecticidal effects of pesticides available from the neem tree, *Azadirachta indica*. Arch Insect Biochem Physiol 22:433–449. https://doi.org/10.1002/arch.940220311

Ashokhan S, Othman R, Rahim MHA, Karsani SA, Yaacob JS (2020) Effect of plant growth regulators on coloured callus formation and accumulation of azadirachtin, an essential biopesticide in *Azadirachta indica*. Plan Theory 9:1–17. https://doi.org/10.3390/plants9030352

Ayoola AA, Efeovbokhan VE, Bafuwa OT, David OT (2014) A search for alternative solvent to hexane during neem oil extraction. Int J Sci Technol 4:66–70. http://www.ejournalofsciences.org

Barceloux DG (2008) Medical toxicology of natural substances: foods, fungi, medicinal herbs, plants, and venomous animals, 1st edn. Wiley, Hoboken, NJ. https://doi.org/10.1002/9780470330319

Benelli G, Murugan K, Panneerselvam C, Madhiyazhagan P, Conti B, Nicoletti M (2015) Old ingredients for a new recipe? Neem cake, a low-cost botanical by-product in the fight against mosquito-borne diseases. Parasitol Res 114:391–397. https://doi.org/10.1007/s00436-014-4286-x

Bernhardt ES, Rosi EJ, Gessner MO (2017) Synthetic chemicals as agents of global change. Front Ecol Environ 15:84–90. https://doi.org/10.1002/fee.1450

Board N (2004) Hand book on neem & allied products, 1st edn. Asia Pacific Business Press, New Delhi. https://www.niir.org/books/book_pdf/49/niir-hand-book-on-neem-allied-products.pdf

Boeke SJ, Boersma MG, Alink GM, van Loon JJAA, van Huis A, Dicke M, Rietjens IMCMCM (2004) Safety evaluation of neem (*Azadirachta indica*) derived pesticides. J Ethnopharmacol 94:25–41. https://doi.org/10.1016/j.jep.2004.05.011

Brotodjojo RR, Arbiwati D (2016) Effect of application of granular organic fertilizer enriched with boiler ash and neem leaves powder on plant resistance against insect pests. Int J Biosci Biochem Bioinforma 6:152–157. https://doi.org/10.17706/ijbbb.2016.6.4.152-157

Campos EVRR, de Oliveira JL, Pascoli M, de Lima R, Fraceto LF (2016) Neem oil and crop protection: from now to the future. Front Plant Sci 7:1–8. https://doi.org/10.3389/fpls.2016.01494

Chaudhary S, Kanwar RK, Sehgal A, Cahill DM, Barrow CJ, Sehgal R, Kanwar JR (2017) Progress on *Azadirachta indica* based biopesticides in replacing synthetic toxic pesticides. Front Plant Sci 8:610. https://doi.org/10.3389/fpls.2017.00610

Coventry E, Allan EJ (2001) Microbiological and chemical analysis of neem (*Azadirachta indica*) extracts: new data on antimicrobial activity. Phytoparasitica 29:441. https://doi.org/10.1007/BF02981863

Dai J, Yaylayan VA, Raghavan GSV, Paré JRJ, Liu Z, Bélanger JMR (2001) Influence of operating parameters on the use of the microwave-assisted process (MAP) for the extraction of azadirachtin-related limonoids from neem (*Azadirachta indica*) under atmospheric pressure conditions. J Agric Food Chem 49:4584–4588. https://doi.org/10.1021/jf010592k

Doshi P, Póss AM, Tóth F, Szalai M, Turóczi G (2018) Effect of neem-derived plant protection products on the isopod species *Porcellionides pruinosus* (Brandt, 1833). Zookeys 2018:415–425. https://doi.org/10.3897/zookeys.801.25510

Doshi P, Nisha N, Yousif AIA, Körösi K, Bán R, Turóczi G (2020) Preliminary investigation of effect of neem-derived pesticides on plasmopara halstedii pathotype 704 in sunflower under in vitro and in vivo conditions. Plan Theory 9:1–14. https://doi.org/10.3390/plants9040535

El-Wakeil NE, Gaafar NM, Vidal S (2006) Side effect of some neem products on natural enemies of Helicoverpa (Trichogramma spp.) and Chrysoperla carnea. Arch Phytopathol Plant Prot 39:445–455. https://doi.org/10.1080/03235400500356160

Feng R, Isman MB (1995) Selection for resistance to azadirachtin in the green peach aphid, *Myzus persicae*. Experientia 51:831–833. https://doi.org/10.1007/BF01922438

Formentini MA, Alves LFA, Schapovaloff ME (2016) Insecticidal activity of neem oil against Gyropsylla spegazziniana (Hemiptera: Psyllidae) nymphs on Paraguay tea seedlings. Braz J Biol 76:951–954. https://doi.org/10.1590/1519-6984.04915

Gbotolorun SC, Osinubi AA, Noronha CC, Okanlawon AO (2008) Antifertility potential of neem flower extract on adult female Sprague-Dawley rats. Afr Health Sci 8:168–173. https://www.ncbi.nlm.nih.gov/pmc/articles/PMC2583274/?report=reader

Giongo AMM, Vendramim JD, Forim MR (2016) Evaluation of neem-based nanoformulations as alternative to control fall armyworm. Cienc e Agrotecnologia 40:26–36. https://doi.org/10.1590/S1413-70542016000100002

Goel N, Paul PK (2015) Plant age affects elicitation of polyphenol oxidase activity by neem extract in *Solanum lycopersicum* against Pseudomonas syringae pv. Tomato. Isr J Plant Sci 62:283–293. https://doi.org/10.1080/07929978.2015.1035606

Govindachari TR, Narasimhan NS, Suresh G, Partho PD, Gopalakrishnan G (1996) Insect antifeedant and growth-regulating activities of Salannin and other c-seco limonoids from neem oil in relation to Azadirachtin. J Chem Ecol 22:1453–1461. https://doi.org/10.1021/jf00096a024

Gribble GW (2012) Progress in the chemistry of organic natural products. Springer Science & Business Media, Berlin. https://doi.org/10.1007/978-3-7091-6971-1

Gupta SC, Prasad S, Tyagi AK, Kunnumakkara AB, Aggarwal BB (2017) Neem (*Azadirachta indica*): an indian traditional panacea with modern molecular basis. Phytomedicine 34:14–20. https://doi.org/10.1016/j.phymed.2017.07.001

Hallur G, Sivramakrishnan A, Bhat SV (2002) Three new tetranortriterpenoids from neem seed oil. J Nat Prod 65:1177–1179. https://doi.org/10.1021/np0105174

Illakwahhi DT, Bali B, Srivastava L (2019) Improving the efficacy of Abamectin using neem oil in controlling tomato Leafminers, Tuta absoluta (Meyrick). Adv J Chem A 2:216–224. https://doi.org/10.33945/sami/ajca.2019.2.216224

Invasive Species Compendium (2019) *Azadirachta indica* (neem tree). https://www.cabi.org/isc/datasheet/8112#0D3CFD2F-B360-4EAF-A39A-12737BB2EAB8. Accessed 1 Jan 2021

Isman MB (2006) Botanical insecticides, deterrents, and repellents in modern agriculture and an increasingly regulated world. Annu Rev Entomol 51:45–66. https://doi.org/10.1146/annurev.ento.51.110104.151146

Jadeja GC, Maheshwari RC, Naik SN (2011) Extraction of natural insecticide azadirachtin from neem (*Azadirachta indica* A. Juss) seed kernels using pressurized hot solvent. J Supercrit Fluids 56:253–258. https://doi.org/10.1016/j.supflu.2011.01.004

Johnson S, Morgan ED (1997) Supercritical fluid extraction of oil and triterpenoids from neem seeds. Phytochem Anal 8:228–232. https://doi.org/10.1002/(SICI)1099-1565(199709/10)8:5<228::AID-PCA367>3.0.CO;2-C

Johnson S, Dureja P (2002) Effect of surfactants on persistence of azadirachtin-a (neem based pesticide). J Environ Sci Heal Part B 37:75–80. https://doi.org/10.1081/PFC-120002899

Johnson S, Dureja P, Dhingra S (2003) Photostabilizers for Azadirachtin-A (a neem-based pesticide). J Environ Sci Health Pt B 38:451–462. https://doi.org/10.1081/PFC-120021665

Kato-Noguchi H, Salam MA, Ohno O, Suenaga K (2014) Nimbolide B and nimbic acid B, phytotoxic substances in neem leaves with allelopathic activity. Molecules 19:6929–6940. https://www.ncbi.nlm.nih.gov/pmc/articles/PMC6290558/

Kayani MZ, Sarwar G, Muhammad S (2001) Control of root-knot nematode (*Meloidogyne incognita*) on tomato plants by using root extracts of plants. J Agric Trop Subtrop 102:143–146. https://www.researchgate.net/publication/236694679_Control_of_root-knot_nematode_Meloidogyne_incognita_on_tomato_plants_by_using_root_extracts_of_plants

Khalil M (2013) Abamectin and Azadirachtin as eco-friendly promising biorational tools in integrated nematodes management programs. J Plant Pathol Microbiol 4:4. https://doi.org/10.4172/2157-7471.1000174

Khan MR, Chonhenchob V, Huang C, Suwanamornlert P (2021) Antifungal activity of propyl disulfide from neem (Azadirachta indica) in vapor and agar diffusion assays against anthracnose pathogens (Colletotrichum gloeosporioides and Colletotrichum acutatum) in mango fruit. Microorganisms 9:839

Kim JH, Jeong D-Y, Jin C-L, Kim W-I, Lim S-J, Choi G-H, Park B-J (2014) Stability of four limonoidal substances of neem extract under controlled aquatic and soil conditions. Korean J Pestic Sci 18:156–160. https://doi.org/10.7585/kjps.2014.18.3.156

Kim JH, Choi G-H, Kang J-E, Park B-J (2015) Stability of representative active compounds on commercial biopesticides based on neem or *Sophora flavescens* extract under controlled temperature. Korean J Pestic Sci 19:88–92. https://doi.org/10.7585/kjps.2015.19.2.88

Koul O, Shankar JS, Kapil RS (1996) The effect of neem allelochemicals on nutritional physiology of larval Spodoptera litura. Entomol Exp Appl 79:43–50. https://doi.org/10.1111/j.1570-7458.1996.tb00807.x

Koul O, Singh G, Singh R, Singh J, Daniewski WM, Berlozecki S (2004) Bioefficacy and mode-of-action of some limonoids of salannin group from *Azadirachta indica* A. Juss and their role in a multicomponent system against lepidopteran larvae. J Biosci 29:409–416. https://doi.org/10.1007/BF02712112

Kumar R, Kranthi S, Nagrare VS, Monga D, Kranthi KR, Rao N, Singh A (2019) Insecticidal activity of botanical oils and other neem-based derivatives against whitefly, *Bemisia tabaci* (Gennadius) (Homoptera: Aleyrodidae) on cotton. Int J Trop Insect Sci 39:203–210. https://doi.org/10.1007/s42690-019-00027-4

Liauw MY, Natan FA, Widiyanti P, Ikasari D, Indraswati N, Soetaredjo FE (2008) Extraction of neem oil (*Azadirachta indica* A. Juss) using n-hexane and ethanol: studies of oil quality, kinetic and thermodynamic. ARPN J Eng Appl Sci 3:49–54. https://citeseerx.ist.psu.edu/viewdoc/download?doi=10.1.1.613.7721&rep=rep1&type=pdf

Lin-Er L, van Loon JJA, Schoonhoven LM (1995) Behavioural and sensory responses to some neem compounds by *Pieris brassicae* larvae. Physiol Entomol 20:134–140. https://doi.org/10.1111/j.1365-3032.1995.tb00809.x

Lokanadhan S, Muthukrishnan P, Jeyaraman S (2012) Neem products and their agricultural applications. J Biopest 5:72–76. http://www.jbiopest.com/users/lw8/efiles/vol_5_0_72_76f.pdf

Lowery DT, Isman MB (1995) Toxicity of neem to natural enemies of aphids. Phytoparasitica 23:297–306. https://doi.org/10.1007/BF02981422

Mamoon-Ur-rashid M, Jilani MS, Khan Q, Hashim MM, Sayal OU, Khan MP, Latif A, Waseem K, Nawaz S (2016) Evaluation of neem (*Azadirachta indica*) derivatives against jassids (Emrasca devastans) and cotton mealybug (Phenacoccus solenopsis), and side effects on the feeding potential of green lacewing (chrysoperla carnea) on cotton aphid (*Aphis gossypii*). Pak J Zool 48:1763–1768. https://www.zsp.com.pk/pdf48/1763-1768%20(21)%20QPJZ-0448-2015-F%20Revied.pdf

Melwita E, Ju Y-H (2010) Separation of azadirachtin and other limonoids from crude neem oil via solvent precipitation. Sep Purif Technol 74:219–224. https://doi.org/10.1016/j.seppur.2010.06.008

Mitchell MJ, Smith SL, Johnson S, Morgan ED (1997) Effects of the neem tree compounds azadirachtin, salannin, nimbin, and 6-desacetylnimbin on ecdysone 20-monooxygenase activity. Arch Insect Biochem Physiol 35:199–209. https://doi.org/10.1002/(SICI)1520-6327(1997)35:1/2<199::AID-ARCH18>3.0.CO;2-6

Moline HE, Locke JC (1993) Comparing neem seed oil with calcium chloride and fungicides for controlling postharvest apple decay. Hort Sci 28:719–720. https://doi.org/10.21273/HORTSCI.28.7.719

Mordue AJ, Morgan ED, Nisbet AJ (2010) Azadirachtin, a natural product in insect control. In: Gilbert LI, Gill SS (eds) Insect control: biological and synthetic agents. Academic Press, London, pp 186–190. https://books.google.co.jp/books?hl=en&lr=&id=nd2euFHjQyQC&oi=fnd&pg=PA185&dq=Azadirachtin#v=onepage&q=Azadirachtin&f=false

Mossa AH, Mohafrash SMM, Chandrasekaran N (2018) Safety of natural insecticides: toxic effects on experimental animals. Biomed Res Int 2018:4308054. https://doi.org/10.1155/2018/4308054

National Research Council (US) Panel on Neem (1992) Neem. In: Ruskin B (ed) A tree for solving problems. National Research Council, Washington, DC, pp 30–39. https://www.nap.edu/read/1924/chapter/1

Ndivo FM, Njeru EM, Birgen J (2018) Efficacy of neem, garlic and Aloe extracts in the management of postharvest potato soft rot caused by *Erwinia carotovora*. Asian J Res Crop Sci 1:1–7. https://doi.org/10.9734/ajrcs/2018/40255

Nicoletti M, Petitto V, Gallo FR, Multari G, Federici E, Palazzino G (2012) The modern analytical determination of botanicals and similar novel natural products by the HPTLC fingerprint approach. Stud Nat Prod Chem 37:217–258. https://doi.org/10.1016/B978-0-444-59514-0.00008-0

Nigam SK, Misra G, Sharma A (1994) Neem: a promising natural insecticide. In: Applied botany abstracts. Economic Botany Information Service, National Botanical Research Institute, Champaign, pp 35–46. https://books.google.co.jp/books/about/Applied_Botany_Abstracts.html?id=XYho849suIEC&redir_esc=y

Nisbet AJ, Woodford JAT, Strang RHC (1994) The effects of azadirachtin-treated diets on the feeding behaviour and fecundity of the peach-potato aphid, *Myzus persicae*. Entomol Exp Appl 71:65–72. https://doi.org/10.1111/j.1570-7458.1994.tb01770.x

Nurmayulis, Eris FR, Hastuti D, Utami RT, Denny YR, Firmansyah T (2019) Effects of neem plant extract (Azadirachta indica A. Juss) and bio-surfactant diethanolamide olien from palm oil to the mortality of cacao moth pest (Conopomorpha cramerella). In: IOP conference series: earth and environmental science. IOP, Bristol, p 12034. https://doi.org/10.1088/1755-1315/383/1/012034

Orhevba BA, Chukwu O, Osunde ZD, Ogwuagwu V (2013) Studies on the effect of pressure on yield of mechanically expressed neem seed kernel oil. Glob J Eng Des Technol 2:20–24. https://www.longdom.org/articles/studies-on-the-effect-of-pressure-on-yield-of-mechanically-expressed-neem-seed-kernel-oil.pdf

Orhevba BA, Chukwu O, Zinash D, Sunmonu MO (2018) Statistical modelling of oil expression from neem seed using a screw press. Arid Zone J Eng Technol Environ 14:450–468. https://

www.academia.edu/37721500/Statistical_Modelling_of_Oil_Expression_from_Neem_Seed_using_a_Screw_Press

Pasquoto-Stigliani T, Campos EVR, Oliveira JL, Silva CMG, Bilesky-José N, Guilger M, Troost J, Oliveira HC, Stolf-Moreira R, Fraceto LF (2017) Nanocapsules containing neem (*Azadirachta indica*) oil: development, characterization, and toxicity evaluation. Sci Rep 7:1–12. https://doi.org/10.1038/s41598-017-06092-4

Pereira VV, Kumar D, Agiwal M, Prasad TG (2019) Stability of azadirachtin: a tetranortriterpenoid from neem tree. Int J Chem Stud 7:412–419. https://www.chemijournal.com/archives/2019/vol7issue6/PartH/7-5-631-904.pdf

Qiao J, Zou X, Lai D, Yan Y, Wang Q, Li W, Deng S, Xu H, Gu H (2014) Azadirachtin blocks the calcium channel and modulates the cholinergic miniature synaptic current in the central nervous system of Drosophila. Pest Manag Sci 70:1041–1047. https://doi.org/10.1002/ps.3644

Raga A, Silva SB, Louzeiro LRF, Sousa EM (2020) Larvicidal and adulticidal activities of neem oil against Anastrepha fraterculus (Wied.) (Tephritidae). Euro J Med Plants 31:44–51. https://doi.org/10.9734/ejmp/2020/v31i1030280

Raguraman S, Kannan M (2014) Non-target effects of botanicals on beneficial arthropods with special reference to *Azadirachta indica*. Springer, New Delhi. https://doi.org/10.1007/978-81-322-2006-0_10

Roy A, Saraf S (2006) Limonoids: overview of significant bioactive triterpenes distributed in plants kingdom. Biol Pharm Bull 29:191–201. https://doi.org/10.1248/bpb.29.191

Sadeghian MM, Mortazaienezhad F (2007) Investigation of compounds from *Azadirachta indica* (neem). Asian J Plant Sci 6:444–445. https://doi.org/10.3923/ajps.2007.444.445

Sarah R, Tabassum B, Idrees N, Hussain MK (2019) Bio-active compounds isolated from neem tree and their applications. In: Natural bio-active compounds. Springer, Berlin, pp 509–528. https://doi.org/10.1007/978-981-13-7154-7_17

Saxena RC (2014) Neem for sustainable pest and vector management and environmental conservation. Adv Biotechnol Patenting 241. https://www.echocommunity.org/en/resources/d7cd259f-eb9e-4a04-bc17-b26bdaf7e42d

Schmutterer H (1995) The neem tree: source of unique natural products for integrated pest management, medicine, industry and other purposes. VCH, Weinheim, pp 1–696. https://doi.org/10.1002/3527603980

Schumacher M, Cerella C, Reuter S, Dicato M, Diederich M (2011) Anti-inflammatory, pro-apoptotic, and anti-proliferative effects of a methanolic neem (*Azadirachta indica*) leaf extract are mediated via modulation of the nuclear factor-κB pathway. Genes Nutr 6:149–160. https://doi.org/10.1007/s12263-010-0194-6

Schwinger M, Ehhammer B, Kraus W (1984) Methodology of the *Epilachna varivestis* bioassay of antifeedants demonstrated with some compounds from Azadirachta indica and *Melia azedarach*. In: Natural pesticides from the neem tree and other tropical plants. Proceedings of the 2nd International Neem Conference Deutsche Gesellschaft fur Technische Zusammenarbeit (GTZ), pp 111–125. https://ci.nii.ac.jp/naid/10009708667/

Scudeler EL, Silvia A, Garcia G, Padovani CR, Fernanda P, Pinheiro F, Carvalho D (2016) Are the biopesticide neem oil and the predator Ceraeochrysa claveri (Navás, 1911) compatible? J Entomol Zool Stud 4:340–346. https://www.entomoljournal.com/archives/2016/vol4issue2/PartE/4-3-27.1.pdf

Schluter U, Schulz WD (1983) Structural damages caused by neem in Epilachna verivestis: a summary of histological and ultrastructural damage. 1. Tissues affected in larvae. Proc. 2nd Int. neem conf. Rottach. Egern, p 294

Shafie R, Kheder A, Farghaly A (2017) Induction of resistance in pepper plants against potato virus Y (PVY) NTN by two medicinal and aromatic plant essential oils and their major components. Egypt J Phytopathol 45:1–15. https://doi.org/10.21608/ejp.2017.89458

Shah FM, Razaq M, Ali A, Han P, Chen J (2017) Comparative role of neem seed extract, moringa leaf extract and imidacloprid in the management of wheat aphids in relation to yield losses in Pakistan. PLoS One 12:1–14. https://doi.org/10.1371/journal.pone.0184639

Shannag HS, Capinera JL, Freihat NM (2014) Efficacy of different neem-based biopesticides against green peach aphid, *Myzus persicae* (Hemiptera: Aphididae). Int J Agric Policy Res 2:61–68. http://www.journalissues.org/journals-home.php?id=1

Shao X, Lai D, Zhang L, Xu H (2016) Induction of autophagy and apoptosis via PI3K/AKT/TOR pathways by azadirachtin a in spodoptera litura cells. Sci Rep 6:1–12. https://doi.org/10.1038/srep35482

Siddiqui BS, Afshan F, Gulzar T, Hanif M (2004) Tetracyclic triterpenoids from the leaves of *Azadirachta indica*. Phytochemistry 65:2363–2367. https://doi.org/10.1248/cpb.51.415

Simmonds MSJ, Jarvis AP, Johnson S, Jones GR, Morgan ED (2004) Comparison of anti-feedant and insecticidal activity of nimbin and salannin photo-oxidation products with neem (*Azadirachta indica*) limonoids. Pest Manag Sci 60:459–464. https://doi.org/10.1002/ps.834

Singh RP, Raheja AK (1996) Strategies in Management of Insect Pests with Neem (Azadirachta indica A. Juss). In: Singh RP, Chari MS, Raheja K, Kraus W (eds) Neem and environment, vol I. Oxford and IBH Publishing Co. Pvt. Ltd, New Delhi

Sirohi A, Tandon P (2014) Insecticidal effects of various parts of *Azadirachta indica* on adults of Aulacophora foveicollis (Lucas)(Coleoptera: Chrysomelidae). Trends Biosci 7:3947–3949. http://www.indianjournals.com/ijor.as

Srivastava S, Srivastava AK (2012) Azadirachtin production by hairy root cultivation of Azadirachta indica in a modified stirred tank reactor. Bioprocess Biosyst Eng 35:1549–1553. https://link.springer.com/article/10.1007/s00449-012-0745-x

Sivakumar M, Gunasekaran K (2011) Management of root knot nematodes in tomato, chilli and brinjal by neem oil formulations. J Biopest 4:198–200. http://www.jbiopest.com/users/LW8/efiles/Vol_4_2_264C.pdf

Soetaredjo FE, Budijanto GM, Prasetyo RI, Indraswati N (2008) Effects of pre-treatment condition on the yield and quality of neem oil obtained by mechanical pressing. ARPN J Eng Appl Sci 3:45–49. http://www.arpnjournals.com/jeas/research_papers/rp_2008/jeas_1008_130.pdf

Sujanya S, Devi BP, Sai I (2008) In vitro production of azadirachtin from cell suspension cultures of Azadirachta indica. J Biosci 33:113–120. https://link.springer.com/article/10.1007/s12038-008-0027-6

Tavares WS, Costa MA, Cruz I, Silveira RD, Serrao JE, Zanuncio JC (2010) Selective effects of natural and synthetic insecticides on mortality of *Spodoptera frugiperda* (Lepidoptera: Noctuidae) and its predator Eriopis connexa (Coleoptera: Coccinellidae). J Environ Sci Health Pt B 45:557–561. https://doi.org/10.1080/03601234.2010.493493

Tinghui X, Wegener M, Shea OM, Deling M, Shea WMO (2001) World distribution and trade in neem products with reference to their potential in China. In: Conference of Australian Agricultural and Resource Economics Society, Adelaide, pp 1–15. https://econpapers.repec.org/paper/agsaare01/126066.htm

Tonthubthimthong P, Chuaprasert S, Douglas P, Luewisutthichat W (2001) Supercritical CO2 extraction of nimbin from neem seeds—an experimental study. J Food Eng 47:289–293. https://doi.org/10.1016/S0260-8774(00)00131-X

U.S. Environmental Protection Agency (2007) Cold pressed neem oil (025006) Fact Sheet. 2. https://www.yumpu.com/en/document/view/7300723/cold-pressed-neem-oil-025006-fact-sheet-us-environmental-

Van der Nat JM, Van der Sluis WG, De Silva KTD, Labadie RP (1991) Ethnopharmacognostical survey of *Azadirachta indica* A. Juss (Meliaceae). J Ethnopharmacol 35:1–24. https://doi.org/10.1016/0378-8741(91)90131-V

Venzon M, Togni PHB, Perez AL, Oliveira JM (2020) Control of two-spotted spider mites with neem-based products on a leafy vegetable. Crop Prot 128:1–13. https://doi.org/10.1016/j.cropro.2019.105006

Verma VS (1974) Chemical compounds from *Azadirachta indica* as inhibitors of potato virus X. Acta Microbiol Pol B 6:9. https://pubmed.ncbi.nlm.nih.gov/4829363/

Wang J, Li J, Cao J, Jiang W (2010) Antifungal activities of neem (*Azadirachta indica*) seed kernel extracts on postharvest diseases in fruits. Afr J Microbiol Res 4:1100–1104

World Health Organization and Food and Agriculture Organization of the United Nations (2019) Preventing suicide: a resource for pesticide registrars and regulators. World Health Organization, Geneva. https://www.paho.org/en/node/64630

Xuan TD, Tsuzuki E, Hiroyuki T, Mitsuhiro M, Khanh TD, Chung I-M (2004) Evaluation on phytotoxicity of neem (*Azadirachta indica* A. Juss) to crops and weeds. Crop Prot 23:335–345. https://doi.org/10.1016/j.cropro.2003.09.004

Yehia HM (2016) Methanolic extract of neem leaf (*Azadirachta indica*) and its antibacterial activity against foodborne and contaminated bacteria on sodium dodecyl sulfate-polyacrylamide gel electrophoresis (SDS-PAGE). Am J Agric Environ Sci 16:598–604. https://doi.org/10.5829/idosi.aejaes.2016.16.3.1037

Zanuncio JC, Mourão SA, Martínez LC, Wilcken CF, Ramalho FS, Plata-rueda A, Soares MA, Serrão JE (2016) Toxic effects of the neem oil (*Azadirachta indica*) formulation on the stink bug predator, Podisus nigrispinus (Heteroptera: Pentatomidae). Sci Rep 6:30261. https://doi.org/10.1038/srep30261

Zuleta-Castro C, Rios D, Hoyos R, Orozco-Sánchez F (2017) First formulation of a botanical active substance extracted from neem cell culture for controlling the armyworm. Agron Sustain Dev 37:2–8. https://doi.org/10.1007/s13593-017-0448-4

Chapter 6
Role of Plant Essential Oils in Pest Management

Lizzy A. Mwamburi

Abstract The growing concern over potential hazards from chemical pesticide safety among consumers and potential harm to the environment has culminated in consideration of natural management strategies of pests. Because they are complementary to most crop production systems, biopesticides based on plants can be integrated into pest management systems. Plant essential oils (EOs) can replace the more persistent non-natural pesticides in protecting the environment from the accumulation of chemicals reduce resistance and increase crop productivity. In addition, they possess low mammalian toxicity, broad-spectrum activity, and degrade rapidly in foodstuffs. In addition to exhibiting distinctive properties compared with synthetic pesticides, including high levels of pest toxicity and reduced toxicity toward non-target organisms, EOs possess contact, feeding deterrence, fumigant toxicity, oviposition, and repellent properties. In this chapter, we review the sources of EOs, their insecticidal activities, constituents, and mode of action and discuss their synergism and formulation with encapsulation for producing nanoinsecticidal products.

Keywords Essential oils · Arthropod pests · Insecticidal activity

6.1 Introduction

In spite of its importance in the region, agricultural yields in sub-Saharan Africa (SSA) are generally quite low, leading to food insecurity in the region. One major hindrance to food security in Africa is the insurgence of arthropod pests that are responsible for a huge magnitude of agricultural economic losses both in the field and in storage. Losses as a result of insect pests in SSA countries may result to about 10–88% (Kfir et al. 2002; Ogendo et al. 2004; Ojo and Omoloye 2012; Midega et al.

L. A. Mwamburi (✉)
Department of Biological Sciences, University of Eldoret, Eldoret, Kenya
e-mail: lizzy.mwamburi@uoeld.ac.ke

S. De Mandal et al. (eds.), *New and Future Development in Biopesticide Research: Biotechnological Exploration*, https://doi.org/10.1007/978-981-16-3989-0_6

2016). Losses in the field and during storage result from direct feeding and reproduction, indirectly by insects acting as carriers of other pathogens or raising the humidity and stimulating fungal growth in storage (Tefera et al. 2010; Midega et al. 2016). Furthermore, favorable tropical climatic conditions in the region often favor the rapid population growth of these pests (Bekele et al. 1997; Midega et al. 2016).

In order to overcome food insecurity, there has been need to increase crop production, resulting in escalated and intensified pesticide applications in the last decade. Furthermore, the use of non-moderated applications of pesticides has led to residues in foods above approved limits resulting in detrimental effects on human health. In addition to having widespread insecticide resistance in the field and storage (Georghiou 1990), synthetic insecticides have non-selective action resulting in accumulation and persistence in the environment and food chains, posing risks to human health and imbalance of ecosystems.

Considering that most chemicals are banned, some control methods are either not available or are too costly for most farmers and as the basic requirement to achieve food security, there is an urgent requirement for simple, affordable and effective pest management for the smallholder farmers in SSA who are the majority producers.

The growing concern over potential hazards from chemical pesticide safety among consumers and the possibility of environmental harm has amounted in much deliberation being shifted to the use of natural products for the management of pests in agriculture. Hence, there is a need to seek an array of safe and long-term alternatives to synthetic pesticides that can increase horticultural crop productivity, decrease resistance and protect the environment from insecticidal pollution.

6.2 Plant-Based Biopesticides

Plants are furnished with possible substitutes for insect-control because they contain copius amounts of a wide array of biological compounds, among which are essential oils (EOs) making the application of plant EOs for biological control of economically important insect a subject of interest. In addition, EOs are considered safer than other plant-derived chemicals. Contrary to the problems that arise as a result of using synthetic pesticides, EOs are biodegradable and non-pollutive to the environment, easily accessible, inexpensive, and have appropriately found a promising role as biopesticides in pest management. Furthermore, the utilization of EOs to manage arthropod pests has been used traditionally to protect stored cereals from insect pests and are, therefore, culturally acceptable (Koul et al. 2008). In addition, EOs may be used further to develop pesticidal molecules to target specific insects (Rattan 2010). Thus, the inquisitiveness in the EOs has been revived with recent observations of their bioactivities to a diversity of pests (Isman 2006). Examples of some of these bioactivities are summarized in Table 6.1. One of the advantages of EOs includes the fact that they degrade rapidly in the environment and are more specific, and therefore favor beneficial insects. There are approximately 2000 plant species from Anacardiaceae, Annonaceae, Apiaceae, Araliaceae, Asteraceae, Cannabinaceae, Chenopodiaceae, Cupressaceae, Dipsacaceae, Ericaceae, Euphorbiaceae, Fabaceae,

Table 6.1 Examples of essential oils and insects that are affected by these oils

Plant	EOs	Insect	Mechanism	Reference
Achillea biebersteinii	Cis-Ascaridol, *p*-cymene, camphor, 1,8-cineole	*Tribolium castaneum*	Contact toxicity, inhibits growth	Nenaah (2014)
Agastache foeniculum	Estragole, 1,8-cineole, 1-octen-3-ol, Germacrene D	*Tribolium castaneum* *Rhyzopertha dominica*	Fumigant toxicity	Ebadollahi (2011b)
Allium sativum	Methyl allyl disulfide Diallyltrisulfide	*Sitophilus zeamais* (Motschulsky) *Tribolium castaneum* (Herbst)	Antifeedant activity, contact toxicity, decreased growth, fumigant toxicity, reduced oviposition	Huang et al. (2000a)
Anethum graveolense	Carvone, limonene	*Callosobruchus chinensis* (L.)	Disturb oviposition, hatching, pupal formation, adult emergence	Sefidkon (2001) and Chaubey (2008)
Artemisia herba-alba	1,8-cineol, camphene, α-pinene, borneol	*Tribolium castaneum* *Oryzaephilus surinamensis* L.	Contact toxicity, fumigant toxicity	Bachrouch et al. (2015)
Artemisia nilagirica	Camphor, β-farnesene, β-bisabolene, caryophyllene oxide	*Rhynchophorus ferrugineus* (Oliver)	Antifeedant activity against adults	Shukla et al. (2012)
Artemisia princeps	Bornane, camazulene, 3-methyl-6-(1-methylethyl)-1,2-cyclohexanediol	*Sitophilus oryzae* L.	Repellent activity against adults	Liu et al. (2001, 2006)
Artemesia sieberi	Camphor, camphene, 1,8-cineol, thujone	*Callosobruchus maculatus* F. *Sitophilus oryzae* L.	Adulticidal	Negahban et al. (2007)
Baccharis salicifolia	β–pinene, α-pinene, Sabinene, α-thujene	*Tribolium castaneum* (Herbst)	Contact toxicity, repellency	Garcia et al. (2005)
Callistemon citrinus	1,8-cineole, α-pinene, α-terpineol	*Callosobruchus maculatus* F.	Insecticidal, repellency against adults	Zandi-Sohani et al. (2013)
Callistemon sieberi	1,8-cineole, α-terpineol, α-pinene	*Callosobruchus chinensis* L.	Contact toxicity, disturbance on oviposition, feeding behavior	Lee et al. (2002) and Shukla et al. (2011)

(continued)

Table 6.1 (continued)

Plant	EOs	Insect	Mechanism	Reference
Carum copticum	Thymol, α-terpinolene, *p*-cymene	*Callosobruchus maculatus* F.	Ovicidal, larvicidal, adulticidal	Sahaf et al. (2007) and Sahaf and Moharramipour (2008)
Cinnamomum osmophloem	Cinnamaldehyde, β-cubebene, Linalool	*Meligethes aeneus* (Fabricius)	Repellency, adulticidal	Pavela (2011)
Citrus aurantium	*Meligethes aeneus*	Linalool, limonene, linalool acetate β-pinene	Repellency toxicity	Pavela (2011)
Clausena anisata	α-Pinene, *trans*-β-ocimene, estragole, β-elemene	*Acanthoscelides obtectus* (Say)	Toxicity, decrease in F1 progeny	Ndomo et al. (2008) and Usman et al. (2010)
Cuminum cyminum	Caryophyllene oxide, acaryphyllene α-pinene, geranylacetate	*Callosobruchus chinensis* L.	Egg hatching, disturbance of oviposition, emergence of adults Pupa formation,	Chaubey (2008) and Romeilah et al. (2010)
Cupressus sempervirens	α-Pinene, terpinene, α-terpinene, Sabinene	*Callosobruchus maculatus* F.	Contact toxicity, negatively affects longevity and fecundity	Hedjah-Chehheb et al. (2013)
Cymbopogon citratus	Geranial, neral, neryl acetate	*Tribolium castaneum* (Herbst) *Sitophilus oryzae* L.	Contact toxicity, repellent, feeding deterrent	Stefanazzi et al. (2011)
Cymbopogon schoenanthus	Limonene, β-phellandrene, δ-terpinene	*Callosobruchus maculatus* F.	Inhibits development	Ketoh et al. (2005) and Khadria et al. (2008)
Drimys winteri	α-Pinene, β-pinene, germacrene, safrole	*Tribolium castaneum* (Herbst)	Repellent, contact toxicity	Zapata and Smagghe (2010) and Muñoz et al. (2011)
Etlingera yunnanensis	1,8-cineole, estragole,, α-pinene, β-caryophyllene, limonene	*Liposcelis bostrychophila* (Badonnel) *Tribolium castaneum* (Herbst)	Contact toxicity, repellency	Shan-Shan et al. (2015)
Eucalyptus benthamii	α-Pinene, viridiflorol, 1,8-cineole	*Sitophilus zeamais* (Motschulsky)	Insecticidal, repellent activity	Mossi et al. (2011)

(continued)

Table 6.1 (continued)

Plant	EOs	Insect	Mechanism	Reference
Eucalyptus globulus	1,8-cineole, 3,7-dimethyl-2-octen-1-ol, trans-3-caren-2-ol	*Lasioderma serricorne* (F.)	Contact toxicity, repellent activities on adult	Ebadollahi et al. (2010)
Eucalyptus staigerana	1,8-limonene, z-citral, E-citral	*Zabrotes subfasciantus* (Boheman) *Callosobruchus maculatus* F.	Disturbs oviposition, reduction in number of emerged insects	Brito et al. (2006) and Maciel et al. (2010)
Eugenia caryophyllus	Eugenol, eugenol acetate, methyl chavicol	*Leptinotarsa decemlineata* (Say)	Contact toxicity, Antifeedant activity	Taghizadeh-Saroukolai et al. (2014)
Foeniculum vulgare	Anethole, limonene, α-fenchone	*Sitophilus zeamais* (Motschulsky) *Tenebrio molitor* (L.)	Repellant activity on adults	Cosimi et al. (2009) and Ebadollahi et al. (2014)
Gomortega keule	Limonene, α-pinene, 1,8-cineol, α-terpinene,	*Acanthoscelides obtectus* (Say)	Contact toxicity on adults	Bittner et al. (2008)
Hypericum scabrum	α-Pinene, camphor, terpinen-4-ol, and δ-cadinene	*Bruchus dentipes* (Baudi)	Contact toxicity on adults	Tozlu et al. (2011)
Illicium fargesii	Carvone, α-terpineol, *trans*-carveol, D-limonene	*Sitophilus zeamais*	Contact, fumigant toxicity	Wang et al. (2011)
Juniperus oxycedrus	Germacrene-D, α-pinene, Myrcene	*Sitophilus oryzae* L.	Toxicity	Athanassiou et al. (2012)
Lantana camara	α-Curcumene, α-acoradiene, β-caryophyllene	*Trogoderma granarium* (Everts)	Repellent activity	Tripathi and Kumar (2007) and Zoubiri and Baaliouamer (2012)
Laurus nobilis	α-Terpinyl acetate, β-pinene 1,8-cineole, Sabinene	*Trobilium confusum*	Toxicity	Isikber et al. (2006) and Cosimi et al. (2009)
Lavandula angustifolia	Linalool, 1,8-cineole, 1-borneol	*Meligethes aeneus* (Fabricius)	Repellent activity, mortality of adults	Pavela (2011) and Ebadollahi et al. (2014)
Litsea cubeba	*E*-citral, D-limonene, neral	*Lasioderma serricorne*	Contact toxicity in adults	Yang et al. (2014)
Mentha piperita	Menthol, Menthofuran, Menthone	*Callosobruchus maculatus* F.	Fumigant toxicity, affects mating, oviposition	Bassole et al. (2010) and El Nagar et al. (2012)

(continued)

Table 6.1 (continued)

Plant	EOs	Insect	Mechanism	Reference
Mentha pulegium	Decane, Pulegone, Limonene, Piperitenone	*Alphis gossypii*	Contact toxicity on adults	Ebadollahi et al. (2017)
Mentha spicata	Carvone, *cis*-Dihydrocarvone, *trans*-piperitone epoxide	*Leptinotarsa decemlineata* (Say)	Antifeedant activity, contact toxicity	Taghizadeh-Saroukolai et al. (2014)
Micromelum minutum	9-epi-β-caryophyllene, Bicyclogermacrene, 1,8-cineole, Tricyclene	*Callosobruchus maculatus* F.	Contact toxicity, fumigant toxicity, repellent activity	Paranagama and Gunasekera (2011)
Nardostachys chinensis	β-Gurjunene, Jatamansome, Aristolemone	*Tribolium castaneum* (Herbst)	Repellent on adults	Paudyal et al. (2012) and Liang et al. (2013)
Nigella sativa	Carvone, trans-anethone, limonene, *p*-cymene	*Sitophilus oryzae* L.	Contact toxicity, repellent activity	Chaubey (2012)
Ocimum basilicum	Linalool, Euginol, α-cadinol, β-ocimene	*Meligethes aeneus* (Fabricius)	Repellent activity, Adulticidal	Pavela (2011)
Origanum vulgare	Thymol, *p*-cymene, Carvacol	*Anobium punctatum*	Toxicity	Palla et al. (2020)
Piper sarmentosum. Roxb	Myristine	*Brontispa longissima* (Gesturo)	Antifeedant activity, Contact toxicity, Fumigation toxicity, Growth and development inhibition	Qin et al. (2010)
Rosmarinus officinalis	α-Pinene, 1,8-cineole, limonene, camphene	*Supella longipalpa*	Contact, fumigant toxicity, Repellency activity	Caballero-Gallardo et al. (2011) and Sharififard et al. (2016)
Salvia leucantha	Bornyl acetate, Caryophyllene oxide, Spathulenol, Caryophyllene	*Aedes aegypti* *A. quadrimaculatus*	Larvicidal activity	Ali et al. (2015)
Tagetes lucida	Methylchavicol, β-myrcene, β-ocimene, linalool	*Sitophilus zeamais*	Repellent activity	Nerio et al. (2009) and Caballero-Gallardo et al. (2011)

(continued)

Table 6.1 (continued)

Plant	EOs	Insect	Mechanism	Reference
Tagetes ternifora	*Cis*-ocimene, Ocimene Tagetone	*Tribolium castaneum* (Herbst) *Sitophilus oryzae* L.	Toxic activity, repellent activity, antifeedant	Stefanazzi et al. (2011)
Thymus satureidoides	Borneol, α-terpineol, camphene, α-pinene	*Varroa destructor*	Acaricidal activity	Ramzi et al. (2017)
Zataria multifora Boiss.	Thymol, ???-cymene, Carvacrol, Lonalool	*Tribolium castaneum*, *Callosobruchus maculatus* F., *Trogoderma granarium*	Fumigant toxicity on adults	Saei-Dehkordi et al. (2010) and Mahmoudvand et al. (2011)
Zingiber zerumbet	Camphene, a-humulene, camphor, 1,8-cineole	*Sitophilus zeamais Tribolium castaneum*	Fumigant toxicity	Suthisut et al. (2011)

Illiciaceae, Lamiaceae, Lauraceae, Meliaceae, Myrtaceae, Papaveraceae, Pedaliaceae, Piperaceae, Poaceae, Rutaceae, Schisandraceae, Scrophulariaceae, Verbenaceae, Vitaceae and Zingiberaceae plant families have been investigated for the insecticidal potential of their EOs have been found to exhibit lethal and sub-lethal effects such as adulticidal, feeding deterrent, growth and development inhibition, larvicidal, ovicidal, oviposition, progeny production, pupicidal and repellent (Grainge and Ahmed 1988).

6.2.1 Essential Oils

EOs are natural compounds that are volatile in nature and, have aromatic constituents characteristic in plants for various functions. They are synthesized via a combination of secondary metabolic pathways in plants and have a distinctive odour (Ebadollahi et al. 2020), may be composed of complex mixtures of aromatic compounds (Bakkali et al. 2008; Rajendran and Sriranjini 2008) and are present as droplets of fluid in the bark, flowers, fruits, leaves, stems and roots in different plants. Many EOs contain natural antioxidants and natural antimicrobial agents (Dorman et al. 2000). In Lamiaceae, they are produced by glandular trichomes, secretory cavities in Myrtaceae and Rutaceae and resin ducts in Asteraceae, Apiaceae (Fahn 1988). These structures burst open and the compounds are let out in copious amounts when herbivores feed or move on the surface of the plants (Duke et al. 2000).

In addition to their role in triggering the revitalization process for the plant through reproduction processes as attracts of pollinators and seed disseminators, and plant thermotolerance (Zhang et al. 2016), EOs also take part either directly or indirectly in plant defenses against arthropod pests (War et al. 2012).

Direct defense responses against insect pests target the biological systems for example the digestive and nervous systems, the endocrine organs of the insects and may be toxic and repellent, result in antinutrition and reduced digestibility, slowed growth and reduced reproduction (War et al. 2012). While indirect responses are insect-specific, and their compositions vary with the attacking insect. They may also involve the release of chemicals that lure the natural enemies of the herbivore by releasing aromatic compounds that lure or favour another organism(s) that reduce herbivore populations (War et al. 2012; Scholz et al. 2016).

6.2.2 Components of Essential Oils

The volatile compounds of EOs may be grouped into four: benzene derivatives, hydrocarbons, terpenes and other compounds (Haagen-Smit 1949; Ngoh et al. 1998). Terpenes and terpenoids are characterized by low molecular weight terpenes form the main group; C% hemiterpenes, the C10 monoterpenes, C15 sesquiterpenes, C20 diterpenes, C30 triterpenes and C40 tetraterpenes. Monoterpenoids constitute about 90% of the total EOs with a wide variety of functions and structures. Other related compounds are acids (e.g. chrysanthemic acid), acyclic alcohols (e.g. citronellol, geraniol), aldehydes (e.g. citronellal), bicyclic alcohols such as verbenol, cyclic alcohols such as menthol, ketones such as menthone, phenols such as thymol, and oxides (cineole) (Koul et al. 2008).

The chemical compounds of EOs vary within different species of the same genus and may also vary in various plant parts, geographical factors, time of harvest, season, climate and extraction method (Rocha et al. 2014). For instance, the concentration of 1,8-cineole was found to vary in the EOs of *Eucalyptus citriodora* (18.9%) (Karemu et al. 2013), *E. globulus* (31%) (Ebadollahi et al. 2010), *E. radiata* (63.3%) (Toudert-Taleb et al. 2014), and *E. saligna* (45.2) (Mossi et al. 2011). Similarly, limonene concentrations were reported to vary in *Citrus bergamia* (38.4%) (Cosimi et al. 2009), *C. limonum* (54.6%), (Bertuzzi et al. 2013), *C. reticulata* (64.1%) and *C. sinensis* (72.7%) (Kamal et al. 2011).

Camphor is well documented for its insecticidal properties (Singh et al. 2014; Tembo et al. 2018). In the same way, camphor concentrations were found to vary in different species of *Artemisia* (Kordali et al. 2006; Negahban et al. 2007; Shukla et al. 2012).

6.3 Pesticidal Properties of Essential Oils

EOs display a wide range of biopesticidal activities ranging from lethal to sublethal effects against Coleoptera, Diptera, Hemiptera Isoptera and Lepidoptera (Regnault-Roger et al. 2012; Pavela and Benelli 2016; Campos et al. 2019). Table 6.1 shows the pesticidal properties of some EOs from various plants. The differences in constituents of various EOs account for various mechanisms of action that range from antinutritional, developmental inhibitory, acute toxicity to repellency effects (Isman 2006; Pavela 2008; Hernández-Carlos and Gamboa-Angulo 2019). Other than the various patterns of phytochemical activties, toxic effects of EOs have been attributed to several other factors. One of which is the point at which the toxin penetrates the insect. The conventional modes of entry are through inhalation, ingestation or through skin absorption by the insect (Ozols and Bicevskis 1979).

EOs of *Artemisia spp.* are known to possess repellent and toxicity properties against coleopteran beetles. Examples of which include *Sitophilus spp.*, *Tribolium castaneum,* and *Callosobruchus maculatus*. In a similar manner, Nyamador et al. (2010) found that the EOs of *Cinnamomum spp* possessed contact, fumigant and repellent activity against *C. maculatus*. The authors reported on adulticidal, antifeed, deterrent, ovicidal and oviposition activities towards *C. maculatus* and *C. subinnotatus* as a result of exposure of EO of *Cymbopogon giganteus* and *C. nardus* Similarly, Ketoh et al. (2005) reported on development inhibition towards *C. maculatus* using *C. schoenanthus*.

EOs of *Eucalyptus spp* were found to exhibit adulticidal, repellency, oviposition, contact toxicity and fumigant toxicity against coleopteran beetles (Mohan et al. 2011). For instance, *Eucalyptus* EOs with large amounts of cineole were shown to be insecticidal towards *Varroa jacobsoni* that is parasitic towards the honeybee (Calderone and Spivak 1995), *Tetranychus urticae* and *Phytoseiulus persimilis* (Choi et al. 2004) and *Dermatophagoides pteronyssinus* (El-Zemity et al. 2006). A similar study by Chagas et al. (2002) reported insecticidal activity against the tick *Boophilus microplus* using EOs from three *Eucalyptus* spp., *E. citriodora, E. globulus* and *E. staigeriana*.

Taking into consideration the various activities of the EOs against pests of agriculture, and the fact that plant extracts contain compounds that exhibit various bioactivities including ovicidal, repellent, and antifeedant properties, it is feasible to combine the EOs with methods such as gamma radiation (Ahmadi et al. 2008a, b). Monoterpenoids account for a large percentage of the constituents of many plant extracts that display bioinsecticidal activities and the EOs. Citronella, camphor, citral, camphene, geraniol, methyi acetate, linalool, thymol, limonene, eugenol, menthone, carvacrol, trans- anethole 1,8-cineol and α -pinene, are well-known examples of biopesticide compounds (Phillips et al. 2010; Negahban et al. 2007; Isman and Machial 2006; Isman 2006). Furthermore, the fact that monoterpenoids possess antifeedant properties (Sbeghen-Loss et al. 2011; Shukla et al. 2012) acute toxicity repellent (Mediouni-Ben and Tersim 2011; Kim et al. 2010), larvicidal, adulticidal, ovicidal, and pupicidal activities (Yang et al. 2014; Waliwitiya et al. 2009; Murugan et al. 2012) make them potential pest control agents.

6.4 Mode of Activity of Essential Oils

Understanding the mode of activity of available EOs is of importance as it helps qualify the chemical characteristics of novel compounds that may be appropriate for insect pest control and dosage that can be safe and economical in agriculture (Haynes 1988). Given the encouraging results observed with EOs against insect pests, there has been rapid development to evaluate the appropriateness of the formulations of their active ingredient for application in integrated pest control programs.

Being distinctively lipophilic and volatile, EOs can permeate the insects' cuticle and disrupt their physiological processes (Lee et al. 2002) cause biochemical dysfunction and mortality. Furthermore, this fast action is a demonstration of the neurotoxicity nature of some EOs against some pests (Kostyukovsky et al. 2002). The mechanism of action, lethal doses, time taken to achieve lethal effects and site for bioactivities from plant EOs has been widely studied.

Plant EOs act at multiple levels of insects as fumigants, insect growth regulators, toxicants, repellents, phagodeterrents and synergists (Table 6.1). Neurotoxicity as a result of exposure to EOs in insects is characterized by hyperactivity, hyperexcitation and finally knockdown and immobilization (Enan 2001).

6.4.1 Fumigant Properties of Essential Oils

Current research has established that active ingredients from EOs may be possible alternatives to prevailing fumigants since they are easily changed to vapour at room temperature, including having various activities against a diversity of insects and fast penetrating. Various studies have investigated the probability of the use of components of plant EOs as insect fumigants. EOs of *Artemisia* spp., *Citrus* spp., *Eucalyptus* spp. *Lavalandula* spp., *Mentha* spp., have been well documented as fumigants. Table 6.1 shows fumigant activities of various plants.

The action of EOs as fumigants against stored product beetles *Sitophilus* spp. and *T. castaneum* has been a subject of immense interest (Fang et al. 2010; Ebadollahi et al. 2012; Franca et al. 2012; Germinara et al. 2017; Salem et al. 2017; Idouaarame et al. 2018; Devi et al. 2020). Findings of the various studies demonstrate that the mechanism of action for the oils is predominantly in the gaseous phase and through the respiratory system. Since most insects respire through the trachea, the vapour causes the spiracles to open. Suffocation occurs due of obstruction the tracheal respiration (Schoonhoven 1978) resulting in death of the insect (Pugazhvendan et al. 2012; Wafaa et al. 2017).

Studies by Huang et al. (2000a) reported that diallyl trisulfide and methyl allyl disulfide from garlic possessed fumigant and toxicity properties against the two beetles *Sitophilus zeamais* and *Tribolium castaneum* Likewise, Sahaf et al. (2007, 2008) demonstrated that the EOs of *Carum copticum* possessed fumigant properties against *S. zeamais* and *T. castaneum* The fumigant toxicity was accredited to the presence of monoterpenoids especially thymol. Monoterpenoids are basically volatile and induce toxic effects such as fumigants as a result of their ability to penetrate the insect cuticles.

Plant EOs obtained from Cymbopogon (Stefanazzi et al. 2011), *Myrtus communis* (Bertoli et al. 2012) anise, eucalyptus, cumin, rosemary and oregano were also demonstrated to have fumigant effects resulting in total mortality of the eggs of *Tribolium confusum* and *Ephestia kuehniella* (Tunç et al. 2000). *Ocimum* spp. extracts, and their active ingredients were found to possess insecticidal effects against a diversity of insects (Ebadollahi et al. 2020).

In a separate study, Singh and Pandey (2018) found that linalool, pulegone, limonene, linalayl acetate found in *Mentha* induced fumigant toxicity to *S. oryzae*. The apiaceae family have been found to have potential as fumigants agents for insects of stored products. For instance, Kim et al. (2003) reported significant mortalities using *Foeniculum vulgare* against *S. oryzae* and *Callosobruchus chinensis* using. While Chaubey (2008) reported fumigant toxicity using *Apium graveolens* and *Cuminum cyminum* against *C. chinensis*, Park et al. (2006) attributed toxicity of larvae of *Lycoriella ingenua* to the activie ingredient limonene, menthone and pulegone of *Schizonepeta tenuifolia*.

6.4.2 Antifeedant Properties

Antifeedant chemicals may deter feeding after contact or act as repellents without making direct contact with the insects (Koul et al. 2008). A significant aspect of the antifeed properties of the EOs of plants is that they have found use in pest management (Table 6.1). Nevertheless, their action on insects is varied and are generally not harmful to the environment .

Indices such as feeding deterrence index (FDI), efficiency of conversion of ingested food (ECI), relative growth rate (RGR) and relative consumption rate (RCR) and are used to determine feeding deterrence. For instance, while the EOs of *Artemisia sieberi* and *A. scoparia* displayed antifeeding activity against *Tribolium castaneum*, EOs of *A. sieberi* oil had higher efficacy in comparison to those from *A. scoparia* and significantly decreased RGR and RCR. *A. sieberi* oil displayed higher efficacy in terms FDI compared to oils from *A. scoparia* (Negahban et al. 2007). In a separate study by Sahaf and Moharramipour (2008), the efficacy of *Carum copticum* and *Vitex pseudonegundo* EOs against *C. maculatus* was found to increase FDI.

EOs in various plants may disrupt or hinder feeding by making the plant matter unappealing or unappetizing (Talukder 2006; Rajashekar et al. 2012). The insects linger on the plants and ultimately die from starvation. Ebadollahi (2011a) demonstrated antifeed activities of *Lavandula* against adults of *T. castaneum*. *Melaleuca alternifolia* and its constituent compounds manifested antifeedant activities against *Helicoverpa armigera* (Liao et al. 2017). Similarly, Taghizadeh-Sarikolaei et al. (2014) reported antifeed activities of *Thymus daenensis* towards *Leptinotarsa decemlineata* and the prominent constituents as thymol, ρ-cymene and γ-terpinene. Shukla et al. (2012) reported antifeed efficacy of oils of *Eupatorium adenophorum* aerial parts and the florescence of *Artemisia nilagirica* against adults of *Rhynchophorus ferrugineus*. The authors reported significantly higher antifeed activity from *E. adenophorum* and *A. nilagirica* from the aerial parts compared to those from *E. adenophorum* leaves. The differences in activity was attributed to the difference in chemical composition, with the major components in the oils from the florescence and leaves of *E. adenophorum* showing approximately 41% oxygenated sesquiterpenes and 64% sesquiterpene hydrocarbons, respectively. The principal class of compounds in EOs *A. nilagirica* aerial parts were composed of monoterpenes (32.92%) and sesquiterpenes (37.02%).

Similarly, the EO constituents citronellal, thymol and α –terpineol were reported to result in feeding deterrence in tobacco cutworm, *Spodoptera litura* (Hummelbrunner and Isman 2001). *Dictamnus dasycarpus* rootbark demonstrated feeding inhibition against *T. castaneum* and *S. zeamais* (Liu et al. 2002). The authors established that fraxinellone resulted in feeding deterrence in the adults and larvae of *T. castaneum* and adults of *S. zeamais*, while dictamnine was responsible for feeding deterrence in adults and larvae of *T. castaneum* and *S. zeamais*.

EOs of *Salvia mirzayanii* displayed a strong feeding deterrence activity towards adults of *T. confusum* (Soleimannejad et al. 2011). The authors observed an increase in the concentration of the EO, RGR, RCR, whereas ECI were reduced significantly. Therefore, nutritional indices may have been influenced by the EO by interfering with the pre-ingestive and post-ingestive process. Consequently, feeding behavior of an insect may result in a reduction in the consumption and consequently growth rate of the insect.

6.4.3 Repellent Properties

Several studies report on the activities of various plants' EOs as repellents (Table 6.1). Repellents provide plants with protection gainst insect pests with minimal harm to the ecosystem.

For example, the oils of *Laureliopsis philippiana* manifested repellent against *Sitophilus* weevils (Norambuena et al. 2016). Methyleugenol and safrole were established to the compounds responsible for the repellency. Other activities observed during the study were contact toxicity and reduced emergence.

Akrami et al. (2011) reported repellency of *Mentha longifolia* on *C. maculatus* and *T. castaneum*. The authors found significantly more repellency of the EO of *A. sieberi* at 1.5 ppm towards *T. castaneum* compared to *C. maculatus* and *S. oryzae*

(Negahban et al. 2007), and *A. scoparia* significantly repelled *T. castaneum* and *S. oryzae* compared to *C. maculatus* (Negahban et al. 2006). Strong repellency (100%) was also observed in *Anethum graveolens* and *T. vulgaris*, towards *P. interpunctella* (Rafiei et al. 2009). Repellency may differ between arthropods. For instance, Taghizadeh-Saroukolai et al. (2009) showed that EO of *P. acaulis* varied in its degrees of repellency against *S. oryzae* (83.6%), *C. maculatus* (71.6) and *T. castaneum* (63.6%).

Nerio et al. (2010) demonstrated that the composition of active components influenced the repellency of the EOs. For instance, feeding deterrence, repellency and toxic activities were exhibited against *T. castaneum* larvae and adults using EOs extracted from fruits and leaves of *Schinus areira* (Descamps et al. 2011). The authors reported repellency from the oils obtained from the leaves. The compostions of the EOs of the leaves were predominantly camphene, monoterpenoids, α-phellandrene and 3-carene, whereas 3-carene, α-phellandrene, and β-myrcene were the principal oils obtained from the fruits. All the oils caused mortality of larvae in fumigant and topical bioassays. However, the former was not observed in the adults. Furthermore, both EOs influenced the nutritional index.

6.4.4 Toxicants

The red flour beetle (*Tribolium species*), rice weevil (*S. oryzae*) and the maize weevil (*S. zeamais*) account for more than 60% losses of cereals and pulses during storage in tropical countries (Singh et al. 2012). The use of EOs from plants as toxicants is an attractive option as these are effective and have been used traditionally. Furthermore, studies on plant derivatives have demonstrated that many plant products are toxic to insects that infest stored products (Table 6.1).

Research shows that the efficacy of EOs towards most insects is associated to terpenes. Monoterpenoids and sesquiterpenes account for the major proportion of the major essential constituents (Table 6.1). For instance, carvacrol, 1,8-cineol, thymol, eugenol, α-pinene and limonene, have been reported to have toxic effects against storage insects. While linalool EO of coriander seed was reported to be toxic towards *S. oryzae* (Knio et al. 2008), limonene, carvone, and (E)-anethole were the principal active components found in the EO of caraway (Fang et al. 2010). High insecticidal toxicity of *Carum carvi* and *Coriandum sativum* against *Cryptolestes pusillus* and *Rhyzopertha dominica* has been attributed to linalool and camphor-rich fractions (Lopez et al. 2008).

Insecticidal toxicity against *Aphis craccivora* was observed when faba beans were treated with a neem oil formulation (neemix®) from *Azadirachta indica* and *Ocimum basilicum* (Sammour et al. 2011). In addition, the authors also reported cumulative adult mortality of up to 100% after 7 days. Geranial, linalool and methyl chavicol were established as the components for insecticidal activities in basil oil. Similarly, Aslan et al. (2004) reported that geranial, linalool and methyl chavicol were insecticidal towards *Tetranychus urticae* and *Bemisia tabaci*.

Insecticidal activity of terpenes (carvone, linalool, terpeniol, phellandrine and citronellol) from garlic and mint were observed on different growth stages of *Agrostis ipsilon* (Sharaby and El-Nujiban 2015). Similarly, high insecticidal efficacy of *Salvia officinalis* towards *A. ipsilon* was reported to result from the presence of sesquiterpenes and terpenes (Sharaby and Al-Dosary 2014). In a parallel study, Sharaby et al. (2012) demonstrated that garlic, eucalyptus, and mint EOs caused toxicity to grasshopper (*Heteracris littoralis*).

Similarly, toxic effects were observed when EOs from *Triaenops persicus* were assessed against adults *S. oryzae* and *T. castaneum* (Koul et al. 2008). Neurotoxic effects as a result of thymol in thyme were observed when *Thymus vulgaris* was assayed against *Nezara viridula* (Koul et al. 2008). Similarly, thymol induced high toxicity to *Lipaphis pseudobrasicae* (Sampson et al. 2005), *Spodoptera litura* (Hummelbrunner and Isman 2001) and *S. oryzae* (Rozman et al. 2006).

6.4.5 Growth Retardants and Inhibitors of Development

Several studies (Table 6.1) have reported effects of plant EOs and their components that disrupt the development and growth of insects, reducing the weight at various stages of growth prolonging the developmental stages (Talukder 2006; Athanassiou et al. 2014; Aziza et al. 2014). The survival rates of larvae, pupae, and adult emergence may also be affected (Koul et al. 2008).

Studies using EOs from azadirachtin and neem seed were reported to increase nymphal mortality of aphids at 80 and 77%, respectively, resulting in prolonged maturation time to adulthood (Kraiss and Cullen 2008). In a similar manner, some botanical biopesticides have been found to especially have dramatized effects during the development and maturation periods, including emergence of adults (Shaalan et al. 2005). Chaubey (2008) found that EOs from *Piper nigrum*, *Myristica Nigella sativa*, *fragrans,* and *Trachyspermum ammi* influenced changes in the reproduction and growth of *C. chinensis*. In a separate study, Abbas et al. (2012) found that *Citrus reticulata* EOs inhibited growth and caused decline in population of *Rhyzopertha domonica*. In a similar manner, EOs from citrus peels resulted in reduced oviposition of *C. maculatus* (Elhag 2000). Likewise, *Elettaria cardamomum* EOs were reported to exhibit deter the oviposition of *C. maculatus* (Abbasipour et al. 2011).

EOs have also been reported to prolong growth stages. For example, basil oil prolonged the duration of the nymph stage of *Aphis craccivora* causing a reduction in number of adults (Sammour et al. 2011). Similarly, Anshul et al. (2014) showed that *Artemisia annua* EOs reduced the weights of *Helicoverpa armigera* larvae while prolonging the larval stage.

6.4.6 Sterility/Reproduction Inhibitors

Sterility may happen as a consequence of induced insect sterility technique or by the use of a chemosterilant that hinders reproduction (Morrison et al. 2010). Chemosterilants may cause permanent or temporary sterility of either male or female insects or interfere with the development of sexual stages from the young to the adults (Wilke et al. 2009; Navarro-Llopis et al. 2011). Asawalam and Adesiyan (2001) and Shaalan et al. (2005) drew attention to the fact that grains mixed with different parts of plants, extracts, oils or powder had the effect of reducing insect eggs hatchability oviposition, postembryonic or progeny development. For example, Elango et al. (2009) reported on the ovicidal effects against *Anopheles subpictus* using extracts of *Andrographis paniculata, A. lineat,* and *Tagetes erecta.*

Use botanical insecticides as chemosterilants may be at the physiological level for instance azadirachtin has been found to interfere with the synthesis of hormones responsible for molting and release of the same from the prothoracic gland, resulting in incomplete ecdysis in young insects, and sterility in adult insects (Isman 2006).

Constituent compounds from garlic; diallyl disulfide and methyl allyl have been found to display toxicity towards *T. castaneum* and *S. zeamais* (Ho et al. 1996; Huang et al. 2000a) at various stages of development. Egg hatching was totally suppressed at 0.32 mg/cm^2 using diallyl trisulfide, while at 0.08 mg/cm^2 larval and adult emergence were repressed. The food consumption, food utilization and growth rate were significantly reduced by methyl allyl disulfide for adults in both insect species, with feeding deterrence indices of 1.52 mg/g food for *T. castaneum* and 44% at 6.08 mg/g food for *S. zeamais* (Huang et al. 2000b). Similarly, Plata-Rueda et al. (2017) demonstrated that the pupal stages of *Tenebrio molitor* were more susceptible to diallyl disulfide and diallyl sulfide compared to larvae and adult stages. The authors attributed the difference in the developmental stages to the fact that efficacy may have been influenced by the way garlic compounds penetrated of the insect body and the capability of the insect to break down these compounds. Furthermore, the insects exhited changes in movement, muscle contraction and paralysis were they came into contact to the EOs of garlic Muscle contractions and paralysis could have been been as a result of neutotoxicity, coupled with hyperextension and hyperactivity of the abdomen and legs and resulting in an instant knockdown effect or immobilization (Prowse et al. 2006; Zhao et al. 2013).

6.5 Synergistic Action of Essential Oils

Various studies have revealed that mixtures or combinations of various EOs compounds exhibit additive, synergistic, and/or antagonist toxicity effects in different groups of insects (Ntalli et al. 2011; Gallardo et al. 2015; Tak et al. 2016; Wu et al. 2017; Gaire et al. 2020). The synergy observed in EOs may be as a result of the different mechanisms of action of their chemical constituents. For example, using synergistically interacting groups of monoterpenoids found in products allows for the achievement of higher insecticidal activity by using smaller amounts of the active constituents (Tak and Isman 2015; Tak et al. 2016).

The rationale for using combinations of EOs is to produce a superior product with multiple mechanisms of action, bearing in mind that the product has a significant effect than the total effects of the known and unknown chemical components of the individual EOs. Earlier studies proposed that the synergistic role of constituents in the EOs with high camphor content (Gonzalez-Coloma et al. 2006; Nerio et al. 2010). For instance, Tak and Isman (2015) showed that a combination of camphor and 1,8-cineole displayed hightened penetration of the cuticle, resulting in a synergy toxic effect towards the larvae of the cabbage looper. The authors found that these changes increased the ability of the mixture of two oils to penetrate the cuticle, resulting in reduced surface tension and increased solubility. Similarly, Abbassy et al. (2009) demonstrated a higher synergistic insecticidal effect of terpien-4-ol and c-terpinene from EOs of *Majorana hortensis* against larvae of *Spodoptera littoralis* than either of the individual compounds.

Faraone et al. (2015) reported increased toxicity (16–20-fold) against *Myzus persicae* as a result of the synergistic action of imidacloprid and two EOs linalool and thymol of *Lavendula angustifolia* and *Thymus vulgaris*, respectively. Mixtures of EO ingredients particularly monoterpenoids exhibited toxic synergy effects against insects as a result of increased ability to penetrate the cuticle (Gaire et al. 2020). The authors observed heightened synergistic effect in toxicity against bed bugs using a mixture of eugenol, thymol and carvacrol. The authors further reported that the synergistic interaction displayed by the mixture was most likely influenced by factors associated to the target site. Including the capability of the monoterpenoids to be operate on various sites within the nervous system of insect.

6.6 Nanoencapsulation

Despite their promising properties, EOs possess problems related to potential for oxidation, solubility in water, volatility, that should be rectified before they can be used effectively (Martin et al. 2010). Turek and Stintzing (2013) explored the factors that influenced EO stability. Besides their being highly volatile, EOs easily decompose in direct heat, exposure to high humidity, light, and/or oxygen. Degradation of the constituents may be as a result of cyclization, oxidation, dehydrogenation or isomerization reactions stimulated chemically or enzymatically (Scott 2005) and may be affected by the conditions during distillation, processing, storage of the plant material, and handling of the final product (Schweiggert et al. 2007).

To achieve high efficacy and stability, EOs are encapsulated and used to deliver EOs in insect pest management programs. Nanoencapsulation uses an approach of encapsulating the active agent in a thin layer of protective membrane in order to cushion it from extreme environmental effects. Nanocapsules consist of a shell, active ingredients that may be adsorbed on the surface or dissolved in the inner core

(Khoee and Yaghoobian 2009). The carrier or envelope may be made up of natural polymers such as proteins or polysaccharides or synthetic polymers such as polyamides or melamineformaldehyde, lipids, phospholipids, or inorganic materials such as SiO_2 (Nagpal et al. 2001; Kumari et al. 2010), giving EOs high efficacy while cushioning them from the likelihood of evaporation or degradation.

An important characteristic of nanoencapsulated EOs is the controlled release, that is characterized by a a prolonged release that follows a preliminary burst (São Pedro et al. 2013). In addition to minimized evaporation and exposure to extreme environmental conditions, nanoencapsulation of EOs represents a practicable and logical approach that modulates drug release, increases the stability of the active ingredients, decreases their volatility, enhances their bioactivity, and reduces toxicity (Ravi Kumar 2000).

6.7 Essential Oil Nanoformulations and Insect Pest Control

Nanoformulated EOs exhibit distinctive properties including higher pest toxicity. For instance, nanopermethrin had higher larvicidal efficacy towards *Culex quinquefasciatus* compared to the non-formulated form of permethrin (Anjali et al. 2010). Studies also reveal that when transformed into nanoparticles novel non-precise and biological properties become part of EOs. They gain entry into epithelial and endothelial cells of the pest and move from one cell to another by transcytosis along the axons and dendrites, blood, and lymph, triggering oxidative stress and other reactions (Devi and Maji 2011). For example, geranium oil used as mosquito repellent when transformed into high-quality solid lipid nanoparticle-loading geranium oil (Asnawi et al. 2008). In separate studies by Yang et al. (2009) and Werdin-Gonzalez et al. (2014), EOs of garlic and geranium incorporated into nanoparticles of polyethylene glycol and tested against *Tribolium castaneum* and *Rhyzopertha dominica* produced an increase in contact toxicity as a result of the slow and sustained dissemination of the effective terpenes. Furthermore, the nanoformulations increased the ability of the EO contact toxicity and changed the feeding ability of both pests. While the nanoemulsion of EO citronella caused a higher release rate against mosquito (Nuchuchua et al. 2009; Solomon et al. 2012).

Encapsulation of EOs enhances their bioactivity. For instance, Ferreira et al. (2019) demonstrated the efficiency and prolonged activity of chitosan encapsulated EO of *Siparuna guianensis* against *Aedes aegypti* larvae as a result of increased contact and slow and controlled release conferred by chitosan nanoparticles. Similarly, chitosan and angico gum nanoparticles containing EOs of *Lippia sidoides* caused 92% mortality of larvae of the mosquito *Aedes aegypti* (Paula et al. 2010). Likewise, chitosan and cashew gum nanoparticles containing EO *L. sidoides* caused 75–100% mortality of *A. aegypti* after 48 and 72 h respectively (Paula et al. 2011). In a separate study, Christofoli et al. (2015) found that nanoencapsulated EOs from *Zanthoxylum rhoifolium* displayed great efficacy in reducing the egg numbers and nymphs of *Bemisia tabaci* populations. The *in vitro* release was chararacterized by an initial fast release, followed by second sustained slow release. Nanoencapsulated oils display more chemical activity compared to non-encapsulated material, more mobility, allowing entry into the tissues of the insect. This can be achieved by feeding and entry through digestive tract or contact through the insect's cuticle. The pest cells, Entry into the endothelial and epithelial, is by transcytosis (Devi and Maji 2011).

Campolo et al. (2017) demonstrated significant insecticidal activity against the invasive tomato pest *Tuta absoluta* using nanoformulations of EOs of citrus peel with polyethylene glycol. Khoobdel et al. (2017) demonstrated insecticidal activity of nanoformulations of EOs of *Rosmarinus officinalis* towards the red flour beetle, *Tribolium castaneum*. Louni et al. (2018), showed that nanoemulsion formulation of *Mentha longifolia* had enhanced contact toxicity on *Ephestia kuehniella*. Nanoencapsulation mechanism can, therefore serve as novel formulations for the establishment of the EOs and their chemical derivatives with improved functions.

6.8 Conclusion

The use of plant EOs as insecticides offers several advantages over synthetic chemicals. Moreover, their individual components have been determined to have potential for insecticidal activity against several arthropods of economic importance in agriculture. Furthermore, studies have shown that mixtures or combinations of several EO active ingredients exhibit additive, synergistic,and/or antagonist toxicity in various arthropod species. However, despite their promising properties, EOs face obstacles related to solubility in water, possibility of oxidation and volatility, due to exposure air, direct sunlight, high temperatures and moisture, resulting in possible degradation and evaporation of some active components. These setbacks have been solved by encapsulation of EOs resulting in controlled release characterized by a two-phase release; an first burst, then by a prolonged release, minimized evaporation and exposure to extreme environmental conditions, increased stability of the active ingredients, decreased volatility, and enhanced bioactivity.

References

Abbas SK, Ahmad F, Sagheer M et al (2012) Insecticidal and growth inhibition activities of *Citrus paradisi* and *Citrus reticulata* essential oils against lesser grain borer, *Rhyzopertha dominica* (F.) (Coleoptera: Bostrichidae). World J Zool 7:289–294

Abbasipour H, Mahmoudvand M, Fahimeh R, Hosseinpour MH (2011) Fumigant toxicity and oviposition deterrency of the essential oil from cardamom, *Elettaria cardamomum*, against three stored-product insects. J Insect Sci 11:165. https://doi.org/10.1673/031.011.16501

Abbassy MA, Abdelgaleil SAM, Rabie RYA (2009) Insecticidal and synergistic effects of *Majorana hortensis* essential oil and some of its major constituents. Entomol Exp Appl 131:225–232

Ahmadi M, Moharramipour S, Mozdarani H, Negahban M (2008a) Combination of medicinal plant essential oils with gamma radiation in management of *Tribolium castaneum*. In: The 1st international symposium on medicinal plants, their cultivation and aspects of uses, Jordan, Petra, 15–16 Oct 2008, M.A. Ateyyat, p 50

Ahmadi M, Moharamipour S, Zolfagharieh HR (2008b) Comparative fumigant toxicity of *Rosmarinus officinalis* and *Artemisia sieberi* against *Tribolium castaneum*. Integr Prot Stored Prod IOBC/WPRS Bull 40:243–247

Akrami H, Moharramipour S, Imani S (2011) Comparative effect of *Thymus kotschyanus* and *Mentha longifolia* essential oils on oviposition deterrence and repellency of *Callosobruchus maculatus* F. Iran J Med Aromat Plants 27:1–10

Ali A, Tabanca N, Demirci B, Blythe EK, Ali Z et al (2015) Chemical composition and biological activity of four *Salvia* essential oils and individual compounds against two species of mosquitoes. J Agric Food Chem 63:447–456

Anjali CH, Sudheer Khan S, Margulis-Goshen K et al (2010) Formulation of water-dispersible nanopermethrin for larvicidal applications. Ecotox Environ Safe 73:1932–1936

Anshul N, Kalra A, Singh D (2014) Biological effect of sweet wormwood *Artemisia annua* methanol extracts and essential oil against *Helicoverpa armigera* hub. (Lepidoptera: Noctuidae). J Entomol Zool Stud 2:304–307

Asawalam E, Adesiyan S (2001) Potential of *Ocimum basilicum* (Linn) for the control of maize weevil *Sitophilus zeamais* (Motsch). Nig Agric J 32:195–201

Aslan I, Özbek H, Çalmaşurand Ö, Şahin F (2004) Toxicity of essential oil vapoura to two greenhouse pests, *Tetranychus orticae* Koch and *Bemisia tabaci* Genn. Ind Crop Prod 19:167–173

Asnawi S, Abd Aziz A, Abd Aziz R, Khamis AK (2008) Formulation of geranium oil loaded solid lipid nanoparticles formosquito repellent application. J Chem Nat Resour Eng 2:90–99

Athanassiou CG, Kavallieratos NG, Evergetis E, Kasoula A-M, Haroutounian SA (2012) Insecticidal efficacy of silica gel with *Juniperus oxycedrus* ssp. *oxycedrus* (Pinales: Cupressaceae) essential oil against *Sitophilus oryzae* (Coleoptera: Curculionidae) and *Tribolium confusum* (Coleoptera: Tenebrionidae). J Econ Entomol 106:1902–1910

Athanassiou CG, Kavallieratos NG, Throne JE, Nakas CT (2014) Competition among species of stored product psocids in stored grains (Psocoptera). PLoS One 9:e102867

Aziza S, Rahman HA, Abdel-Aziz S, Moawad S (2014) Natural plant oils and terpenes as protectors for the potato tubers against *Phthorimaea operculella* (Zeller) infestation by different application methods. Egypt J Biol Pest Control 24:265–274

Bachrouch O, Ferjani N, Haouel S, Ben Jemâa JM (2015) Major compounds and insecticidal activities of two Tunisian *Artemisia* essential oils toward two major coleopteran pests. Ind Crop Prod 65:127–133

Bakkali F, Averbeck S, Averbeck D, Idaomar M (2008) Biological effects of essential oils—a review. Food Chem Toxicol 46:446–475

Bassole IHN, Lamien-Meda A, Bayala B et al (2010) Composition and antimicrobial activities of *Lippia multiflora* Moldenke, *Mentha piperita* L and *Ocimum basilicum* L essential oils and their major monoterpene alcohols alone and in combination. Molecules 15:7825–7839

Bekele AJ, Obeng-Ofori D, Hassanali A (1997) Evaluation of *Ocimum kenyense* (Ayobangira) as a source of repellents toxicants and protectants in storage against three major stored product insect pests. J Appl Entomol 121:169e173

Bertoli A, Conti B, Mazzoni V, Meini L, Pistelli L (2012) Volatile chemical composition and bioactivity of six essential oils against the stored food insect *Sitophilus zeamais* Motsch. (Coleoptera: Dryophthoridae). Nat Prod Res 26:2063–2071

Bertuzzi G, Tirillini B, Angelini P, Venanzoni R (2013) Antioxidative action of *Citrus limonum* essential oil on skin. Eur J Med Plant 3:1–9

Bittner ML, Casanueva ME, Arbert CC et al (2008) Effect of essential oils from plant species against the granary weevils, *Sitophilus zeamais* and *Acanthoscelides obtectus* (Coleoptera). Chil J Agric Res 53:1455–1459

Brito JP, Baptistussi RC, Funichelo M, Oliveira JEM, Bortoli SA (2006) Effect of essential oils of *Eucalyptus* spp. under *Zabrotes subfasciatus* (both. 1833) (Coleoptera: Bruchidae) and *Callosobruchus maculatus* (Fabr. 1775) (Coleoptera: Bruchidae) in two beans species. Bol Sanidad Veg Plagas 32:573–580

Caballero-Gallardo K, Olivero-Verbel J, Stashenko EE (2011) Repellent activity of essential oils and some of their individual constituents against *Tribolium castaneum* Herbst. J Agric Food Chem 59:690–696

Calderone NW, Spivak M (1995) Plant extracts for control of the parasitic mite *Varroa jacobsoni* (Acari: Varroidae) in colonies of the western honeybee (Hymenoptera: Apidae). J Econ Entomol 88:1211–1215

Campolo O, Cherif A, Ricupero M, Siscaro G et al (2017) Citrus peel essential oil nanoformulations to control the tomato borer, *Tuta absoluta*: chemical properties and biological activity. Sci Rep 7:13036

Campos EVR, Proença PLF, Oliveira JL et al (2019) Use of botanical insecticides for sustainable agriculture: future perspectives. Ecol Indic 105:483–495

Chagas ACS, Passos WM, Prates HT et al (2002) Acaricide effect of *Eucalyptus* spp. essential oils and concentrated emulsion on *Boophilus microplus*. Braz J Vet Res Anim Sci 39:247–253

Chaubey MK (2008) Fumigant toxicity of essential oil from some common species against pulse beetle, *Callosobruchus chinensis* (Coleoptera: Bruchidae). J Oleo Sci 57:171–179

Chaubey MK (2012) Biological effects of essential oils against rice weevil *Sitophilus oryzae* L. (Coleoptera: Curculionidae). J Essent Oil Bear Pl 15:809–815

Choi W, Lee SG, Park HM, Ahn YJ (2004) Toxicity of plant essential oils to *Tetranychus urticae* (Acari: Tetranychidae) and *Phytoseiulus persimilis* (Acari: Phytoseiidae). J Econ Entomol 97:553–558

Christofoli M, Costaa ECC, Bicalhoc KU et al (2015) Insecticidal effect of nanoencapsulated essential oils from *Zanthoxylum rhoifolium* (Rutaceae) in *Bemisia tabaci* populations. Ind Crop Prod 70:301–308

Cosimi S, Rossi E, Cioni PL, Canale A (2009) Bioactivity and qualitative analysis of some essential oils from Mediterranean plants against stored-product pests: evaluation of repellency against *Sitophilus zeamais* Motschulsky, *Cryptolestes ferrugineus* (Stephens) and *Tenebrio molitor* (L.). J Stored Prod Res 45:125–132

Descamps LR, Sanchez Chopa C, Ferrero AA (2011) Activity of *Schinus areira* (Anacardiaceae) essential oils against the grain storage pest *Tribolium castaneum*. Nat Prod Commun 6:887–891

Devi N, Maji TK (2011) Study of complex coacervation of gelatin a with sodium carboxymethyl cellulose: microencapsulation of neem (*Azadirachta indica* a. Juss.) seed oil (NSO). Int J Polym Mater Polym Biomater 60:1091–1105

Devi MA, Sahoo D, Singh TB, Rajashekar Y (2020) Toxicity, repellency and chemical composition of essential oils from *Cymbopogon* species against red flour beetle *Tribolium castaneum* Herbst (Coleoptera: Tenebrionidae). J Consum Prot Food Saf 15:181–191

Dorman HJD, Surai P, Dean SG (2000) *In vitro* antioxidant of a number of plant essential oils and phytoconstituents. J Essent Oil Res 12:241–248

Duke SO, Canel C, Rimando AM et al (2000) Current and potential exploitation of plant glandular trichome productivity. Adv Bot Res Incorp Adv Plant Pathol 31:121–151

Ebadollahi A (2011a) Antifeedant activity of essential oils from *Eucalyptus globulus* Labill and *Lavandula stoechas* L. on *Tribolium castaneum* Herbst (Coleoptera: Tenebrionidae). Biharean Biol 5:8–10

Ebadollahi A (2011b) Chemical constituents and toxicity of essential oil from *Agastache foeniculum* (Pursh) Kuntze against two stored-product insect pests. Chil J Agric Res 71:212–217

Ebadollahi A, Safaralizadeh MH, Pourmirza AA (2010) Fumigant toxicity of *Lavandula stoechas* L. oil against three insect pests attacking stored products. J Plant Prot Res 50:56–60

Ebadollahi A, Nouri-Ganbalani G, Hoseini SA, Sadeghi GR (2012) Insecticidal activity of essential oils of five aromatic plants against *Callosobruchus maculatus* F. (Coleoptera: Bruchidae) under laboratory conditions. J Essent Oil Bear Pl 15:256–262

Ebadollahi A, Jalali Sendi J, Aliakbar A, Razmjou J (2014) Chemical composition and acaricidal effects of essential oils of *Foeniculum vulgare* mill. (Apiales: Apiaceae) and *Lavandula angustifolia* miller (Lamiales: Lamiaceae) against *Tetranychus urticae* Koch (Acari: Tetranychidae). Psyche 2014:427078. https://doi.org/10.1155/2014/424078

Ebadollahi A, Davari M, Razmjou J, Naseri B (2017) Separate and combined effects of *Mentha piperata* and *Mentha pulegium* essential oils and a pathogenic fungus *Lecanicillium muscarium* against *Aphis gossypii* (Hemiptera: Aphididae). J Econ Entomol 110:1025–1030

Ebadollahi A, Ziaee M, Palla F (2020) Essential oils extracted from different species of the lamiaceae plant family as prospective bioagents against several detrimental pests. Molecules 25:1556

El Nagar TFK, Abdel Fattah HM, Khaled AS, Aly SA (2012) Efficiency of peppermint oil fumigant on controlling *Callosobruchus maculatus* F. infesting cowpea seeds. Life Sci J 9:375–383

Elango G, Rahuman AA, Bagavan A et al (2009) Laboratory study on larvicidal activity of indigenous plant extracts against *Anopheles subpictus* and *Culex tritaeniorhynchus*. Parasitol Res 104:1381–1388

Elhag EA (2000) Deterrent effects of some botanical products on oviposition of the cowpea bruchid *Callosobruchus maculatus* (F.) (Coleoptera: Bruchidae). Int J Pest Manag 46:109–113

El-Zemity S, Hussien R, Saher F, Ahmed Z (2006) Acaricidal activities of some essential oils and their monoterpenoidal constituents against house dustmite, *Dermatophagoides pteronyssinus* (Acari: Pyroglyphidae). J Zhejiang Univ Sci B 7:957–962

Enan E (2001) Insecticidal activity of essential oils: octopaminergic sites of action. Comp Biochem Physiol C 130:325–337

Fahn A (1988) Secretory tissues in vascular plants. New Phytol 108:229–257

Fang R, Jiang CH, Wang XY et al (2010) Insecticidal activity of essential oil of *Carum carvi* fruits from China and its main components against two grain storage insects. Molecules 15:9391–9402

Faraone N, Hiller NK, Cutler GC (2015) Plant essential oils synergize and antagonize toxicity of different conventional insecticides against *Myzus persicae* (Hemiptera: Aphididae). PLoS One 10:e0127774. https://doi.org/10.1371/journal.Pone.0127774

Ferreira PT, Haddi K, Corrêa FTR et al (2019) Prolonged mosquitocidal activity of *Siparuna guianensis* essential oil encapsulated in chitosan nanoparticles. PLoS Negl Trop Dis 13(8): e0007624

Franca SM, Oliveira JV, Esteves Filho AB, Oliveira CM (2012) Toxicity and repellency of essential oils to *Zabrotes subfasciatus* (Boheman) (Coleoptera, Chrysomelidae, Bruchinae) in *Phaseolus vulgaris* L. Acta Amazon 42:381–386

Gaire S, Scharf ME, Gondhalekar AD (2020) Synergistic toxicity interactions between plant essential oil components against the common bedbug (*Cimex lectularius* L.). Insects 11:133

Gallardo A, Picollo MI, Mougabure-Cueto G (2015) Lethal activity of individual and mixed monoterpenoids of geranium essential oil on *Musca domestica*. Parasitol Res 114:1229–1232

Garcia M, Donadel OJ, Ardanaz CE, Tonn CE, Sosa ME (2005) Toxic and repellent effects of *Baccharis salicifolia* essential oil on *Tribolium castaneum*. Pest Manag Sci 61:612–618

Georghiou GP (1990) Overview of insecticide resistance. In: Green MB, Lebaron HM, Moberg WK (eds) ACS symposium series 421. American Chemical Society, Washington, DC, pp 19–41

Germinara G, Stefano M, Acutis L, Pati S et al (2017) Bioactivities of *Lavandula angustifolia* essential oil against the stored grain pest *Sitophilus granarius*. Bull Insectol 70:129–138

Gonzalez-Coloma A, Martın-Benito D, Mohamed N, Garcia-Vallejo MC, Soria AC (2006) Antifeedant effects and chemical composition of essential oils from different populations of *Lavandula luisieri* L. Biochem Syst Ecol 34:609–616

Grainge M, Ahmed S (1988) Handbook of plants with pest-control properties. Wiley-Interscience, Hoboken, NJ

Haagen-Smit AJ (1949) Essential oils-a brief survey of their chemistry and production in the United States. Econ Bot 3:71–83
Haynes KF (1988) Sublethal effects of neurotoxic insecticides on insect behaviour. Annu Rev Entomol 33:149–168
Hedjah-Chehheb M, Toudert-Taleb K, Khoudja ML et al (2013) Essential oils compositions of six conifers and their biological activity against the cowpea weevil, *Callosobruchus maculatus* Fabricius, 1775 (Coleoptera: Bruchidae) and *Vigna unguiculata* seeds. Afr Entomol 21:243–254
Hernández-Carlos B, Gamboa-Angulo M (2019) Insecticidal and nematicidal contributions of Mexican flora in the search for safer biopesticides. Molecules 24:897
Ho SH, Koh L, Ma Y, Huang Y, Sim KY (1996) The oil of garlic, *Allium sativum* L. (Amaryllidaceae), as apotential grain protectant against *Tribolium castaneum* (Herbst) and *Sitophilus zeamais* Motsch. Postharvest Biol Technol 9:41–48
Huang Y, Chen SX, Ho SH (2000a) Bioactivities of methyl allyl disulfide and diallyl trisulfide from essential oil of garlic to two species of stored-product pests, *Sitophilus Zeamais* (Coleoptera: Curculionidae) and *Tribolium Castaneum* (Coleoptera: Tenebrionidae). J Econ Entomol 93:537–543
Huang Y, Lam SL, Ho SH (2000b) Bioactivities of essential oils from *Elletaria cardamomum* (L.) Maton. To *Sitophilus zeamais* Motschulsky and *Tribolium castaneum* (Herbst). J Stored Prod Res 36:107–117
Hummelbrunner LA, Isman MB (2001) Acute sublethal, antifeedant and synergistic effects of monoterpenoid essential oil compounds on the tobacco cutworm, *Spodoptera litura* (Lep. Noctuidae). J Agric Food Chem 49:715–720
Idouaarame S, Abdel-hamid AA, Elfarnini M, Filali OA, Blaghen M (2018) Insecticidal activity of essential oils from five Moroccan plants on three insect pests of stored cereals. GSC Biol Pharm Sci 4:052–057
Isikber AA, Alma MH, Kanat M, Karci A (2006) Fumigant toxicity of essential oils from *Laurus nobilis* and *Rosmarinus officinalis* against all life stages of *Tribolium confusum*. Phytoparasitica 34:167–177
Isman MB (2006) Botanical insecticides, deterrents and repellents in modern agriculture and an increasingly regulated world. Annu Rev Entomol 51:45–66
Isman MB, Machial CM (2006) Pesticides based on plant essential oils: from traditional practice to commercialization. In: Rai M, Carpinella MC (eds) Naturally occurring bioactive compounds. Elsevier BV, Amsterdam, pp 29–44
Kamal GM, Anwar F, Hussain AI, Sarri N, Ashraf MY (2011) Yield and chemical composition of Citrus essential oils as affected by drying pretreatment of peels. Int Food Res J 18:1275–1282
Karemu CK, Ndung'u MW, Githua M (2013) Repellent effects of essential oils from selected eucalyptus species and their major constitutnets against *Sitophilus zeamais* (coleopteran: Curculionidae). Int J Trop Insect Sci 33:188–194
Ketoh GK, Koumaglo HK, Glitho IA (2005) Inhibition of *Callosobruchus maculatus* (F.) (Coleoptera: Bruchidae) development with essential oil extracted from *Cymbopogon schoenanthus* L. Spreng. (Poaceae), and the wasp *Dinarmus basalis* (Rondani) (Hymenoptera: Pteromalidae). J Stored Prod Res 41:363–371
Kfir R, Overholt WA, Khan ZR, Polaszek A (2002) Biology and management of economically important lepidopteran cereal stem borers in Africa. Annu Rev Entomol 47:701–731
Khadria A, Serralheirob MLM, Nogueirab JMF et al (2008) Antioxidant and antiacetylcholinesterase activities of essential oils from *Cymbopogon schoenanthus* L. Spreng. Determination of chemical composition by GC–mass spectrometry and 13C NMR. Food Chem 109:630–637
Khoee S, Yaghoobian M (2009) An investigation into the role of surfactants in controlling particle size of polymeric nanocapsules containing penicillin-G in double emulsion. Eur J Med Chem 44:2392–2399

Khoobdel M, Ahsaei SA, Farzaneh M (2017) Insecticidal activity of polycaprolactone nanocapsules loaded with *Rosmarinus officinalis* essential oil in *Tribolium castaneum* (Herbst). Entomol Res 47:175–184

Kim SII, Park C, Ohh MH, Cho HC, Ahn YJ (2003) Contact and fumigant activities of aromatic plant extracts and essential oils against *Lasioderma serricorne* (Coleoptera: Anobiidae). J Stored Prod Res 39:11–19

Kim S, Yoon JS, Jung JW et al (2010) Toxicity and repellency of *Origanum* essential oil and its components against *Tribolium castaneum* (Coleoptera: Tenebrionidae) adults. J Asia Pac Entomol 13:369–373

Knio KM, Usta J, Dagher S, Zournajian H, Kreydiyyeh S (2008) Larvicidal activity of essential oils extracted from commonly used herbs in Lebanon against the seaside mosquito, *Ochlerotatus caspius*. Bioresour Technol 99:763–768

Kordali S, Aslan I, Calmasur O, Cakir A (2006) Toxicity of essential oils isolated from three *Artemisia* species and some of their major components to granary weevil, *Sitophilus granarius* (L.) (Coleoptera: Curculionidae). Ind Crop Prod 23:162–170

Kostyukovsky M, Rafaeli A, Gileadi C, Demchenko N, Shaaya E (2002) Activation of octopaminergic receptors by essential oil constituents isolated from aromatic plants: possible mode of action against insect pests. Pest Manag Sci 587:1101–1116

Koul O, Walia S, Dhaliwal GS (2008) Essential oils as green pesticides: potential and constraints. Biopestic Int 4:63–84

Kraiss H, Cullen EM (2008) Insect growth regulator effects of azadirachtin and neem oil on survivorship, development and fecundity of *Aphis glycines* (Homoptera: Aphididae) and its predator, *Harmonia axyridis* (Coleoptera: Coccinellidae). Pest Manag Sci 64:660–668

Kumari A, Yadav SK, Yadav SC (2010) Biodegradable polymeric nanoparticles based drug delivery systems. Colloids Surf 75:1–18

Lee S, Peterson CJ, Coats JR (2002) Fumigation toxicity of monoterpenoids to several stored product insects. J Stored Prod Res 39:77–85

Liang Y, Li JL, Xu S et al (2013) Evaluation of repellency of some Chinese medicinal herbs essential oils against *Liposcelis bostrychophila* (Psocoptera: Liposcelidae) and *Tribolium castaneum* (Coleoptera: Tenebrionidae). J Econ Entomol 106:513–519

Liao M, Xiao JJ, Zhou LJ et al (2017) Chemical composition, insecticidal and biochemical effects of *Melaleuca alternifolia* essential oil on the *Helicoverpa armigera*. J Appl Entomol 141:721–728

Liu CH, Mishra AK, He B, Tan RX (2001) Composition and antifungal activity of essential oils from *Artemisia princeps* and *Cinnamomum camphora*. Int Pest Control 43:72–74

Liu ZL, Xu YJ, Wu J, Goh SH, Ho SH (2002) Feeding deterrents from *Dictamnus dasycarpus* Turcz against two stored-product insects. J Agric Food Chem 50:1447–1450

Liu CH, Mishra AK, Tan RX et al (2006) Repellent and insecticidal activities of essential oils from *Artemisia pinceps* and *Cinnamomum camphora* and their effect on seed germination of wheat and broad bean. Bioresour Technol 97:1969–1973

Lopez MD, Jordan MJ, Pascual-Villalobus MJ (2008) Toxic compounds in essential oils of coriander, caraway and basil active against stored rice pests. J Stored Prod Res 44:273–278

Louni M, Shakarami J, Negahban M (2018) Insecticidal efficacy of nanoemulsion containing *Mentha longifolia* essential oil against *Ephestia kuehniella* (Lepidoptera: Pyralidae). J Crop Prot 7:171–182

Maciel MV, Morais SM, Bevilaqua CML et al (2010) Chemical composition of *Eucalyptus* spp. essential oils and their insecticidal effects on *Lutzomyia longipalpis*. Vet Parasitol 167:1–7

Mahmoudvand M, Abbasipour H, Basij M et al (2011) Fumigant toxicity of some essential oils on adults of some stored-product pests. Chil J Agric Res 71:83–89

Martin AS, Varona S, Navarrete A, Cocero MJ (2010) Encapsulation and co-precipitation processes with supercritical fluids: applications with essential oils. Open Chem Eng J 4:31–41

Mediouni-Ben JJ, Tersim N (2011) Composition and repellent efficacy of essential oil from *Laurus nobilis* against adults of cigarette beetle *Lasioderma serricorne* (Coleoptera: Anobiidae). Tunisian J Plant Prot 6:29–42

Midega CAO, Murage AW, Pittchar JO, Khan ZR (2016) Managing storage pests of maize: Farmers' knowledge, perceptions and practices in western Kenya. Crop Prot 90:142–149

Mohan M, Haider SZ, Andola HC, Purohit VK (2011) Essential oils as green pesticides: for sustainable agriculture. Res J Pharm Biol Chem Sci 2:100–105

Morrison NI, Franz G, Koukidou M et al (2010) Genetic improvements to the sterile insect technique for agricultural pests. Asia Pac J Mol Biol 18:275–295

Mossi AJ, Astolfi V, Kubiak G et al (2011) Insecticidal and repellency activity of essential oil of Eucalyptus sp. against *Sitophilus zeamais* Motschulsky (Coleoptera, Curculionidae). J Sci Food Agric 91:273–277

Muñoz O, Christen P, Cretton S et al (2011) Comparison of the essential oils of leaves and stem bark from two different populations of *Drimys winteri* a Chilean herbal medicine. Nat Prod Commun 6:879–882

Murugan K, Kumar PM, Kovendan K et al (2012) Larvicidal, pupicidal, repellent and adulticidal activity of *Citrus sinensis* orange peel extract against *Anopheles stephensi*, *Aedes aegypti* and *Culex quinquefasciatus* (Diptera: Culicidae). Parasitol Res 111:1757–1769

Nagpal BN, Srivastava A, Valecha NA, Sharma VP (2001) Repellent action of neem cream against *an. Culicifacies* and *cx. Quinquefasciatus*. Curr Sci 80:1270–1271

Navarro-Llopis V, Vacas S, Sanchis J, Primo J, Alfaro C (2011) Chemosterilant bait stations coupled with sterile insect technique: an integrated strategy to control the mediterranean fruit fly (Diptera: Tephritidae). J Econ Entomol 104:1647–1655

Ndomo AF, Ngamo LT, Tapondjou LA et al (2008) Insecticidal effects of the powdery formulation based on clay and essential oil from leaves of *Clausena anisata* (Willd) J.D. ex. Benth (Rutaceae) against *Acanthoscelides obtectus* (say) hook (Coleoptera; Bruchidae). J Pestic Sci 81:227–234

Negahban M, Moharramipour S, Sefidkon F (2006) Chemical composition and insecticidal activity of *Artemisia scoparia* essential oil against three coleopteran stored-product insects. J Asia Pac Entomol 9:381–388

Negahban M, Moharamipour S, Sefidkon F (2007) Fumigant toxicity of essential oil from *Artemisia sieberi* Besser against three storedproduct insects. J Stored Prod Res 43:123–128

Nenaah GE (2014) Chemical composition, toxicity and growth inhibitory activities of essential oils of three *Achillea* species and their nanoemulsions against *Tribolium castaneum* (Herbst). Ind Crop Prod 53:252–260

Nerio LS, Olivero-Verbel J, Stashenko EE (2009) Repellent activity of essential oils from seven aromatic plants grown in Colombia against *Sitophilus zeamais* Motschulsky (Coleoptera). J Stored Prod Res 45:212–214

Nerio LS, Oliver-Verbel J, Stashenko E (2010) Repellent activity of essential oils: a review. Bioresour Technol 101:372–378

Ngoh SP, Cho LEW, Pang FY et al (1998) Insecticidal and repellent properties of nine volatile constituents of essential oils against the American cockroach, *Periplaneta americana* (L.). J Pestic Sci 54:261–268

Norambuena C, Silva G, Urbina A, Figueroa I, Rodrı́guez-Maciel JC (2016) Insecticidal activity of *Laureliopsis philippiana* (looser) Schodde (Atherospermataceae) essential oil against *Sitophilus* spp. (Coleoptera Curculionidae). Chilean J Agric Res 76:330–336

Ntalli NG, Ferrari F, Giannakou I, Menkissoglu-Spiroudi U (2011) Synergistic and antagonistic interactions of terpenes against *Meloidogyne incognita* and the nematicidal activity of essential oils from seven plants indigenous to Greece. Pest Manag Sci 67:341–351

Nuchuchua O, Sakulku U, Uawongyart N et al (2009) *In vitro* characterization and mosquito (*Aedes aegypti*) repellent activity of essential-oils-loaded nanoemulsions. AAPS Pharm Sci Technol 10:1234–1242

Nyamador WS, Ketoh GK, Amevoin K et al (2010) Variation in the susceptibility of two *Callosobruchus* species to essential oils. J Stored Prod Res 46:48–51
Ogendo JO, Deng AL, Belmain SR et al (2004) Pest status of *Sitophilus zeamais* Motschulsky, control methods and constraints to safe maize grain storage in western Kenya. Egert J Sci Tech 5:175–193
Ojo JA, Omoloye AA (2012) Rearing the maize weevil, *Sitophilus zeamais*, on an artificial maize-cassava diet. J Insect Sci 12:1–9
Ozols G, Bicevskis M (1979) Respects for the use of *Ips tyrographus* attractant. In: Shumakov EM, Chekmenev SY, Ivanova TV (eds) Biologia Aktualis Veshchestva Zashchiva Rastenij. Izd. Kolos, Moscow, pp 49–51
Palla F, Bruno M, Mercurio F, Tantillo A, Rotolo V (2020) Essential oil as natural biocides in conservation of cultural heritage. Molecules 25:730
Paranagama PA, Gunasekera JJ (2011) The efficacy of the essential oils of Sri Lankan *Cinnamomum zeylanicum* fruit and *Micromelum minutum* leaf against *Callosobruchus maculatus* (F.) (*Coleoptera*: Bruchidae). J Essent Oil Res 23:75–82
Park IK, Choi KS, Kim DH et al (2006) Fumigant activity of plant essential oils and components from horseradish (*Armoracia rusticana*), anise (*Pimpinella anisum*) and garlic (*Allium sativum*) oils against *Lycoriella ingenua* (Diptera: Sciaridae). Pest Manag Sci 62:723–728
Paudyal M, Rajbhandari M, Basnet P, Yahara S, Gewali M (2012) Quality assessment of the essential oils from *Nardostachys jatamansi* (d. Don) dc and *Nardostachys chinensis* batal obtained from Kathmandu valley market. Sci World 10:13–16
Paula HCB, Sombra FM, Abre FOMS, Paul RC (2010) *Lippia sidoides* essential oil encapsulationby angico gum/chitosan nanoparticles. J Braz Chem Soc 21:2359–2366
Paula HCB, Sombra FM, Cavalcante RF, Abreu FOMS, Paula RCM (2011) Preparation and characterization of chitosan/cashew gum beads loaded with *Lippia sidoides* essential oil. Mater Sci Eng C 31:173–178
Pavela R (2008) Insecticidal properties of several essential oils on the house fly (*Musca domestica* L.). Phytother Res 22:274–278
Pavela R (2011) Insecticidal and repellent activity of selected essential oils against of the pollen beetle, *Meligethes aeneus* (Fabricius) adults. Ind Crop Prod 34:888–892
Pavela R, Benelli G (2016) Essential oils as ecofriendly biopesticides? Challenges and constraints. Trends Plant Sci 21:1000–1007
Phillips AK, Appel AG, Sims SR (2010) Topical toxicity of essential oils to the German cockroach (Dictyoptera: Blattellidae). J Econ Entomol 103:448–459
Plata-Rueda A, Martinez LC, Santos MHD et al (2017) Insecticidal activity of garlic essential oil and their constituents against the mealworm beetle, *Tenebrio molitor* Linnaeus (Coleoptera: Tenebrionidae). Sci Rep 7:46406
Prowse MG, Galloway TS, Foggo A (2006) Insecticidal activity of garlic juice in two dipteran pests. Agr Forest Entomol 8:1–6
Pugazhvendan SR, Ross PR, Elumalai K (2012) Insecticidal and repellent activities of four indigenous medicinal plants against stored grain pest, *Tribolium castaneum* (Herbst) (Coleoptera: Tenebrionidae). Asian Pac J Trop Dis 2:S16–S20
Qin W, Huang S, Li C et al (2010) Biological activity of the essential oil from the leaves of *Piper sarmentosum* Roxb. (Piperaceae) and its chemical constituents on *Brontispa longissima* (Gestro) (Coleoptera: Hispidae). Pestic Biochem Phys 96:132–139
Rafiei KZ, Moharramipour S, Rahbarpour A (2009) Investigated repellency effect of some essential oils of 17 native medicinal plants on adults *Plodia interpunctella*. Am Eurasian J Sustain Agric 3:181–184
Rajashekar Y, Rao LJM, Shivanandappa T (2012) Decaleside: a new class of natural insecticides targeting tarsal gustatory sites. Naturwissenschaften 99:843–852
Rajendran S, Sriranjini V (2008) Plant products as fumigants for stored product insect control. J. Stored Prod Res 44:126–135

Ramzi H, Ismaili MR, Aberchane M, Zaanoun S (2017) Chemical characterization and acaricidal activity of *Thymus satureioides* C. & B. and *Origanum elongatum* E. & M. (Lamiaceae) essential oils against *Varroa destructor* Anderson & Trueman (Acari: Varroidae). Ind Crop Prod 108:201–207

Rattan RS (2010) Mechanism of action of insecticidal secondary metabolites of plant origin. Crop Prot 29:913–920

Ravi Kumar MN (2000) Nano and microparticles as controlled drug delivery devices. J Pharm Pharm Sci 3:234–258

Regnault-Roger C, Vincent C, Arnason JT (2012) Essential oils in insect control: low-risk products in a high-stakes world. Annu Rev Entomol 57:405–424

Rocha RP, Melo EDC, Barbosa LCA et al (2014) Influence of plant age on the content and composition of essential oil of *Cymbopogon citratus* (DC.) Stapf. J Med Plant Res 8:1121–1126

Romeilah RM, Fayed SA, Mahmoud GI (2010) Chemical compositions, antiviral and antioxidant activities of seven essential oils. J Appl Sci Res 6:50–62

Rozman V, Kalinovic I, Liska A (2006) Insecticidal activity of some aromatic plants from Croatia against granary weevil (*Sitophilus granarius* L) on stored wheat. Cereal Res Commun 34:705–708

Saei-Dehkordi SS, Tajik H, Moradi M, Khalighi-Sigaroodi F (2010) Chemical composition of essential oils in *Zataria multiflora* Boiss. From different parts of Iran and their radical scavenging and antimicrobial activity. Food Chem Toxicol 48:1562–1567

Sahaf BZ, Moharramipour S (2008) Fumigant toxicity of *Carum copticum* and *Vitex pseudo-negundo* essential oils against eggs, larvae and adults of *Callosobruchus maculatus*. J Pestic Sci 81:213–220

Sahaf BZ, Moharamipour S, Meshkatassadat MH (2007) Chemical constituents and fumigant toxicity of essential oil from *Carum copticum* against two stored product beetles. Insect Sci 14:213–218

Sahaf BZ, Moharramipour S, Meshkatalsadat MH (2008) Fumigant toxicity of essential oil from *Vitex pseudo- negundo* against *Tribolium castaneum* (Herbst) and *Sitophilus oryzae* (L). J Asia Pac Entomol 11:175–179

Salem N, Olfa Bachrouch O, Sriti J, Msaada K et al (2017) Fumigant and repellent potentials of *Ricinus communis* and *Mentha pulegium* essential oils against *Tribolium castaneum* and *Lasioderma serricorne*. Int J Food Prop 20:S2899–S2913

Sammour EA, El Hawary FMA, Abdel-Aziz NF (2011) Comparative study on the efficacy of neemix and basil oil formulations on the cowpea aphid *Aphis craccivora* Koch. Arch Phytopathol Plant Prot 44:655–670

Sampson BJ, Tabanaca N, Kirimer N et al (2005) Insecticidal activity of 23 essential oils and their major compounds against adults *Lipaphis pseudobrassicae* (Davis) (Aphididae: Homoptera). Pest Manag Sci 61:1122–1128

São Pedro A, Santo IE, Silva C, Detoni C, Albuquerque E (2013) The use of nanotechnology as an approach for essential oil-based formulations with antimicrobial activity. In: Méndez-Vilas A (ed) Microbial pathogens and strategies for combating them: science, technology and education. Formatex Research Center, Zurbaran, Badajoz, pp 1364–1374

Sbeghen-Loss AC, Mato M, Cesio MV, Frizzo C et al (2011) Antifeedant activity of Citrus waste wax and its fractions against the dry wood termites *Cryptotermes brevis*. J Insect Sci 11:1–7

Scholz SS, Heyer M, Vadassery J, Mithöfer A (2016) A role for calmodulin-like proteins in herbivore defense pathways in plants. J Endocytobiosis Cell Res 27:1–12

Schoonhoven AV (1978) The use of vegetable oils to protect stored beans from bruchid attack. J Econ Entomol 71:254–256

Schweiggert U, Carle R, Schieber A (2007) Conventional and alternative processes for spice production-a review. Trends Food Sci Technol 18:260–268

Scott RPW (2005) Essential oils. In: Worsfold P, Townshend A, Poole C (eds) Encyclopedia of analytical science, 2nd edn. Elsevier, London, pp 554–561

Sefidkon F (2001) Essential oil composition of *Anethum graveolens* L. Iranian J Med Arom Plant 8:45–62

Shaalan EAS, Canyon D, Younes MWF, Abdel-Wahab H, Mansour AH (2005) A review of botanical phytochemicals with mosquitocidal potential. Environ Int 31:1149–1166

Shan-Shan G, Chun-Xue Y, Jun-Yu L et al (2015) Chemical composition and bio-activities of the essential oil from *Etlingera yunnanensis* against two stored product insects. Molecules 20:15735–15747

Sharaby A, Al-Dosary M (2014) An electric air flow olfactometer and the olfactory response of *Rhynchophorous ferrugineus* weevil to some volatile compounds. J Agric Ecol Res Int 1:40–50

Sharaby A, El-Nujiban A (2015) Evaluation of some plant essential oils against the black cutworm *Agrotis ipsilon*. Global J Adv Res 2:701–711

Sharaby A, Montasser SA, Mahmoud YA, Ibrahim SA (2012) Natural plant essential oils for controlling the grasshopper *(Hetracis littoralis)* and their pathological effects on the alimentary canal. Ecol Balk 4:39–52

Sharififard M, Sharififard F, Safdari A, Siahpoush H, Kassiri A (2016) Evaluation of some plant essential oils against the brown-banded cockroach, *Supella longipalpa* (Blattaria: Ectobiidae): a mechanical vector of human pathogens. J Arthropod Borne Dis 10:528–537

Shukla R, Singh P, Prakash B et al (2011) Efficacy of essential oils of *Lippia alba* (mill.) N.E. Brown and *Callistemon lanceolatus* (Sm.) sweet and their major constituents on mortality, oviposition and feeding behaviour of pulse beetle, *Callosobruchus chinensis* L. J Sci Food Agric 91:2277–2283

Shukla P, Vidyasagar PSPV, Aldosari SA, Abdel-Azim M (2012) Antifeedant activity of three essential oils against the red palm weevil, Rhynchophorus *ferrugineus*. Bull Insectol 65:71–76

Singh P, Pandey AK (2018) Prospective of essential oils of the genus *Mentha* as biopesticides: a review. Front Plant Sci 9:1295

Singh P, Pandey AK, Tripathi NN (2012) Essential oils: a renewable source for the management of stored product insects—a review. Agric Rev 33:226–236

Singh P, Jayaramaiah RH, Sarate P et al (2014) Insecticidal potential of defense metabolites from *Ocimum kilimandscharicum* against *Helicoverpa armigera*. PLoS One 9:e104377

Soleimannejad S, Moharramipour S, Fathipour Y, Nikooei M (2011) Efficiency of essential oil from *Salvia mirzayanii* against nutritional indices of *Tribolium confusum*. Integr Prot Stored Prod IOBC/WPRS Bull 69:299–302

Solomon B, Sahle FF, Gebre-Mariam T, Asres K, Neubert RHH (2012) Microencapsulation of citronella oil for mosquito-repellent application: formulation and *in vitro* permeation studies. Eur J Pharm Biopharm 80:61–66

Stefanazzi N, Stadler T, Ferrero A (2011) Composition and toxic, repellent and feeding deterrent activity of essential oils against the stored-grain pests *Tribolium castaneum* (Coleoptera: Tenebrionidae) and *Sitophilus oryzae* (Coleoptera: Curculionidae). Pest Manag Sci 67:639–646

Suthisut D, Fields PG, Chandrapatya A (2011) Fumigant toxicity of essential oils from three Thai plants (Zingiberaceae) and their majorcompounds against *Sitophilus zeamais,Tribolium castaneum* and two parasitoids. J Stored Prod Res 47:222–230

Taghizadeh-Saroukolai A, Moharramipour S, Meshkatalsadat MH, Fathipour Y, Talebi AA (2009) Repellent activity and persistence of essential oil extracted from *Prangos acaulis* to three stored-product beetles. Am Eurasian J Sustain Agric 3:202–204

Taghizadeh-Saroukolai A, Nouri-Ganbalani G, Rafiee-Dastjerdi H, Hadian J (2014) Antifeedant activity and toxicity of some plant essential oils to Colorado potato beetle, *Leptinotarsa decemlineata* say (Coleoptera: Chrysomelidae). Plant Protect Sci 50:207–216

Tak JH, Isman MB (2015) Enhanced cuticular penetration as the mechanism for synergy of insecticidal constituents of rosemary essential oil in *Trichoplusia ni*. Sci Rep 5:12690

Tak JH, Jovel E, Isman MB (2016) Comparative and synergistic activity of *Rosmarinus officinalis* L. essential oil constituents against the larvae and an ovarian cell line of the cabbage looper, *Trichoplusia ni* (Lepidoptera: Noctuidae). Pest Manag Sci 72:474–480

Talukder FA (2006) Plant products as potential storedproduct insect management agents—a mini review. Emirates J Food Agric 18:17–32

Tefera T, Kanampiu F, Groote HD et al (2010) The metal silo: an effective grain storage technology for reducing post-harvest insect and pathogen losses in maize while improving smallholder farmers' food security in developing countries. Crop Prot 30:240–245

Tembo Y, Mkindi AG, Mkenda PA et al (2018) Pesticidal plant extracts improve yield and reduce insect pests on legume crops without harming beneficial arthropods. Front Plant Sci 9:1425

Toudert-Taleb K, Hedjal-Chehheb M, Hami H, Debras JF, Kellouche A (2014) Composition of essential oils extracted from six aromatic plants of Kabylian origin (Algeria) and evaluation of their bioactivity on *Callosobruchus maculatus* (Fabricius, 1775) (Coleoptera: Bruchidae). Afr Entomol 22:417–427

Tozlu E, Cakir A, Kordali S et al (2011) Chemical compositions and insecticidal effects of essential oils isolated from Achilleagypsicola, Saturejahortensis, Origanumacutidens and Hypericumscabrum against broadbean weevil (*Bruchus dentipes*). Sci Hortic 130:9–17

Tripathi NN, Kumar N (2007) *Putranjiva roxburghii* oil—a potential herbal preservative for peanuts during storage. J Stored Prod Res 43:435–442

Tunç İ, Berger BM, Erler F, Dağlı F (2000) Ovicidal activity of essential oils from five plants against two stored-product insects. J Stored Prod Res 36:161–168

Turek C, Stintzing FC (2013) Stability of essential oils: a review. Compr Rev Food Sci F 12:40–53

Usman LA, Hamid AA, Olawore NO et al (2010) Chemical composition of leaf essential oil of *Clausena anisata* growing in north-Central Nigeria. J Appl Sci Res 6:891–894

Wafaa MH, Rowida SB, Hussein AHS-AA (2017) Botanical insecticide as simple extractives for pest control. Cogent Biol 3:1404274

Waliwitiya R, Kennedy CJ, Lowenberger CA (2009) Larvicidal and oviposition-altering activity of monoterpenoids, trans-anethole and rosemary oil to the yellow fever mosquito *Aedes aegypti* (Diptera: Culicidae). Pest Manag Sci 65:241–248

Wang CF, Liu P, Yang K et al (2011) Chemical composition and toxicities of essential oil of *Illicium fragesii* fruits against *Sitophilus zeamais*. Afr J Biotechnol 10:18179–18184

War AR, Paulraj MG, Ahmad T et al (2012) Mechanisms of plant defense against insect herbivores. Plant Signal Behav 7:1306–1320

Werdin-Gonzalez JO, Gutiérrez MM, Ferrero AA, Band BF (2014) Essential oils nanoformulations for stored-product pest control—characterization and biological properties. Chemosphere 100:130–138

Wilke ABB, Nimmo DD, John O et al (2009) Mini-review: genetic enhancements to the sterile insect technique to control mosquito populations. Asia Pac J Mol Biol Biotechnol 17:65–74

Wu L, Huo X, Zhou X et al (2017) Acaricidal activity and synergistic effect of thyme oil constituents against carmine spider mite (*Tetranychus cinnabarinus* (Boisduval)). Molecules 22:1873

Yang FL, Li SG, Zhu F, Lei CL (2009) Structural characterization of nanoparticles loaded with garlic essential oil and their insecticidal activity against *Tribolium castaneum* (Herbst) (Coleoptera: Tenebrionidae). J Agric Food Chem 57:10156–10162

Yang K, Wang CF, You CX et al (2014) Bioactivity of essential oil of *Litsea cubeba* from China and its main compounds against two stored product insects. J Asia Pac Entomol 17:459–466

Zandi-Sohani N, Hojjati M, Carbonell-Barrachina AA (2013) Insecticidal and repellent activities of the essential oil of *Callistemon citrinus* (Myrtaceae) against *Callosobruchus maculatus* (F.) (*Coleoptera:* Bruchidae). Neotrop Entomol 42:89–94

Zapata N, Smagghe G (2010) Repellency and toxicity of essential oils from the leaves and bark of *Laurelia sempervirens* and *Drimys winteri* against *Tribolium castaneum*. Ind Crop Prod 32:405–410

Zhang F-P, Yang Q-Y, Wang G, Zhang S-B (2016) Multiple functions of volatiles in flowers and leaves of *Elsholtzia rugulosa* (Lamiaceae) from southwestern China. Sci Rep 6:27616

Zhao NN, Zhang H, Zhang XC et al (2013) Evaluation of acute toxicity of essential of garlic (*Allium sativum*) and its selected major constituent compounds against overwintering *Cacopsylla chinensis* (Hemiptera: Psyllidae). J Econ Entomol 106:1349–1354

Zoubiri S, Baaliouamer A (2012) Chemical composition and insecticidal properties of *Lantana camara* L. leaf essential oils from Algeria. J Essent Oil Res 24:377–383

Chapter 7
Plant Secondary Metabolites: Emerging Trends in Agricultural Pests Control

Abid Hussain and Ahmed Mohammed AlJabr

Abstract Plants are producing enormous chemical diversity of plant secondary metabolites as a self-defense against biotic and abiotic stresses. With the advancement in analytical chemistry, different groups of plant secondary metabolites including nitrogen containing compounds, phenolic compounds, and terpenes were evolved as candidate bioactive substances to control the infestations of agricultural pests. Recent efficacy results of plant secondary metabolites against major agricultural pests involved in pre-harvest and post-harvest losses revealed that plant secondary metabolites have tremendous potential to be incorporated into the Integrated Pest Management (IPM) strategy of agricultural pests. However, extensive research is needed to overcome the challenges through scientific knowledge in order to develop eco-friendly formulations of plant secondary metabolites against agricultural pests adversely threatening sustainable global food production.

Keywords Agricultural pests · Biopesticides · Botanicals · Pest management · Secondary metabolites

A. Hussain (✉)
Laboratory of Bio-Control and Molecular Biology, Department of Arid Land Agriculture, College of Agricultural and Food Sciences, King Faisal University, Hofuf, Al-Ahsa, Saudi Arabia

Institute of Research and Consultancy, King Faisal University, Hofuf, Al-Ahsa, Saudi Arabia

Ministry of Environment, Water and Agriculture, Riyadh, Saudi Arabia
e-mail: abhussain@kfu.edu.sa

A. M. AlJabr
Laboratory of Bio-Control and Molecular Biology, Department of Arid Land Agriculture, College of Agricultural and Food Sciences, King Faisal University, Hofuf, Al-Ahsa, Saudi Arabia

S. De Mandal et al. (eds.), *New and Future Development in Biopesticide Research: Biotechnological Exploration*, https://doi.org/10.1007/978-981-16-3989-0_7

7.1 Introduction

Sustainable food production and nutrition to fulfill the emerging global food demands for the rapidly growing human population in the context of depleting natural resources is posing a serious challenge for the policy makers, public and private sector, academia, researchers and farmers. In this regard, the idea of crop intensification was flourished over the period of time to cope food security challenges. However, crop intensification leads to increase the number of arthropod pests attacking various crops. Their infestations impart huge economic losses. According to the Swedish Institute for Food and Biotechnology estimations, which was reported by the FAO that the global edible food losses and waste reaches up to one third of total production, which translates into wasting of 1.3 billion tonnes of food per annum (FAO, IFAD, UNICEF, WFP, WHO 2019). However, roughly half of this wasting is mainly because of the infestations of arthropod pests. In the past, the spraying of broad spectrum pesticides due to their presumed effectiveness have played a major role to protect crops against pests. However, indiscriminate use of pesticides leads to the development of pesticide resistance (Al-Ayedh et al. 2016), pest resurgence (Dutcher 2007), environmental pollution, applicator's safety and persistence of residues. Furthermore, non-specific nature of pesticides results in the killing of non-target beneficial organisms especially parasitoids and predators that are found to be highly susceptible against pesticides (Theiling and Croft 1988; Desneux et al. 2007; Martinou et al. 2014). These concerns raised the dire need to research on the eco-friendly alternate pest management measures. Consequently, use of biological control agents has become acceptable technique and tried against various pest species with variable effectiveness (AlJabr et al. 2018). With the advancement in the field of analytical chemistry, great diversity of plant secondary metabolites belonging to various classes have been discovered. Currently, the use of plant secondary metabolites is gaining popularity due to the issue of food safety. Therefore, plant secondary metabolites are establishing as key component in the fight against various agricultural pest species, and accordingly resulted in the rise of global market about plant secondary metabolites as effective pest management chemicals to protect crops. The chapter will highlight the research developments and future prospects about the potential utility of plant secondary metabolites as alternate eco-friendly pest management option to protect crops against agricultural pests.

7.2 Current Market Value

The economics plays a key role for the development of any plant protectant. The damages caused by the agricultural pests not only make them unfit for human consumption but also reduce the market value along with the reduction in crop productivity. Due to the widespread adoption of the concept of organic farming, the

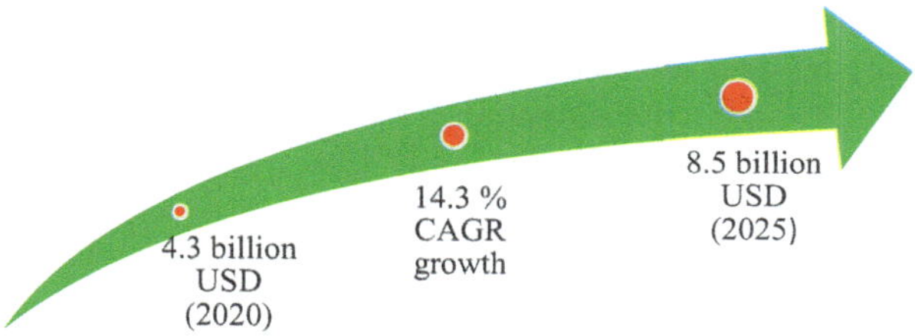

Fig. 7.1 Growing trends in Global Biopesticides market (2020–2025) (Markets and Markets 2020)

Table 7.1 Global top key market players in biopesticides

No	Name	Country	Website link
1	BASF SE	Germany	https://www.basf.com/global/en.html
2	Syngenta AG	Switzerland	https://www.syngenta.com/en
3	Marrone Bio Innovations	US	https://marronebio.com/
4	Bayer AG	Germany	https://www.bayer.com/en/
5	Novozymes A/S	Denmark	https://www.novozymes.com/en

use of biopesticides in which plant secondary metabolites as a key component is encouraged among farmers. According to the latest Market Research Report released during June 2020 revealed that global biopesticides market is forecasted to reach 8.5 billion USD (2025) from 4.3 billion USD (2020) with a significant growth at a 14.7% Compound Annual Growth Rate (CAGR) as shown in Fig. 7.1 (Markets and Markets 2020).

The strategic developments in terms of collaboration and agreements, consumer awareness about the pesticide residues in their food, and phasing-out of key pesticides from the market due to their deleterious health and environmental impacts lead to the expansion of biopesticides market. Among all the regions, until now North America witnessed the dominance of biopesticides market compared with Europe, Asia pacific, South America, and rest of the world. In this potential strong market, top five key global players involved in the manufacturing and marketing of biopesticides are mentioned below (Table 7.1).

7.3 Classification of Plant Secondary Metabolites

The metabolic pathways within the plant cells generated the plant secondary metabolites as an important defense line in order to cope with biotic and abiotic stresses. Historically, Nobel Laurate Albrecht Kossel (1853–1927) for the first time during 1910 defined the concept of plant secondary metabolites (Jones 1953). While the advances in the analytical chemistry especially chromatography laid the foundation of photochemistry discipline in the mid-twentieth century. Previous studies have revealed that plants produce a complex mixture of plant secondary metabolites belonging to various structural classes. However, it is almost undeterminable to count the exact number of bioactive plant secondary metabolites. A through literature review complied one and half decade before revealed that between 150,000 to 200,000 bioactive compounds out of one million natural compounds extracted only from the Plant Kingdom (Bérdy 2005). The plant secondary metabolites are mainly classified into three major groups such as (1) Nitrogen containing compounds, (2) Phenolic compounds, and (3) Terpenes as mentioned their prominent compounds in the Table 7.2.

7.4 Plant Secondary Metabolites as Biopesticides Against Agricultural Pests

7.4.1 Current Status as Management Option

The plant in the form of plant secondary metabolites are providing untapped reservoir of chemicals with numerous potential uses with most importantly low environment risk. In nature, plant secondary metabolites are imparted by the plants themselves spontaneously to act as defender by minimizing feeding injury occurred in response to pest attack. These compounds are concentrated within different plant parts including seeds, bark, leaves, roots and fruits (Bérdy 2005; Wink and Schimmer 2018). Therefore, plants are being utilized to extract plant secondary metabolites to kill or repel the agricultural pests throughout the world as eco-management strategy. Over the period of time, the toxicity of plant-based products or their secondary metabolites imparted various responses such as abnormal growth, feeding deterrence, low fecundity, respiratory failure, molting inhibition, feeding inhibition, suppression of calling behavior, oviposition deterrence and repellency among agricultural pests as shown in Fig. 7.2 (Zhao et al. 1998; Isman 2000; AlJabr et al. 2017a, b; Hussain et al. 2019).

The infestations of agricultural pests are posing serious threats to global food security. Few years before CABI (Centre for Agriculture and Biosciences International) in their Compendia published full database of 1187 arthropod pests. Based on the arthropods database, Royal Botanical Garden published a report estimating a potential loss of 540 billion USD/annum in case the arthropod pests and pathogens especially invasive species were not stopped to spread their infestations. In this regard, plant secondary metabolites deemed fit to stop the infestations of agricultural pests due to their wider range of modes of actions. Pest management potential of various plant secondary metabolites belonging to different groups (Table 7.2) have been evaluated in the past, which mainly aimed to screen and then subsequently scaled up for commercialization. Until now, the status of top 19 agricultural pests belonging to various taxa tested against the application of plant secondary metabolites by various methodologies including fumigation (Prates et al. 1998; Cavalcanti et al. 2010; de Souza Born et al. 2018), topical application (Faraone et al. 2015; Giongo et al. 2016), substrate treatment (Faraone et al. 2015), contact/residual

Table 7.2 Major groups of plant secondary metabolites

Groups	Key compounds
1. *Nitrogen containing compounds*	
a. Alkaloids	Atropine, Berberine, Caffeine, Camptothecin, Cocaine, Codeine, Morphine, Nicotine, Quinine, Reserpine, Vinblastine (Richard et al. 2013)
b. Cyanogenic glucosides (CNglcs)	Amygdalin, Dhurrin, Linamarin, Lotaustralin, Prunasin (Ben-Yehoshua and Conn 1964; Zagrobelny et al. 2004)
c. Non-protein amino acids	3,4-dihydroxyphenylalanine (L-DOPA), Azatyrosine, Canavanine, Mimosine, Hypoglycine (Yamane et al. 2010)
2. *Phenolic compounds*	
a. Coumarin	Coumarin, Esculetin, Scopoletin, Umbelliferone
b. Furano-coumarins	Angelicin, Psoralen
c. Tannins	Ellagitannins, Gallotannin, Flavan-3.4-diol, Proanthocyanidins (Hussein and El-Anssary 2019)
d. Lignin	Softwood lignin (coniferyl alcohol), hardwood lignin (coniferyl alcohol and sinapyl alcohol), grass lignin (coniferyl, sinapyl and *p*-coumaryl alcohol), caffeyl lignin (C-lignin) (Liu et al. 2018)
e. Flavonoids	Pinocembrin, quercetin, sakuranetin, 6-isopentenylnaringenin I, luteone, wighteone
3. *Terpenes*	
a. Hemiterpenes	Isovaleric acid, Isoprene, trans-2-Methyl-2-butenal (Tiglic aldehyde), 2-Methyl-3-buten-2-ol, 3-Methyl-3-buten-1-ol, 3-Methyl-2-buten-1-ol, 3-Methyl-3-buten-2-one, 3-Methylcrotonic Acid, Methyl 3,3-Dimethylacrylate, Tiglic acid ((E)-2-Methyl-2-butenoic Acid)
b. Acyclic monoterpenes	Citral (3,7-Dimethyl-2,6-octadienal,), Citral dimethyl acetal, (S)-(−)-Citronellal, Citronellic acid, (S)-(−)-β-Citronellol, β-Citronellol, Citronellyl Acetate, R-(−)-2,6-Dimethyloctane, 3,7-Dimethyl-1-octanol, 2,6-Dimethyl-2,4,6-octatriene, Geraniol, Geranyl Acetate, Geranylacetone, Geranyl formate, 3,7-Dimethyl-2,6-octadienenitrile (Geranyl Nitrile), Linalool, Linalyl acetate, Linalyl butyrate, Linalyl propionate, Myrcene, Nerol, Neryl acetate
c. Monocyclic monoterpenes	(1R,3S)-(+)-Camphoric acid, Carvacrol, (R)-(−)-Carvone, (S)-(+)-Carvone, Cuminaldehyde, *o*-Cymene, *m*-Cymene, *p*-Cymene, Ethyl chrysanthemate, N-Ethyl-2-isopropyl-5-methylcyclohexanecarboxamide, β-Thujaplicin, (±)-Limonene, (R)-(+)-Limonene, (S)-(−)-Limonene, Linalool oxide, DL-Menthol, (1S,2R,5S)-(+)-Menthol, (−)-Menthol, (−)-Menthone, (−)-Menthoxyacetic Acid, (−)-Menthoxyacetyl chloride, (1S)-(+)-Menthyl acetate, (1R)-(−)-Menthyl acetate, (−)-Menthyl chloride, (1S)-(+)-Menthyl chloroformate, (1R)-(−)-Menthyl chloroformate, L-Menthyl 2,2-dihydroxyacetate, L-Menthyl lactate, (1R,2S,5R)-(−)-Menthyl (S)-p-toluenesulfinate, (1S,2R,5S)-(+)-Menthyl (R)-p-toluenesulfinate, 8-Mercaptomenthone, (1S,2S,5R)-(+)-Neomenthol, (S)-(−)-Perillaldehyde, (R)-(−)-α-Phellandrene, (R)-(+)-Pulegone, cis-Terpin monohydrate, α-Terpinene, γ-Terpinene, 4-Carvomenthenol, (+)-Terpinen-4-ol, (−)-Terpinen-4-ol, α-Terpineol, Terpinolene, Terpinyl acetate, Thymol
d. Bicyclic monoterpenes	(−)-Borneol, (−)-Bornyl acetate, [(1R)-(endo,anti)]-(+)-3-Bromocamphor-8-sulfonic acid ammonium salt, (1S)-(−)-

(continued)

Table 7.2 (continued)

Groups	Key compounds
	Camphanic acid, (1S)-(−)-Camphanic chloride, Camphene, (±)-Camphor, (1R)-(+)-Camphor, (1S)-(−)-Camphor, (1R)-Camphor oxime, Camphorquinone, (1R)-(−)-Camphorquinone, (1S)-(+)-Camphorquinone, (1R,E)-(+)-Camphorquinone 3-oxime, Camphor-10-sulfonic acid (β), (1S)-(+)-10-Camphorsulfonic acid, (1R)-(−)-10-Camphorsulfonic acid, (1S)-(−)-Camphorsulfonylimine, (1S)-(+)-10-Camphorsulfonyl chloride, (1R)-(−)-10-Camphorsulfonyl chloride, (1R)-(+)-2,10-Camphorsultam, (1S)-(−)-2,10-Camphorsultam, (1S)-(+)-(10-Camphorsulfonyl)oxaziridine, (1R)-(−)-(10-Camphorsulfonyl)oxaziridine, 3-Carene, Eucalyptol (1,8-Cineole), (+)-Fenchone, (1R)-(−)-Fenchone, Genipin, (1R,2R,5R)-(+)-2-Hydroxy-3-pinanone, (1S,2S,5S)-(−)-2-Hydroxy-3-pinanone, Isoborneol, Isobornyl acetate, Isobornyl acrylate, Isobornyl methacrylate, (1S)-(+)-Ketopinic acid, Paeoniflorin, (1S,2S,3R,5S)-(+)-Pinanediol, (1R)-(+)-α-Pinene, (−)-α-Pinene, (−)-β-Pinene, α-Pinene oxide, (1R,4R,5R)-4,7,7-Trimethyl-6-thiabicyclo[3.2.1]octane, (1S)-(−)-Verbenone
e. Sesquiterpenes	(+)-Abscisic acid, (±)-Abscisic acid, Artemether, Artemisinin, Artesunate, (−)-Bilobalide, α-Caryophyllene, β-Caryophyllene, (+)-Cedrol, Farnesol, Farnesyl acetate, Guaiazulene, Nerolidol, (+)-Nootkatone, Parthenolide, Picrotoxin, (−)-α-Santonin (Hussain et al. 2019)
f. Diterpenes	Abietic acid, Carnosic acid, 10-Deacetylbaccatin III, (+)-Dehydroabietylamine, Docetaxel, Forskolin, Gibberellic acid, Isophytol, Paclitaxel, Phytol, 13-cis-Retinoic acid (Isotretinoin), Sclareol, (3aR)-(+)-Sclareolide, Stevioside hydrate, Triptolide
g. Sesterpenes	Geranyl farnesol
h. Triterpenes	Betulinic acid, Betulin, Celastrol, 18β-Glycyrrhetinic acid, Glycyrrhizic acid, Glycyrrhizic acid ammonium salt, Limonin, Oleanolic acid, Squalane, Squalene, Ursolic acid

contact (Prates et al. 1998; Khan et al. 2014; de Souza Born et al. 2018), diet incorporation (Silva et al. 2016; AlJabr et al. 2017a, b; Hussain et al. 2017, 2019; Su et al. 2018), ingestion (Prates et al. 1998; Giongo et al. 2016), and choice assays (Marazzi et al. 2004; Juan Hikawczuk et al. 2006), have been summarized below in the Table 7.3. Their finding revealed that plant secondary metabolites have the full potential to be incorporated into the Integrated Pest Management (IPM) strategy for the control of agricultural pests due to the issue of food safety that raised mainly due to indiscriminate use of synthetic pesticides.

The addition of these compounds from different plant parts into the integrated management strategy of agricultural pests is advantageous because they have secondary metabolites that are known for broad spectra of activity. Numerous deleterious effects of plant extracts have been manifested including (1) low fecundity, (2) molting inhibition, (3) toxicity, (4) respiratory disturbances, (5) feeding inhibition, (6) oviposition deterrence, (7) reduced fecundity, (8) suppression of calling behavior, and (9) repellency (Khan and Saxena 1986; Zhao et al. 1998; Isman 2000).

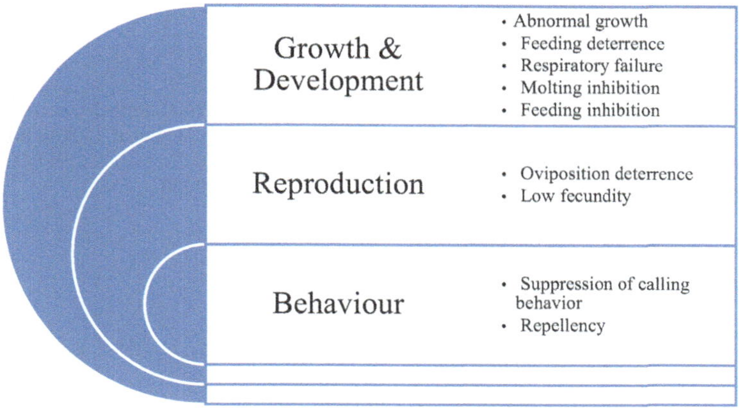

Fig. 7.2 Responses of Agricultural pests against the toxicity of plant-based products or their secondary metabolites

7.4.2 Challenges

The main challenges to the commercialization of plant secondary metabolites as biopesticides, which have to be tackled are the following:

- Lack of interdisciplinary collaboration ultimately leads to failure in deep technical knowledge understanding dilemma that found to be the primary root cause for large scale commercialization
- Little interest of agro-industry due to plant genetic variability and high cost of research and development to commercialize plant secondary metabolites
- Tough patent and registration mechanism for plant secondary metabolites
- Lack of awareness about the availability, application advantages and safety benefits of plant secondary metabolites among the dealers and ultimately the farmers
- Low efficacy due to the instability of the active ingredients under direct sunlight compared with the synthetic pesticides
- Low adoption by the farmers due to the high price of the plant secondary metabolites

7.5 Future Prospects and Research Directions

The use of plant secondary metabolites-based biopesticides are utmost important for biotic stress management due to the emerging global demands of organically produced food that recently crossed 100 billion USD during 2018. Plant secondary metabolites have tremendous potential to support the organic farming as their farmland expand up to 186 countries with 71.5 million ha of land globally (Willer

Table 7.3 Status of plant secondary metabolites against top 19 agricultural pests

Agricultural pests	Plant secondary metabolites	References
Helicoverpa armigera (Lepidoptera: Noctuidae)	Quercitin, cinnamic acid, caffeic acid, chlorogenic acid, catechin, trihydroxyflavone, gentisic acid, ferulic acid, protocatechuic acid, umbelliferone, phenylethylamine A, 3-(4-hydroxyphenyl)-N-[2-(4-hydroxyphenyl)-2-methoxyethyl]-acrylamide (2), 24 N-trans-*p*-coumaroyl octopamine (3), N-trans-*p*-coumaroyl-3′,4-′-dihydroxyphenylethylamine (4), trans-N-coumaroyltyramine (5), N-cis-p-coumaroyloctopamine (6), cis-N-phydroxycinnamoyl-7-′-methoxyltyramine (7), cis-N-coumaroyltyramine, Azadirectin, gossypol, camphor, limonene, b-caryophyllene	de la Paz Celorio-Mancera et al. (2011), War et al. (2013), Singh et al. (2014), Chen et al. (2017) and Liu et al. (2020)
Bemisia tabaci (Hemipteran: Aleyrodidae)	Nicotine, Azadirectin	Kumar et al. (2005), Bezerra-Silva et al. (2012) and Kliot et al. (2014)
Tetranychus urticae (Acari: Tetranychidae)	Thymola, p-Cymenea, b-Caryophyllene, Carvacrol, β-Pinene, Limonene, 1,8-Cineole, Terpinolene, Thymol, Eugenol, Azamax (Azadirectin)	Cavalcanti et al. (2010) and de Souza Born et al. (2018)
Plutella xylostella (Lepidoptera: Plutellidae)	Allyl isothiocyanate, Z-3-hexen-1-ol, 2,5-hexanediol, Z-3-hexenyl isovalerate, Z-3-hexenyl acetate, D-terpinene, Caffeic acid	Baoyu et al. (2001), Marazzi et al. (2004) and Peres et al. (2017)
Spodoptera litura (Lepidoptera: Noctuidae)	Rutin, chlorogenic acid, quinic acid, caffeic acid, naringenin, quercitin, kaempferol, myricetin, catechin, and ferulic acid	Su et al. (2018) and Kundu et al. (2018)
Tribolium castaneum (Tenebrionidae: Coleoptera)	neo-clerodane diterpenes, triterpenoid 2α,3β,21β,23,28-penta hydroxyl 12-oleanene, 1,8-cineole, R-(+)-limonene, Cineole	Prates et al. (1998), Juan Hikawczuk et al. (2006) and Khan et al. (2014)
Myzus persicae (Hemiptera: Aphididae)	Linalool, Thymol	Faraone et al. (2015)
Spodoptera frugiperda (Lepidoptera: Noctuidae)	Cedrelone, Scopoletin, (+/−)-Catechin, Triglyceride, Dammaradienol, rutin	Silva et al. (2016) and Giongo et al. (2016)
Aphis gossypii (Hemiptera: Aphididae)	Tannic acid, gossypol, cucurbitacin B	Yousaf et al. (2018) and Ma et al. (2019)

(continued)

Table 7.3 (continued)

Agricultural pests	Plant secondary metabolites	References
Nilaparvata lugens (Hemiptera: Delphacidae)	Gramine, quercetin, spermidine, azadirachtin, Schaftoside	Senthil-Nathan et al. (2009), Sun et al. (2013), Hao et al. (2018) and Kang et al. (2019)
Frankliniella occidentalis (Thysanoptera: Thripidae)	Retrorsine, retrorsine N-oxide, senecionine, retrorsine, jacobine, erucifoline, monocrotaline	Liu et al. (2017, 2019)
Ceratitis capitata (Diptera; Tephritidae)	3-methyl-1-butanal; decanal; 3-methyl-1-butanol; (Z)-2-pentenol; (E)-2-hexenol; 2-heptanone; 2-heptanone, limonene, beta caryophyllene	Light et al. (1988), Prokopy et al. (1998) and Fombong et al. (2016)
Cydia pomonella (Lepidoptera: Tortricidae)	Juglone	Piskorski and Dorn (2011)
Callosobruchus maculatus (Coleoptera: Chrysomelidae)	Extracts from *Hemizygia welwitschii*	Fotso et al. (2019)
Spodoptera Littoralis (Lepidoptera: Noctuidae)	Cinnamic acid, salicylic acid, xanthotoxin, quercetin, flavone, coumarin	Wang et al. (2012)
Acyrthosiphon pisum (Homoptera: Aphididae)	Quercetin; kaempferol+RCO−; kaempferol; tricin; apigenin+RCO−; apigenin; *Medicago sativa* leaves fractions (Protocatechuic acid; Chlorogenic Acid; 4-Hydroxycinnamic Acid; Caffeic Acid; Syringic Acid; Rutosid; Genistin; Ferulic Acid; Genistein	Goławska et al. (2008) and Yuan et al. (2019)
Diaphorina citri (Hemiptera: Liviidae)	Formic acid	George et al. (2019)
Tuta absoluta (Lepidoptera: Gelechiidae)	Azadirachtin; spilanthol; (E)-N-isobutylundeca-2-en-8,10-diynamide; (R, E)-N-(2- methylbutyl)undeca-2-en-8,10-diynamide	Terzidis et al. (2014) and Ndereyimana et al. (2019)
Thrips tabaci (Thysanoptera: Thripidae)	*Artemisia arborescens; Origanum majorana; Ocimum gratissimum; Melaleuca alternifolia*; iso-nicotinate	van Tol et al. (2007)

Ranking is based on the abstracts in the CABI Compendia on crop protection, forestry and invasive species

and Lernoud 2018). However, the full potential of plant secondary metabolites yet remains to be seen. Therefore, it is of prime importance to focus our future research to tackle the challenges that hinder their commercialization.

7.5.1 Interdisciplinary Collaboration

The production of plant secondary metabolites on commercial scale required various steps including extraction of compounds, characterization of compounds, toxicity evaluation of compounds (lab and field), large scale production of compounds, formulation, testing (lab and field), which involves various disciplines. A strong interdisciplinary collaboration is the key to link interdisciplinary research and technology to generate deep understanding and develop effective plant secondary metabolites-based eco-friendly products.

7.5.2 Efficacy

There is a general perception and also to some extent reality that biopesticides have low efficacy especially under field conditions due to various reasons including high degradation of active ingredient, and instability under direct sunlight, etc. Therefore, it is very important to divert research interests to develop stable formulations of plant secondary metabolites that can withstand under challenging environmental conditions. In this regard, nanoformulations is an emerging avenue of research that has great potential in encapsulating plant secondary metabolites to achieve great performance in terms of efficacy against the infestations of agricultural pests.

7.5.3 Plant Genetic Variability

Genetic variability among the same species is posing a serious threat that demands a production system that could grow plants resulting homogeneity, and uniformity. This production system could help to stabilize this industry. However, tissue culture is pertinent to produce and propagate genetically homogeneous plants for constant supply of target plant secondary metabolites.

7.5.4 Public-Private Partnership (PPP)

The development of plant secondary metabolites as commercial products involves a multidisciplinary approach. Currently, a lot of research have been done on the extraction and characterization of plant secondary metabolites. Instead of wasting time in reinventing the same idea, as it is the time to work hand in hand through Partnership with best players. In this regard, it is very important for public sector to develop partnership with best private sector players to develop economically viable stable products from plant secondary metabolites to control agricultural pests.

7.5.5 *CRISPR/Cas9 System for Plant Secondary Metabolites*

Plant secondary metabolites in spite of their usage as agro-chemicals are also being utilized as flavors, fragrances, coloring, and pharmaceutical ingredients. However, their production from plants in small quantities demands an active and efficient system for their sustainable production. Over the last period of time, metabolic engineering and synthetic system was found to be quite promising for the production of plant secondary metabolites (Verpoorte et al. 1999; Marchev et al. 2020; Birchfield and McIntosh 2020). However, the recently introduced precise genome editing CRISPR/Cas9 technology that revolutionized genome engineering technology has sparked extraordinary interest for the biosynthesis of plant secondary metabolites due to the ease of targeting sequence insertions, deletions and substitutions. The future research to manipulate plant secondary metabolism through CRISPR/CAS9 technology will help to produce cheap, fast, and effective plant secondary metabolites.

7.6 Conclusion

Securing global food production is critically important to reduce the increasing number of hungry peoples in the world. In this regard, agricultural pests are posing severe threat by reducing crop yield and productivity. The plant secondary metabolites are a potential biopesticides candidates to replace the environment deteriorating synthetic pesticides by their wider modes of actions to overcome the host detoxification defense mechanism. However, interdisciplinary collaboration is critically important for deep understanding and incorporation of modern technologies to develop economically viable products for commercialization of plant secondary metabolites-based eco-friendly products through Public-Private-Partnership.

References

Al-Ayedh H, Hussain A, Rizwan-ul-Haq M, Al-Jabr AM (2016) Status of insecticide resistance in field-collected populations of *Rhynchophorus ferrugineus* (Olivier) (Coleoptera: Curculionidae). Int J Agric Biol 18:103–110. https://doi.org/10.17957/IJAB/15.0070

AlJabr AM, Hussain A, Rizwan-ul-haq M, Al-Ayedh H (2017a) Method for controlling growth of red palm weevil. US patent no US009596851B1. pp 1–6

AlJabr AM, Hussain A, Rizwan-ul-Haq M, Al-Ayedh H (2017b) Toxicity of plant secondary metabolites modulating detoxification genes expression for natural red palm weevil pesticide development. Molecules 22:169. https://doi.org/10.3390/molecules22010169

AlJabr A, Hussain A, Rizwan-ul-haq M (2018) Toxin-pathogen synergy reshaping detoxification and antioxidant defense mechanism of *Oligonychus afrasiaticus* (McGregor). Molecules 23:1978. https://doi.org/10.3390/molecules23081978

Baoyu H, Zhongning Z, Yuling F (2001) Electrophysiology and behavior feedback of diamondback moth, *Plutella xylostella*, to volatile secondary metabolites emitted by Chinese cabbage. Chin Sci Bull 46:2086–2088. https://doi.org/10.1007/BF02901138

Ben-Yehoshua S, Conn EE (1964) Biosynthesis of Prunasin, the cyanogenic glucoside of peach. Plant Physiol 39:331–333. https://doi.org/10.1104/pp.39.3.331

Bérdy J (2005) Bioactive microbial metabolites. J Antibiot (Tokyo) 58:1–26. https://doi.org/10.1038/ja.2005.1

Bezerra-Silva GCD, Silva MA, Vendramim JD, Dias CTDS (2012) Insecticidal and behavioral effects of secondary metabolites from meliaceae on *Bemisia tabaci* (Hemiptera: Aleyrodidae). Florida Entomol 95:743–751. https://doi.org/10.1653/024.095.0325

Birchfield AS, McIntosh CA (2020) Metabolic engineering and synthetic biology of plant natural products—a mini review. Curr Plant Biol 24:100163. https://doi.org/10.1016/j.cpb.2020.100163

Cavalcanti SCH, dos Niculau ES, Blank AF, CAG C, Araújo IN, Alves PB (2010) Composition and acaricidal activity of *Lippia sidoides* essential oil against two-spotted spider mite (*Tetranychus urticae* Koch). Bioresour Technol 101:829–832. https://doi.org/10.1016/j.biortech.2009.08.053

Chen C, Liu Y, Shi X, Desneux N, Han P, Gao X (2017) Elevated carboxylesterase activity contributes to the lambda-cyhalothrin insensitivity in quercetin fed *Helicoverpa armigera* (Hübner). PLoS One 12:e0183111. https://doi.org/10.1371/journal.pone.0183111

de la Paz Celorio-Mancera M, Ahn S-J, Vogel H, Heckel DG (2011) Transcriptional responses underlying the hormetic and detrimental effects of the plant secondary metabolite gossypol on the generalist herbivore *Helicoverpa armigera*. BMC Genomics 12:575. https://doi.org/10.1186/1471-2164-12-575

de Souza Born F, da Camara CAG, de Melo JPR, de Moraes MM (2018) Acaricidal property of the essential oil from *Lippia gracilis* against *Tetranychus urticae* and a natural enemy, *Neoseiulus californicus*, under greenhouse conditions. Exp Appl Acarol 75:491–502. https://doi.org/10.1007/s10493-018-0286-3

Desneux N, Decourtye A, Delpuech J-M (2007) The sublethal effects of pesticides on beneficial arthropods. Annu Rev Entomol 52:81–106. https://doi.org/10.1146/annurev.ento.52.110405.091440

Dutcher JD (2007) A review of resurgence and replacement causing pest outbreaks in IPM. In: General concepts in integrated pest and disease management. Springer, Dordrecht, pp 27–43

FAO, IFAD, UNICEF, WFP, WHO (2019) State of food security and nutrition in the world 2019. Safeguarding against economic slowdowns and downturns. FAO, Rome

Faraone N, Hillier NK, Cutler GC (2015) Plant essential oils synergize and antagonize toxicity of different conventional insecticides against *Myzus persicae* (Hemiptera: Aphididae). PLoS One 10:e0127774. https://doi.org/10.1371/journal.pone.0127774

Fombong AT, Kachigamba DL, Torto B (2016) Chemical ecology of African Tephritid fruit flies. In: Fruit fly research and development in Africa—towards a sustainable management strategy to improve horticulture. Springer International, Cham, pp 163–205

Fotso TG, Tofel HK, Abdou JP, Tchao N, Zourmba CM, Adler C, Nukenine EN (2019) Control of *Callosobruchus maculatus* (Coleoptera: Chrysomelidae) using fractionated extracts from Cameroonian *Hemizygia welwitschii* (Lamiaceae) leaf on stored *Vigna unguiculata* (Fabales: Fabaceae). J Insect Sci 19:22. https://doi.org/10.1093/jisesa/iez029

George J, Shi Q, Stelinski LL, Stover E, Lapointe SL (2019) Host selection, oviposition and feeding by a phytopathogen vector, *Diaphorina citri* (Hemiptera: Liviidae), modulated by plant exposure to formic acid. Front Ecol Evol 7:78. https://doi.org/10.3389/fevo.2019.00078

Giongo AMM, Vendramim JD, Freitas SDL, Silva MFGF (2016) Toxicity of secondary metabolites from Meliaceae against *Spodoptera frugiperda* (J. E. Smith) (Lepidoptera: Noctuidae). Neotrop Entomol 45:725–733. https://doi.org/10.1007/s13744-016-0418-6

Goławska S, Kapusta I, Łukasik I, Wójcicka A (2008) Effect of phenolics on the pea aphid, *Acyrthosiphon pisum* (Harris) population on *Pisum sativum* L. (Fabaceae). Pestycydy/Pesticides 3–4:71–77

Hao P-Y, Feng Y-L, Zhou Y-S, Song X-M, Li H-L, Ma Y, Ye C-L, Yu X-P (2018) Schaftoside interacts with NlCDK1 protein: a mechanism of Rice resistance to Brown Planthopper, *Nilaparvata lugens*. Front Plant Sci 9:710. https://doi.org/10.3389/fpls.2018.00710

Hussain A, Rizwan-ul-Haq M, Al-Ayedh H, Aljabr AM (2017) Toxicity and detoxification mechanism of black pepper and its major constituent in controlling *Rhynchophorus ferrugineus* Olivier (Curculionidae: Coleoptera). Neotrop Entomol 46:685–693. https://doi.org/10.1007/s13744-017-0501-7

Hussain A, Rizwan-ul-haq M, AlJabr AM, Al-Ayedh H (2019) Lethality of Sesquiterpenes reprogramming red palm weevil detoxification mechanism for natural novel biopesticide development. Molecules 24:1648. https://doi.org/10.3390/molecules24091648

Hussein RA, El-Anssary AA (2019) Plants secondary metabolites: the key drivers of the pharmacological actions of medicinal plants. In: Herbal medicine. IntechOpen, London. https://doi.org/10.5772/intechopen.76139

Isman MB (2000) Plant essential oils for pest and disease management. Crop Prot 19:603–608. https://doi.org/10.1016/S0261-2194(00)00079-X

Jones M (1953) Albrecht Kossel, a biographical sketch. Yale J Biol Med 26:80–97

Juan Hikawczuk VE, López Verrilli MA, Borkowski EJ, Sosa ME, Giordano OS, Saad JR, Tonn CE (2006) Antifeedant activity of neo- clerodane diterpenes from *Baccharis flabellata* Hook & Arn var. flabellata toward *Tribolium castaneum* Herbst: structure–activity relationships. Nat Prod Res 20:813–819. https://doi.org/10.1080/14786410500353596

Kang K, Yue L, Xia X, Liu K, Zhang W (2019) Comparative metabolomics analysis of different resistant rice varieties in response to the brown planthopper *Nilaparvata lugens* Hemiptera: Delphacidae. Metabolomics 15:62. https://doi.org/10.1007/s11306-019-1523-4

Khan ZR, Saxena RC (1986) Effect of steam distillate extracts of resistant and susceptible rice cultivars on behavior of *Sogatella furcifera* (Homoptera: Delphacidae). J Econ Entomol 79:928–935. https://doi.org/10.1093/jee/79.4.928

Khan A, Islam M, Rahman M, Zaman T, Haque M (2014) Pesticidal and pest repellency activities of a plant derived triterpenoid 2α,3β,21β,23,28-penta hydroxyl 12-oleanene against *Tribolium castaneum*. Biol Res 47:68. https://doi.org/10.1186/0717-6287-47-68

Kliot A, Kontsedalov S, Ramsey JS, Jander G, Ghanim M (2014) Adaptation to nicotine in the facultative tobacco-feeding hemipteran *Bemisia tabaci*. Pest Manag Sci 70:1595–1603. https://doi.org/10.1002/ps.3739

Kumar P, Poehling H-M, Borgemeister C (2005) Effects of different application methods of azadirachtin against sweetpotato whitefly *Bemisia tabaci* Gennadius (Hom., Aleyrodidae) on tomato plants. J Appl Entomol 129:489–497. https://doi.org/10.1111/j.1439-0418.2005.01009.x

Kundu A, Mishra S, Vadassery J (2018) *Spodoptera litura*-mediated chemical defense is differentially modulated in older and younger systemic leaves of *Solanum lycopersicum*. Planta 248:981–997. https://doi.org/10.1007/s00425-018-2953-3

Light DM, Jang EB, Dickens JC (1988) Electroantennogram responses of the mediterranean fruit fly, *Ceratitis capitata*, to a spectrum of plant volatiles. J Chem Ecol 14:159–180. https://doi.org/10.1007/BF01022539

Liu X, Vrieling K, Klinkhamer PGL (2017) Interactions between plant metabolites affect herbivores: a study with pyrrolizidine alkaloids and Chlorogenic acid. Front Plant Sci 8. https://doi.org/10.3389/fpls.2017.00903

Liu Q, Luo L, Zheng L (2018) Lignins: biosynthesis and biological functions in plants. Int J Mol Sci 19:335. https://doi.org/10.3390/ijms19020335

Liu X, Vrieling K, Klinkhamer PGL (2019) Phytochemical background mediates effects of pyrrolizidine alkaloids on Western flower Thrips. J Chem Ecol 45:116–127. https://doi.org/10.1007/s10886-018-1009-2

Liu C, Tian J, An T, Lyu F, Jia P, Zhou M, Liu Z, Feng Y (2020) Secondary metabolites from *Solanum rostratum* and their antifeedant defense mechanisms against *Helicoverpa armigera*. J Agric Food Chem 68:88–96. https://doi.org/10.1021/acs.jafc.9b06768

Ma K, Tang Q, Liang P, Xia J, Zhang B, Gao X (2019) Toxicity and sublethal effects of two plant allelochemicals on the demographical traits of cotton aphid, *Aphis gossypii* glover (Hemiptera: Aphididae). PLoS One 14:e0221646. https://doi.org/10.1371/journal.pone.0221646

Marazzi C, Patrian B, Stadler E (2004) Secondary metabolites of the leaf surface affected by Sulphur fertilisation and perceived by the diamondback moth. Chemoecology 14:81–86. https://doi.org/10.1007/s00049-003-0264-y

Marchev AS, Yordanova ZP, Georgiev MI (2020) Green (cell) factories for advanced production of plant secondary metabolites. Crit Rev Biotechnol 40:443–458. https://doi.org/10.1080/07388551.2020.1731414

Markets and Markets (2020) Market research report: biopesticides market by type (bioinsecticides, biofungicides, bionematicides, and bioherbicides), source (Microbials, biochemicals, and beneficial insects). Mode of application, formulation, crop application, and region—Global Forecast to 2025

Martinou AF, Seraphides N, Stavrinides MC (2014) Lethal and behavioral effects of pesticides on the insect predator *Macrolophus pygmaeus*. Chemosphere 96:167–173. https://doi.org/10.1016/j.chemosphere.2013.10.024

Ndereyimana A, Nyalala S, Murerwa P, Gaidashova S (2019) Bioactivity of plant extracts against tomato leaf miner, *Tuta absoluta* (Meyrick) (Lepidoptera: Gelechiidae). J Appl Hortic 21:146–150. https://doi.org/10.37855/jah.2019.v21i02.25

Peres L, Sobreiro A, Couto I, Silva R, Pereira F, Heredia-Vieira S, Cardoso C, Mauad M, Scalon S, Verza S, Mussury R (2017) Chemical compounds and bioactivity of aqueous extracts of *Alibertia* spp. in the control of *Plutella xylostella* L. (Lepidoptera: Plutellidae). Insects 8:125. https://doi.org/10.3390/insects8040125

Piskorski R, Dorn S (2011) How the oligophage codling moth *Cydia pomonella* survives on walnut despite its secondary metabolite juglone. J Insect Physiol 57:744–750. https://doi.org/10.1016/j.jinsphys.2011.02.014

Prates HT, Santos JP, Waquil JM, Fabris JD, Oliveira AB, Foster JE (1998) Insecticidal activity of monoterpenes against *Rhyzopertha dominica* (F.) and *Tribolium castaneum* (Herbst). J Stored Prod Res 34:243–249. https://doi.org/10.1016/S0022-474X(98)00005-8

Prokopy RJ, Hu X, Jang EB, Vargas RI, Warthen JD (1998) Attraction of mature Ceratitis capitata females to 2-Heptanone, a component of coffee fruit odor. J Chem Ecol Ecol 24:1293–1304. https://doi.org/10.1023/A:1021218531083

Richard T, Temsamani H, Cantos-Villar E, Monti J-P (2013) Application of LC–MS and LC–NMR techniques for secondary metabolite identification. In: Advances in botanical research, pp 67–98

Senthil-Nathan S, Choi M-Y, Paik C-H, Seo H-Y, Kalaivani K (2009) Toxicity and physiological effects of neem pesticides applied to rice on the *Nilaparvata lugens* Stål, the brown planthopper. Ecotoxicol Environ Saf 72:1707–1713. https://doi.org/10.1016/j.ecoenv.2009.04.024

Silva TRFB, Almeida ACDS, Moura TDL, Da Silva AR, Freitas SDS, Jesus FG (2016) Effect of the flavonoid rutin on the biology of *Spodoptera frugiperda* (Lepidoptera: Noctuidae). Acta Sci Agron 38:165. https://doi.org/10.4025/actasciagron.v38i2.27956

Singh P, Jayaramaiah RH, Sarate P, Thulasiram HV, Kulkarni MJ, Giri AP (2014) Insecticidal potential of defense metabolites from *Ocimum kilimandscharicum* against *Helicoverpa armigera*. PLoS One 9:e104377. https://doi.org/10.1371/journal.pone.0104377

Su Q, Zhou Z, Zhang J, Shi C, Zhang G, Jin Z, Wang W, Li C (2018) Effect of plant secondary metabolites on common cutworm, *Spodoptera litura* (Lepidoptera: Noctuidae). Entomol Res 48:18–26. https://doi.org/10.1111/1748-5967.12238

Sun X-Q, Zhang M-X, Yu J-Y, Jin Y, Ling B, Du J-P, Li G-H, Qin Q-M, Cai Q-N (2013) Glutathione S-transferase of brown planthoppers (*Nilaparvata lugens*) is essential for their adaptation to Gramine-containing host plants. PLoS One 8:e64026. https://doi.org/10.1371/journal.pone.0064026

Terzidis AN, Wilcockson S, Leifert C (2014) The tomato leaf miner (*Tuta absoluta*): conventional pest problem, organic management solutions? Org Agric 4:43–61. https://doi.org/10.1007/s13165-014-0064-4

Theiling KM, Croft BA (1988) Pesticide side-effects on arthropod natural enemies: a database summary. Agric Ecosyst Environ 21:191–218. https://doi.org/10.1016/0167-8809(88)90088-6

van Tol RWHM, James DE, de Kogel WJ, Teulon DAJ (2007) Plant odours with potential for a push-pull strategy to control the onion thrips, *Thrips tabaci*. Entomol Exp Appl 122:69–76. https://doi.org/10.1111/j.1570-7458.2006.00489.x

Verpoorte R, van der Heijden R, Hoopen H, Memelink J (1999) Metabolic engineering of plant secondary metabolite pathways for the production of fine chemicals. Biotechnol Lett 21:467–479. https://doi.org/10.1023/A:1005502632053

Wang R, Sun Y, Liang X, Song Y, Su Y, Salzman K, Zeng R (2012) Effects of six plant secondary metabolites on activities of detoxification enzymes in Spodoptera litura. Acta Ecol Sin 32:5191–5198. https://doi.org/10.5846/stxb201203060302

War AR, Paulraj MG, Hussain B, Buhroo AA, Ignacimuthu S, Sharma HC (2013) Effect of plant secondary metabolites on legume pod borer, *Helicoverpa armigera*. J Pestic Sci 86:399–408. https://doi.org/10.1007/s10340-013-0485-y

Willer H, Lernoud J (eds) (2018) The world of organic agriculture. Research Institute of Organic Agriculture (FiBL) and IFOAM Organics International, Bonn

Wink M, Schimmer O (2018) Modes of action of defensive secondary metabolites. In: Annual plant reviews online. Wiley, Chichester, pp 18–137

Yamane H, Konno K, Sabelis M, Takabayashi J, Sassa T, Oikawa H (2010) Chemical defence and toxins of plants. In: Comprehensive natural products II. Elsevier, Amsterdam, pp 339–385

Yousaf HK, Shan T, Chen X, Ma K, Shi X, Desneux N, Biondi A, Gao X (2018) Impact of the secondary plant metabolite Cucurbitacin B on the demographical traits of the melon aphid, *Aphis gossypii*. Sci Rep 8:16473. https://doi.org/10.1038/s41598-018-34821-w

Yuan E, Yan H, Gao J, Guo H, Ge F, Sun Y (2019) Increases in genistein in *Medicago sativa* confer resistance against the Pisum host race of *Acyrthosiphon pisum*. Insects 10:97. https://doi.org/10.3390/insects10040097

Zagrobelny M, Bak S, Rasmussen AV, Jørgensen B, Naumann CM, Lindberg Møller B (2004) Cyanogenic glucosides and plant–insect interactions. Phytochemistry 65:293–306. https://doi.org/10.1016/j.phytochem.2003.10.016

Zhao B, Grant G, Langevin D, MacDonald L (1998) Deterring and inhibiting effects of quinolizidine alkaloids on spruce budworm (Lepidoptera: Tortricidae) oviposition. Environ Entomol 27:984–992. https://doi.org/10.1093/ee/27.4.984

Chapter 8
Transgenic Plants and Its Role in Insect Control

Joseph Adomako, Stephen Yeboah, Stephen Larbi-Koranteng, Frederick Kankam, Daniel Oppong-Sekyere, Jerry Asalma Nboyine, Yaw Danso, Michael Kwabena Osei, and Patricia Oteng-Darko

Abstract The issue of food security has gained global significance in both political and social discourse due to a projected worldwide population increase by 2050. A major hindrance to achieving food security is the negative effects of insect pests. Insect pests competes with humans at the highest level for agricultural resources and it is estimated that their activities accounts for between 30–40% losses in food crops globally. For decades, numerous policies aimed at ameliorating the impact of insect pests on crops have been implemented. Prominent among these is the development and use of pesticides. Notwithstanding its effectiveness, this strategy is bereft with serious limitations such as poisoning, environmental pollution and insect pest developing resistance to pesticides. A sure way to defeat the food production challenges in a sustainable manner is to explore the use of new engineering techniques to develop superior crop varieties that are high-yielding, environmentally sustainable, cost-effective to produce and resistant to insect pests. Conventional

J. Adomako (✉) · Y. Danso
Plant Health Division, CSIR-Crops Research Institute, Kumasi, Ghana
e-mail: joeadomako@gmail.com

S. Yeboah · P. Oteng-Darko
Resources and Crop Management and Socio-economics Division, CSIR-Crops Research Institute, Kumasi, Ghana

S. Larbi-Koranteng
Crop and Soil Science Department, Akenten Appiah-Menka University of Skills Training and Entrepreneurial Development, Asante-Mampong, Ghana

F. Kankam
Department of Agronomy, Faculty of Agriculture, University for Development Studies, Tamale, Ghana

D. Oppong-Sekyere
Department of Ecological Agriculture, Bolgatanga Technical University, Bolgatanga, Ghana

J. A. Nboyine
CSIR-Savanah Agriculture Research Institute, Nyankpala, Ghana

M. K. Osei
Horticulture Division, CSIR-Crops Research Institute, Kumasi, Ghana

S. De Mandal et al. (eds.), *New and Future Development in Biopesticide Research: Biotechnological Exploration*, https://doi.org/10.1007/978-981-16-3989-0_8

breeding techniques to achieve this may be limited by time and space, hence the need for modern tools for the development of transgenic crops that are resistant to insects attack; these techniques hasten the process of insect pest control strategies in crop husbandry. The adoption of transgenic plants could reduce the usage of pesticides with broad-spectrum effect, in order to reduce the damages they cause. Since its introduction, transgenic plants have been a main tool for managing several insect pests of economic importance successfully. The adoption of such plants will reduce pesticides use. Despite its benefits, there is limited acceptance of transgenic plants globally. Notwithstanding this, the prospects of integrating transgenic plants in crop production to manage the negative effects of insect pests look promising as the demand for safe food and public involvement in evaluating such materials increases.

Keywords Insect pests · Food security · Insect pest management · Transgenic plants · Environmental sustainability

8.1 Introduction

In 2011, it was reported that, the global population was more than seven billion people, with an expected population of up to 9.3 billion by 2050 while per-capita demand for food is likely to increase along with income growth. There is an expected increase in demand for food between 50% and 100% by 2050 (West et al. 2014). The major concern therefore, has been the ability of modern agriculture to support this ever-growing population as it continues to be the concern for policy makers. Meanwhile, crop yield and growth are not sustainable due to reductions in Agricultural research funding, inadequate irrigation systems, and the dependency on rain-fed agriculture. It is therefore imperative that surge in biotic stress, climate change, and human activities pose serious problems to food security (Myers et al. 2017). Hence, tackling hunger remains a major challenge for our generation (Wheeler and Braum 2013). For instance, it is reported that production of cereals globally is to be increased by 56% by the end of 2050 and global production of livestock is also to be up by 90% according to the International Food Policy Research Institute (IFPRI), with developing countries accounting for 93% of cereal and 85% of meat requirement.

Achieving increased food production to meet global demand can materialize by integrating several factors, such as increase in the area of cultivation, use of improved agronomical practices, biocontrol agents, and efficient management of soil and water. In addition to these will be the use of pest-resistant varieties as well as the introduction of transgenic plants that can withstand insect pests and diseases attack (Carvalho 2006). The easiest way to have overcome universal food insecurity was to increase land area for cultivation of crops. This, however, may not be feasible as most arable agricultural lands are unavailable, since they are already in use for different agricultural purposes. The World Bank in 2015, estimated that only 37.3% of the available land worldwide could be used for agricultural production, with 11%

(1.5 billion hectares) considered as arable. Although the Food and Agriculture Organization estimates that about 2.7 billion ha of land area is available to increase crop production, the larger portion of these areas are with harsh geo-climatic conditions; making it difficult to use for farming activities (Tyczewska et al. 2018).

Research indicates that erosion and/or pollution have taken almost a third of arable lands globally over the last four decades (Verheijen et al. 2009), contributing to soil depletion and reduction of soil water retention. (Verheijen et al. 2009). The problem is further exacerbated by the abuse of pesticides by farmers although their use is as a result of increased damages by insect pests. Losses in yield of agricultural productivity due to pests and pathogens is estimated to be between 20 to 40% globally (Oerke 2006; Culliney 2014; Pandey et al. 2017). Also, threats from climate change resulting from increase in temperature and low precipitation have had serious effects on plant growth, increasing availability of carbon dioxide, reducing nutritional levels of our food and at the same time enhancing environmental conditions for the growth of insect pests and disease pathogens.

To overcome global food production challenges in a sustainable manner, it is necessary to explore novel techniques like genetic engineering to develop improved and superior plant varieties such as insect resistant transgenic crops. Genetically engineered crops generally have insect-resistance transgenes incorporated in them that enhance their ability to resist insect pest attacks which has genuinely extended the scope of resistant genes available to plant breeders. Transgenes can be of plant, bacterial, or other origins, and are considered the most ideal since they are highly effective against targeted pest, resilient to adverse environmental conditions, high biodegradability rate, cost effective and less exposure of operator to toxins. Introduction of transgenic plants will therefore reduce the over reliance on broad-spectrum insecticides, and reduce ecological damage these insecticides cause.

Reductions in fertilizer and pesticides usage do not only result in valuable savings for farmers but also improves nutritional quality and health benefits of consumers as no evidence suggest commercial transgenic crops contain allergens other than those in normal foods (Dunn et al. 2017). Therefore, this review examines the importance of insect pests in crop production and the role of transgenic plants in solving the critical challenges of food insecurity resulting from activities of insect pests and the prospects of growing these crops.

8.2 Economic Importance of Insect Pests in Crop Production

For centuries producers have battled with biotics stress more especially with insect pests for food sufficiency to feed both human and animal. The issue of food security has gained global dominance in both political and social agendas due to increasing global population globally projected to be above nine billion by 2050. This has necessitated the development and use of several modern tools to hasten the breeding of superior crop varieties that are high-yielding, and can withstand adverse environmental conditions, and at the same time offer returns on investment. An effective way to achieve this objective, is to reduce losses associated with insect pests in food crop production. As reported by Oerke and Dehne (2004), insect pests are the main competitors with human for agricultural resources, with their activities favoured by monocultures and intense use of fertilizers. The importance of insect pests in food production can never be underestimated as they contribute significantly to global food insecurity and reduced livelihood.

They reduce quality and quantity of crop yield contributing immensely to primary and secondary losses under different (pre and post-harvest) conditions. Primary and secondary pests are the categorization of insects based on their habits and characteristics. The later categorization as primary insects are considered more harmful due to the damage caused to the entire plant and crop derivatives. These insects bore holes into their host plants, which is used as site for oviposition, larval growth and development, as well as serve as avenue for other secondary pathogens to operate. In contrast however, secondary pests are opportunistic agents that feed on processed products or plant tissues previously damaged by primary insects. For centuries, there have been several reports which shows that crop losses globally due to insect pests is immeasurable. Oerke and Dehne (2004) reported that pests account for approximately 10.1% of the total crop losses in potatoes, soybean, rice, barley, maize, sugar beet, notwithstanding the application of control measures. Similarly, the FAO (2019) estimates that between 20–40% of global crop production are lost to pests.

Global loss in food crops is projected to be between 30–40% notwithstanding annual investments on insecticides (Garcı'a-Lara and Saldivar 2016). Despite these direct losses to major crops and their derivatives, they additionally cause indirect losses by contaminating crops or produce with their body parts or exoskeletons, eggs and wastes (Garcı'a-Lara and Saldivar 2016). For years, several insect pests have evolved to biotypes in order to develop resistance to insecticides, host plant or climate change compounding challenges they pose to crop loss. A typical example is the biotype development in aphids, *Aphis craccivora* that causes severe damage in cowpea and groundnuts, especially under drought conditions. Similarly, several species of whiteflies and the cassava hornworm are known to cause enormous damage to cassava production (Bellotti 2008). It has been reported that high population densities of *B. tabaci,* can cause reduction in root yield of cassava as the insect feed directly on the cassava crop (Liu et al. 2007). Similarly, reports indicate that the cassava mealybug, *Phenacoccus manihoti* Matile-Ferrero can cause tuber yield loss by 80% during production (Nwanze 1982). An intensive agricultural production systems will mean the evolvement of biotypes with its attendant effect. Finding eco-friendly means of mitigating their effects are of crucial importance to sustain the environment.

8.3 Strategies for Insect Pest Management

Pest management strategies are mainly established to minimize both field and post-harvest losses in crops. According to Isman (2019), pest management tools can be classified into several categories although none can be used to singly solve insect pest problems despite the success driven by individual tools. According to Oerke and Dehne (2004), crop protection is inefficient in food crops, making insect pests more severe than in cash crop production.

Over dependence on synthetic pesticides to manage the stress of insect pests are environmentally unfriendly, unsustainable, and unprofitable to use by several farmers in developing countries. Apart from these, abuse of pesticides has proven harmful to human beings, beneficial and non-targeted organisms and have led to insect species developing resistance to pesticides. Notwithstanding this, chemical pesticides are the most widely used control strategy. It does not only provide a rapid, effective and dependable pest management approach but also an economical means of controlling insect pest's complex.

To reduce the negative effects of chemical pesticides, research has focused on the development of more selective agents (Eason et al. 2014) that will depart from highly persistent and broad-spectrum products. Also, the search and development of biological insecticides meant to improve environmental safety have been on the ascendancy. Such insecticides act on their target organisms through either direct contact as in aerosol or will require the insect ingesting it during feeding as in the case of *Bacillus thuringiensis*. Apart from these, there are others that are taking up systemically, acting on the vascular systems of plants like in the case of the neonicotinoids and other like the insect growth regulators (IGRs) and ecdysone that disrupt pest development by acting on their endocrine systems (Goldson et al. 2015).

The use of pheromones or other attractants alongside toxicants and a carrier material are used with the objective of attracting insect species to formulations to cause their mortality after eating (El-Sayed et al. 2009). This approach has gained popularity because the toxins on contact kills a few insect species other than the target insects. Another advantage of such insecticides is that they are applied at lower rates than their counterparts that require a spray application. Despite the success in pesticides usage, acquisition of resistance by insect pests remains a source of concern. Resistance acquisition is a natural phenomenon that occur as a result of frequent genetic changes among population over time. However, in a polymorphic population, some individuals are pre-adapted to cope with selective agents, especially when they do not get exposed.

Introduction of natural enemies (insect predators, parasitoids) or biopesticides have been successful in controlling invasive pests (Jackson 2007). According to Hajek and Eilenberg (2018), a significant reduction in populations of pests by biocontrol agents have been achieved. The use of biological control agents requires periodic releases of such organisms and providing a natural habitat for such insects and avoiding practices that will negatively influence their survival to be able to control the endemic pests. Even though biocontrol has been successfully used in greenhouses (Surendra 2019) and sometimes in the field (Zalom et al. 2018), much effort is require to promote the practice. A success story is the control of cassava mealybug using parasitic wasp *Anagyrus lopezi* (Aekthong and Rattanakul 2019).

The employment of cultural control strategies to manage insect pests is environmentally sound and beneficial to farmers. Dara (2019) defined cultural control practices as the manipulation of agronomic practices to minimize effects of pest infestations and damage on crops. This method is also known as the traditional pest management practices due to its long-standing nature, usually by small-scale farmers. This method of pest management relies solely on the characteristics of the cropping systems of a particular environment, use of crop associations, traps, manipulation of planting and harvesting times and field sanitation techniques to reduce insect pest attacks.

Genetic diversity plays an important role in insect pests' management. Growing more than one crop using same piece of land as an improvement on the monocropping system is a sure way of maximizing crop diversity. Cultivating more than one crop is ecologically complex due to the interspecific and intraspecific competition with any prevailing insect pests and natural enemies. Insect pests' population densities are usually low in polycultures due to associational resistance or resource concentration and activities of predators. Varying time for planting and harvesting as a way of avoiding pests during critical stages of crop production is a sure way to avoid damages cause by pests. There are reports on reduction of infestation levels of aphids, thrips and pod bug due to the early planting of cowpea in Uganda. There is a similar report in Nigeria by Asante et al. (2001) which suggested reductions in pod borer (*Maruca vitrata*) flower thrips (*Megaluro thripssjostedti*), and the pod sucking bug (*Clarigrallato mentosicollis*) incidence. The difficulty or poor understanding of the traditional agricultural systems, difficulty in observing and the limited understanding of the insect pest ecology are major constraints to the use of cultural practices to manage insects by farmers (Laizer et al. 2019).

Identification and use of alternative insect control approaches such as resistant host plant has been explored and provided some level of success in insect pest management. For years, the strategy has depended on the use of traditional breeding techniques to develop cultivars that are resistant or tolerant to pests. Some of the characteristics possessed by these cultivars are morphological, physical or biochemical traits that reduce their attractiveness or its suitability for the pest to feed successfully to enable them to develop and reproduce. Resistance of host plants to insects occur naturally which confers on plants to defend themselves against insect pests' attack. This phenomenon helps both the insect and the plant to co-evolute and co-exist (Kalode and Sharma 1995). Numerous sources of resistance have been found and resistant genes introgressed into different high yielding crop cultivars.

Unlike synthetic chemical and biological control strategies, the use of crops varieties that are resistant to insect pest are not affected by environmental conditions. The contribution made by the application of plant varieties with resistance to pests for sustainable crop production cannot be over-emphasized. Developing crop varieties with resistance against insect pests using conventional means is time-consuming. It is reported that developing the midge-sorghum resistant variety, ICSV 745, took not less than 15 years while that of spotted alfalfa-aphid resistant varieties took between three and 5 years (Sharma 1993; Panda and Khush 1995). Traditionally, identifying host-plant resistance plays fundamental roles in insect pests' control, however, its advancement has been slow due to the low yield potential of the developed resistant crop varieties because of a linkage drag (Smith 2005). Again, developing resistant cultivars against insect pests through conventional breeding techniques is limited by time constraints. To overcome these challenges, transgenic breeding strategies have widely been accepted as an important tool in crop improvement to produce crops that can produce insect resistance genes.

8.4 Role of Insect Resistant Transgenic Plants in Crop Production

It is estimated that approximately 40 different insect-resistant genes have been inserted into crops such as broccoli, corn, cotton, rice, soybean and several others (Table 8.1) to control insect pests and increase productivity. Such genes obtained from microorganisms, plants, and animals have proven to be advantageous in controlling pests compared to conventional insecticides. Microbial sources of insecticidal toxins include bacteria and fungi. Remarkably, these secondary metabolites produced by different strains have served as microbial insecticides for several years in IPM package. (Schuler et al. 1998). So far, *Bacillus thuringiensis (Bt)* insecticides have proven to be economically and ecologically successful and provided exceptional achievements in transgenic technology. Crystal protein endotoxic genes produced by different strains of *Bt* are selective and are specific in its action as an insecticide to the lavae of different insect groups: whiles *Cry1* and *Cry2* are specific to lepidopteran, Cry2A *is* toxic to both lepidopterans and dipterans, and *Cry3* is specific to the coleopterans (Malone et al. 2008). These crystal proteins are solubilized and activated by an enzyme, proteinases after a susceptible insect ingest these transgenic plants into the insect midgut. Although the actual process involved in the insect-killing is not fully understood, some studies, such as Schünemann et al. (2014) reports that there is a binding process between activated toxins to the receptors in the epithelium on the insect midgut which is inserted into the membrane of the insect midgut. This triggers disruption of the electrical K1 and pH gradient thereby creating pores which ultimately results in irreversible midgut-wall damage. Similarly, there are also two genes that have been transferred from the bacteria *Agrobacterium tumefaciens* and the genus Streptomyces ie Isopentenyl-transferase gene (*ipt*) and a cholesterol-oxidase gene respectively into tobacco to confer resistance against economically important crop insect pests.

According to Gatehouse (1991), the plant sources exhibiting insecticidal activities, are classified into two groups consisting of (1) protein antimetabolites (example

Table 8.1 Some transgenic crops, inserted genes and their targeted insects

Crop	Inserted gene(s)	Targeted insect(s)	Reference
Broccoli	*CryIAa*	*Plutella xylostella*	Kumar et al. (2018)
Brinjal	*Cry1Ac*	*Leucinodes orbonalis*	Shelton et al. (2018)
Chickpea	*Cry1Ac, cry2Aa*	*Helicoverpa armigera*	Acharjee et al. (2010) and Acharjee and Sarmah (2013)
Chinese cabbage	*Cry1Ac*	*Plutella xylostella, Trichoplusia ni*	Cho et al. (2001)
Cotton	*CryIAc, Cry2Ab*	*Spodoptera frugiperda Anthonomus grandis*	Siddiqui et al. (2019)
Corn	*Cry1Ab, Cry1Ac, Cry1Ah*	*Helicoverpa armigera, Ostrinia furnacalis* and *Chilo suppressalis, O. furnacalis*	Xue et al. (2008) and Sun et al. (2015)
Rice	*Cry1Ab Cry 1Ac Cry2AX1*	*Scirpophaga incertulas, Cnaphalocrocis medinalis, Mythimna separata*	Lu (2010) and Chakraborty et al. (2016)
Sorghum	*Cry1Ac*	*Chilo partellus*	Girijashankar et al. (2005)
Sugar cane	*Cry1Ab cry1Ac*	*Diatraea saccharalis*	Wang et al. (2017) and Dessoky et al. (2021)
Soybean	*CryIAc, CryIAb, CryIF*	Lepidopteran insects	Koch et al. (2015)
Tomato	*Cry2Ab, Cry1Ab*	*Helicoverpa armigera Spodeptera litura*	Saker et al. (2011) and Koul et al. (2014)
Potato	*Cry3A*	*Leptinotarsa decenlineata*	Mi et al. (2015)
Cowpea	*Bt (vip3) Cry1Ab*	*Maruca vitrata*	Bett et al. (2017) and Addae et al. (2020)

proteinase inhibitors, α-amylase inhibitors, lectins and arcelins) and (2) non-protein antimetabolites (alkaloids, non-protein amino acids, terpenoids, retinoids (isoflavonoids), tannins, polysaccharides, glucosinolates, and cyanogenic glycosides). The production of these antimetabolic proteins act on the insect's digestive processes to protect the plants against insects. Proteinase inhibitors from plants constitute integral part of plant's natural defense against insect attacks (Larry and Richard 2002). Second type of enzymes used as inhibitors to modify crop plant is the α-amylases. The pea resistant to the Bruchid beetles, *Callosobruchus maculatus* and *C. chinensis* was as a result of the introduction and expression of the bean a-AI gene under the control of the 5′ and 3′ regions in the bean phytohemagglutinin gene (Shade et al. 1994). Lectin is a carbohydrates-binding protein and is commonly found in storage tissues and seeds of some plants (Babu et al. 2003). Depletion of essential amino acids caused by the presence of inhibitors is as a result of the activity of hypersecretion of the digestive enzymes of the genes from the plant (Gatehouse et al. 1992) or the midgut-epithlial cells binding of lectins of the insect (Gatehouse and Hilder 1994). Wasp, spiders, mammals, and scorpions constitute the sources of resistance genes from animals whilst genes encoding neurotoxins from predatory mites and scorpion (Tomalski and Miller 1991; Stewart et al. 1991) have been used in recombinant baculoviruses to improve their biological activity. Using ribonucleic acid (RNA) interference as a silencer of insect gene have shown potential in improving plant defense system and has proven noble in developing insect resistant transgenic crops. (Baum et al. 2007). Nevertheless, Cry proteins of Bt origin forms the basis of plant defense against insects in most commercially grown transgenic crops (Tabashnik and Carrière 2009).

Cultivation of insect resistant transgenic plants does not need protection with other insects, leading to minimal environmental effect, prevention of the health hazards during the application of insecticides, and developing insecticidal resistance. Research has pointed to the fact that transgenic plants have negligible or no side effect against birds, mammals as well as human beings (Goldberg and Tjaden 1990). The deployment of insect-resistant transgenic crops according Wu and Guo (2003) plays tremendous role in the conservation of biodiversity as has been found in *Bt*-fields compared to fields treated with synthetic insecticides. Again Wu and Guo (2003) found an increase in population of natural predators against aphids in fields cultivated with Bt cotton compared to non-Bt cotton fields. Several improvements have been made in developing and mass cultivation of transgenic crops to minimize pest damage in both food and non-food crops since the first transgenic plant, expressing insecticidal gene was produced in 1987.

A contributory factor to the intense development of transgenic plants for the control of insect pests has been the tremendous resistance of insect pests to chemical pesticides. To overcome this, there have been several Bt toxin genes introduced into crops, such as tobacco, maize, rice, potato, apple, cotton, and tomato, to confer resistance to specific insects. Bt crops initially developed mainly of cotton and maize, producing Cry1Ac and Cry1Ab toxins, respectively (Tabashnik and Carrière 2009), which were biocidal to lepidopteran pests. Following the successful research and limited production in 1987, universal cultivation of *Bt* crops has risen. An area of about 148 million hectares of transgenic crops were cultivated globally in 2010 increasing to 185.1 million ha in 2016 across 26 countries with 19 of these coming from developing countries (Abbas 2018). An average of more than 99.0 million hectares of land was cultivated of transgenic crops in developing countries in 2016 compared to 85.5 million ha in industrialized countries in the same period. Commercialization of transgenic crops resistant to insects actively commenced around the middle of 1990, when transgenic cotton, potato, and maize plants exhibiting the ability to kill insect with toxin d-endotoxin, produced by the *Bacillus thuringiensis* genes (Gatehouse 2008).

Monsanto, in 1996 pioneered the development and commercial production of the first insect-resistant transgenic cotton (highly effective in the control of Lepidoptera pest compared to synthetic insecticides (Betz et al. 2000). Since then, several transgenic crops have been developed to manage significant pests. Recently, approval was given for use in Bangladesh four Bt eggplant resistant to insects as well as a *Bt* soybean variety expressing Cry1Ac + Cry1Ab for the control of the lepidopteran in Latin America (Koch et al. 2015). *Bacillus thuringiensis* sweet corn, the most adapted transgenic vegetable with resistance to the insect pest, *Heliothis zea* according to Shelton et al. (2013) produces cleaner ears comparable to maize cultivars using chemicals. Again, the cultivation of transgenic maize with resistance to the European corn borer (*Ostrinia nubilalis* Hübner) and the western corn rootworm (*Diabrotica virgifera virgifera* LeConte) as well as other coleopteran species are reported to have reduced yield losses caused by these pests without using toxic organophosphate insecticides (DeVilliers and Hoisington 2011). Also, there is a report on the reduction of cotton pests significantly following the cultivation of Bt cotton (Naranjo 2011). Transgenic tomatoes have also been reported to protect either tomato fruit worm (*Heliothis zea*) and tobacco hornworm (*Manduca sexta*) or tomato fruit borer (*Helicoverpa armigera*) (Mandaokar et al. 2000; Kumar and Kumar 2004). Tomato plants containing *Bacillus thuringiensis* subsp. *tenebrionis* (*B.t.t.*) toxin caused a significant insecticidal activity against Colorado potato beetle larvae and under field conditions Kumar (2004) found that *Bt* tomato was effective in managing *Manduca sexta*, *Keiferia lycorersicella*, and *Helicoverpa armigera*. In addition to these, insecticidal activity of transgenic brinjal fruits against larvae of the fruit borer (*Leucinodes orbonalis*) have been documented (Kumar et al. 1998).

Insect-resistant crops developed initially, however, expressed dominant *Bt-Cry* genes, producing single *Bt*-toxin against specific lepidopteran pests, thereby killing a limited set of target pests. This narrow range of action led to the evolution of insect resistance crops, a major setback to the use of this technology. To overcome this limitation, *Bt*-crops producing multiple toxins have been developed by stacking *Cry* genes, targeting multiple receptors in insect pests to provide broad protection to a range of insects, delaying pest resistance development (Christou et al. 2006; Gatehouse 2008). The important role of transgenic crops in sustainable food production systems in sub-Saharan Africa is gradually gaining ground even though their utilization is limited and remains controversial in almost all countries in SSA and other developing countries. Although adopting transgenic crops may not solve all of the continent's food production constraints, adopting the technology will be highly beneficial to crop producers in the region. Currently, few countries in Africa have seen large-scale adoption of transgenic crops, although the continent has witnessed a fair share of insect infestation. According to Adenle (2011) about 15 million farmers estimated to be 90% who planted genetically modified crops were from three developing countries such as Burkina Faso, Egypt, and South Africa, were resource-poor farmers. These three countries together planted about 2.5 million hectares of transgenic crops (James 2011) with Kenya, Nigeria, and Uganda having commenced field trials although Falck-Zepeda et al. (2013) reported that Africa contributed less than 1.6% of the total land area cultivated to transgenic crops in 2011. Since then, the drive-in Africa to grow engineered crops is on the rise with Ethiopia, Nigeria, Kenya, and Malawi approving the planting of insect-resistant cotton (ISAAA 2020).

South Africa is considered the first African country to commercially produce insect's resistant *Bt* cotton to manage insects such as the bollworm infestation, control insect resistance to chemicals and reduce insecticidal use in cotton production (Gouse 2013). Reports indicated that research on *Bt* cotton in South Africa have increased yield and total reduction in chemical expenditure following adoption and commercialization of *Bt* cotton compared to the traditional varieties (Fok et al. 2007). Maize is an important cereal crop in the continent with a high insect infestation rate. The introduction of *Bt* maize with Bt gene (Cry1Ac), has been reported to show great prospects to control *B. fusca* and *C. partellus* in the production of the crop. Like *Bt* cotton, *Bt* maize has recorded higher yields compared to conventional cultivars with lower pesticides cost (Brookes and Barfoot 2008; Gonzales 2002; James 2002). Another challenge to insect damage on crops is the predisposition of their host to secondary infections by other pathogens. Wounds created on the kernels of maize favours fungal colonization thereby exposing them to mycotoxin contaminations. Transgenic insect-resistant (*Bt*) maize has been found as a potential way of reducing fumonisin exposure (Gouse 2013) due to a reduction in fusarium colonization arising from minimal entry points created by insects such as Lepidoptera.

8.5 Limitations to the Adoption and Utilization of Transgenic Plants for Insect Control

Since the advent of using genetic engineering to produce modified crops for human use, public perception of transgenic plants and their recognition in food production has been met with a mixed reaction. The controversy surrounding transgenic products have inspired global public debate on its acceptability and use by farmers in several countries. Advocates of the technology highlights benefits for society through hunger reduction, starvation prevention, and biotic stress management whilst opponents often see it as interference with nature which has dire consequences, disastrous to human genetics and natural ecosystems (Nelson 2001). In addition to these, there is also the fear in developing countries that adoption of transgenic crops, is likely to lead into farmers permanently depending on multinational companies for seed and chemical with the potential of favoring the industrialized countries (Junne 1991).

Despite the entry of transgenic crops into the food system of several countries, public acceptance remains an important factor affecting the future of technology. Notwithstanding, several factors against the adoption of transgenic crops, lack of education and provision of information on risks and benefits associated with the technology can also influence its acceptability. Baker and Burnham (2001) found socioeconomic variables as insignificant, but consumer's cognitive variables as significant determinants for embracing genetically modified food products.

Generally, the adoption of genetically modified crops are widely cited as a solution to combat pest resistance in crop production, yet this technology faces challenges in managing insect pest complexes. Threats of the evolution of resistance to *Bt* in targeted pest lingers on. A resurgence of the pink bollworms in India, and the mutation of corn leaf worms to develop resistance in Brazil are some avenues of concerns and opposition to use of insect-resistant transgenic crops. The long term effects of transgenic plants on agro-ecosystems still needs further understanding. The challenge is that insect species susceptible to expressed toxins, can develop into secondary pests and cause severe damage than initially would. Secondary pests which hitherto were of minor importance might assume major importance. In mid-southern and southeastern cotton-producing regions of the USA, Naranjo (2010) reported higher incidence in minor insect pests such as aphids, leafhoppers, mirid plant bugs, and stinkbugs considered initially as secondary pests in cotton production. Similarly, reduction in insecticide sprays in India has been linked to the prevalence of mealybugs (*Pseudococcus corymbatus*, *Pulvinaria maxima*, and *Saissetia nigra*), thrips (*Thrips tabaci*), and leafhoppers (*Amrasca biguttula biguttula*) in cotton production (Sharma 2005). In effect, reduction in the use of chemical insecticides in Bt cotton production have resulted in upsurge of pests that were not susceptible to *Bt* protein and were initially controlled by pesticides. Reductions in natural enemies' populations and interspecific competition with the target pest according to Mabubu et al. (2016) are some factors that may contribute to the outbreak of secondary pest species with the use of *Bt*-crops.

8.6 Integration of Transgenic Plants into Integrated Pest Management Strategies

As the limitations of completely depending on transgenic plants in insect control continue to expand, utilizing the technology in an IPM) approach will ensure food security, sustainable Agriculture and the protection to the environment. The FAO (2018) defines IPM as "careful use and intergration of all pest control measures that discourage pest population development, reduce pesticides use and other interventions to economically justified levels to reduce human risks as well as the environment. Generally, *Bt* crops have been categorized as either vehicle to deliver selected insecticides or to induce host plant resistance which will affect the growth and development of insects (Naranjo 2010).

Several regulatory agencies, including the United States Environmental Protection Agency (USEPA), as one of the several regulatory bodies, considers *Bt* protein as a "plant-incorporated protectant" (PIP), that regulate transgenic plants with pesticidal properties similar to any synthetic or organic pesticide (Naranjo 2010) sparking the debate as to whether it can fit into the description of host plant resistance, a strategy of integrated pest management. *Bt* crops, according to Naranjo (2010) are considered prophylactic control since *Bt* proteins are continually produced and released by *Bt* crops irrespective of insect infestation. Prophylactic measures are, however, good components of IPM since the concept embodies preventive and prescriptive measures consisting of strategies leading to pest avoidance. Genetic engineering overcomes the limitation of conventional breeding by accelerating host plant resistance breeding processes through recombination of specific genetic material followed by crossing into multiple elite lines. In line with this, genetically engineered crops can be viewed as a form of host plant resistance, highly recognized as a key component of IPM (Kennedy 2008). Host plant resistance developed through traditional breeding process or genetic engineering plays valuable role in IPM and complements different pest management practices.

The principle of IPM hinges on three main control principles viz. biological, cultural control, and host plant resistance. With the biological control, the abundance and the activities of natural enemies are enhanced to suppress pest population. Whilst host plant resistance involves the selection of crop cultivars that have highest pest resistance, cultural control strategy involves all agronomical practices that modify the environment making it less favorable for pest invasion. When there is sufficient combination of all these three control measures, put together, then there could be rational consideration of pesticide use in the IPM strategy (Koul et al. 2004; Romeis et al. 2008).

A well design IPM approach ensures rational use of all approaches that complement each other to eliminate over reliance of any single approach to achieve comprehensive control of pests. Genetically engineered crops are technically resourceful element of IMP that when integrated properly into a cropping system will enhance profit of stakeholders while reducing risks. Integration of transgenes in insect pest management strategies has successfully been implemented in managing insect pests. Integration of refugees, in the cropping system, improves the resilience of *Bt* traits in genetically engineered plant systems. Current technology allows the deployment of sterile insect technique (SIT) and/or pheromone-based mating disruption by refugees crops (Anderson et al. 2019). Cotton growers successfully planted Bt cotton cultivars without planting refuges but supplied them through targeted and proportional release of sterile male pink bollworm moths over *Bt* and non-*Bt* fields throughout Arizona. This process contributed successfully to destruction of the bollworm pest and the lifting of a ban on US cotton (USDA-ARS 2018). Training stakeholders about the contribution of IPM tactics and ways to include genetically engineered crops into agricultural system remains a priority. Knowledge of the socioeconomic factors should be combined with knowledge in agricultural systems to promote strategies that will drive the adoption and acceptance of insect-resistant transgenic plants in IPM.

8.7 Current and Future Research Techniques in Transgenic Crops Development for Insect Pest Management

The creation and commercialization of transgenic plants resistance to major insect pests have been the major achievements of plant biotechnology. Currently, genes that express insect resistance in transgenic plant are not only derived from the bacterium *Bacillus thuringiensis*, but other genes associated with higher plants, particularly genes encoding inhibitors of digestive enzymes and lactin (Schuler et al. 1998). Knowledge into insect resistant transgenic plant technologies keeps increasing very fast, with substantial research in every sector of our economies.

Modeling studies predicted doom for the sustainability of the technology in global agriculture because of the fear of insect developing resistance to single insecticidal gene products. These dreadful predictions notwithstanding, the worse has not yet happened whilst the introduction of transgenic insect resistant crops keeps increasing steadily over the years. To ensure the permanence and ability to sustain resistance, new approaches are being considered. This suggests that the future of transgenic crops for insect crop management in food production, and storage is promising and as new innovative strategies that ensure longevity of the next generation of insect-resistant plants should be in place.

Successful constitutive *Bt* genes expression has been reported, whilst tissues-specific expression has proven a better option in some cases, this is what happens with the epidermal cell which first suffer an attack from sap-sucking insects. Reports show that transcription elements or chemical induction can be used to regulate expression. It is therefore possible to use this technique to create parts of the plant where there will be no expression of genes and therefore the plant acts as a non-genetically refuge (Christou et al. 2006). For example, the chloroplasts where plastid expression occur, could be target and used for future transgenic crops development (Bock 2007). This is because the plastids accumulates high levels of toxins of bacterial origin just as the *B. thuringiensis* genes (McBride et al. 1995). Another way to improve insect resistance is through gene stacking or pyramiding where multiple genes of interest are inserted into a single plant.

Recently, crops expressing several Cry genes to target single insect (Christou et al. 2006) and the development of hybrid Cry proteins to improve toxicity and host range are being evaluated with the aim of slowing down the development of resistance. The combination of the Cry genes and plant lectins to target various pest has also been reported with the snowdrop (*Galanthus nivalis*) lectin for example fused to Cry gene, to deliver protein to heamolymph of lepidopteran larvae.

New engineered transgenic maize with six resistant genes to control corn rootworm and lepidopteran pest and dual herbicide tolerance genes, has been developed to provide a "one-stop solution" to both pest and weed problems through gene stacking (Grainnet 2007).

Transgenic insect-resistant technologies have been a major scientific success in modern plant biotechnology. Notwithstanding this, there are restrictions on these products in many developing countries, due to the lack of understating of the technology and the lack of mechanisms to regulate its deployment. (Paarlberg 2002, 2008). Usually, the problem confronted public institutions of developing world to develop product for farmers, is the insufficient potential gains that will be accrued which eventually make commercialization difficult due to high price. It is for this reason that most of the developed commercial product of genetic engineering are in the hands of big companies.

8.8 Conclusions

Genetic engineering technologies have transformed crop production remarkably with the introduction of insect resistant crop plants that are high in productivity to benefit resource-challenged crop farmers. Introduction of transgenic crop plants has a limited uncontrolled application of chemical insecticides which endanger man, animals, and the environment in some advanced and developing economies. Transgenic crop plants have been employed in crop production to manage many insect pests of economic importance. Research in transgenic crops may offer new means of improving agriculture, especially in Africa and the world in general. However, a major challenge of transgenic research, apart from obtaining transgenic materials or resources, is to adequately understand physiological expression at the plant level of the inserted genes. An all-inclusive approach that integrates genetically engineered crops and other strategies to manage insect pests provides a sure way of producing safe food to feed the growing population. There should be a sustained education and awareness creation targeted at opponents of transgenic engineering for crop improvement so that the technology would be embraced as a whole to benefit mankind.

Conflict of Interest Authors declare no competing interest. All references have been duly cited and authors acknowledged.

Ethics approval and consent to participate.

Not applicable.

References

Abbas MST (2018) Genetically engineered (modified) crops (Bacillus thuringiensis crops) and the world controversy on their safety. Egypt J Biol Pest Control 28:52

Acharjee S, Sarmah BK (2013) Transgenic *Bacillus thuringiensis* (Bt) chickpea: India's most wanted genetically modified (GM) pulse crop. Afr J Biotechnol 12(39):5709–5713

Acharjee S, Sarmah BK, Kumar P, Olsen K, Mahon R, Moar W, Moore A, Higgins T (2010) Transgenic chickpeas (*Cicer arietinum* L.) expressing a sequence-modified cry2Aa gene. Plant Sci 178:333–339

Addae PC, Ishiyaku MF, Tignegre JB, Ba MN, Bationo JB, Atokple IDK, Abudulai M, Dabiré-Binso CL, Traore F, Saba M, Umar ML, Adazebra GA, Onyekachi FN, Nemeth MA, Huesing JE, Beach LR, Higgins TJV, Hellmich RL, Pittendrigh BR (2020) Efficacy of a cry1Ab gene for control of Maruca vitrata (Lepidoptera: Crambidae) in cowpea (Fabales: Fabaceae). J Econ Entomol 113(2):974–979. https://doi.org/10.1093/jee/toz367

Adenle AA (2011) Response to issues on GM agriculture in Africa: are transgenic crops safe? BMC Res Notes 4:388. https://doi.org/10.1186/1756-0500-4-388

Aekthong S, Rattanakul C (2019) Investigating the use of wasps Anagyrus lopezi as a biological control agent for cassava mealybugs: modeling and simulation. Adv Differ Equ 2019:237. https://doi.org/10.1186/s13662-019-2176-3

Anderson JA, Ellsworth PC, Faria JC, Head GP, Owen MDK, Pilcher CD, Shelton AM, Meissle M (2019) Genetically engineered crops: importance of diversified integrated pest management for agricultural sustainability. Front Bioeng Biotechnol 7:24. https://doi.org/10.3389/fbioe.2019.00024

Asante SK, Tamo M, Jackai LEN (2001) Integrated management of cowpea insect pests using elite cultivars, date of planting and minimum insecticide application. Afr Crop Sci J 9:655–665

Babu RM, Sajeena A, Seetharaman K, Reddy MS (2003) Advances in genetically engineered (transgenic) plants in pest management—an overview. Crop Prot 22:1071–1086

Baker GA, Burnham TA (2001) Consumer response to genetically modified foods: market segment analysis and implications for producers and policy makers. J Agric Resour Econ 26(2):387–403

Baum J, Bogaert T, Clinton W et al (2007) Control of coleopteran insect pests through RNA interference. Nat Biotechnol 25:1322–1326. https://doi.org/10.1038/nbt1359

Bellotti AC (2008) Cassava pests and their management. In: Capinera JL (ed) Encyclopedia of entomology, 2nd edn. Springer, Dordrecht

Bett B, Gollasch S, Moore A et al (2017) Transgenic cowpeas (*Vigna unguiculata* L. Walp) expressing *Bacillus thuringiensis* Vip3Ba protein are protected against the Maruca pod borer (Maruca vitrata). Plant Cell Tiss Org Cult 131:335–345. https://doi.org/10.1007/s11240-017-1287-3

Betz FS, Hammond BG, Fuchs RL (2000) Safety and advantages of *Bacillus thuringiensis*-protected plants to control insect pests. Regul Toxicol Pharmacol 32:156–173. https://doi.org/10.1006/rtph.2000.1426

Bock R (2007) Plastid biotechnology: prospects for herbicide and insect resistance, metabolic engineering and molecular farming. Curr Opin Biotechnol 18(2):100–106

Brookes G, Barfoot P (2008) GM crops: global socioeconomic and environmental impacts 1996–2006. PG Economics, Dorchester

Carvalho FP (2006) Agriculture, pesticides, food security and food safety. Environ Sci Policy 9(7–8):685–692

Chakraborty M, Reddy PS, Mustafa G, Rajesh G, Narasu VM, Udayasuriyan V, Rana D (2016) Transgenic rice expressing the cry2AX1 gene confers resistance to multiple lepidopteran pests. Transgenic Res 25(5):665–678. https://doi.org/10.1007/s11248-016-9954-4. Epub 2016 Mar 26

Cho HS, Cao J, Ren JP, Earle ED (2001) Control of lepidopteran insect pests in transgenic Chinese cabbage (*Brassica rapa ssp. pekinensis*) transformed with a synthetic *Bacillus thuringiensis* cry1C gene. Plant Cell Rep 20(1):1–7. https://doi.org/10.1007/s002990000278

Christou P, Capell T, Kohli A, Gatehouse J, Gatehouse A (2006) Recent developments and future prospects in insect pest control in transgenic crops. Trends Plant Sci 11:302–308

Culliney TW (2014) Crop losses to arthropods. In: Pimentel D, Peshin R (eds) Integrated pest management. Springer, Berlin, pp 201–225

Dara SK (2019) The new integrated Pest management paradigm for the modern age. J Integr Pest Manag 10(1):12. https://doi.org/10.1093/jipm/pmz010/Issues

Dessoky ES, Ismail RM, Elarabi NI, Abdelhadi AA, Abdallah NA (2021) Improvement of sugarcane for borer resistance using agrobacterium mediated transformation of cry1Ac gene. GM Crops Food 12(1):47–56. https://doi.org/10.1080/21645698.2020.1809318

DeVilliers SM, Hoisington DA (2011) The trends and future of biotechnology crops for insect pest control. Afr J Biotechnol 10(23):4677–4681. https://doi.org/10.1104/pp.107.111096

Dunn SE, Vicini JL, Glenn KC, Fleischer DM, Greenhawt MJ (2017) The allergenicity of genetically modified foods from genetically engineered crops: a narrative and systemic review. Ann Allergy Asthma Immunol 119:214–222

Eason CT, Miller A, MacMorran DB, Murphy EC (2014) Toxicology and ecotoxicology of Para-aminopropiophenone (PAPP)—a new predator control tool for stoats and feral cats in New Zealand. N Z J Ecol 38:177–188

El-Sayed AM, Suckling DM, Byers JA, Jang EB, Wearing CH (2009) Potential of 'lure and kill' for long-term pest management and eradication of invasive species. J Econ Entomol 102:815–835

Falck-Zepeda J, Guillaume G, Sithole-Niang I (2013) Introduction and background. In: Falck-Zepeda J, Gruère G, Sithole-Niang I (eds) Genetically modified crops in Africa: economic and policy lessons from countries south of the Sahara, pp 1–24. https://doi.org/10.2499/9780896297951

FAO (2018) Five things to know about Africa's latest foe, http://www.fao.org/fao-stories/article/en/c/1104446/. Assessed 3 Sept 2020

FAO (2019) New standards to curb the global spread of plant pests and diseases. http://www.fao.org/news/story/en/item/1187738/icode/. Accessed 6 Jun 2020

Fok M, Hofs JL, Gouse M, Kirsten JF (2007) Contextual appraisal of GM cotton diffusion in South Africa. Life Sci Int J 1(4):468–482

Garcı'a-Lara S, Saldivar SO (2016) Insect science. In: Encyclopedia of food and health, pp 432–436. https://doi.org/10.1016/B978-0-12-384947-2.00396-2

Gatehouse JA (1991) Breeding for resistance to insects. In: Murray DR (ed) Advanced methods in plant breeding and biotechnology. CAB International, Wallingford, pp 250–276

Gatehouse JA (2008) Biotechnological prospects for engineering insect-resistant plants. Plant Physiol 146:881–887. https://doi.org/10.1104/pp.107.111096

Gatehouse AMR, Hilder VA (1994) Genetic manipulation of crops for insect resistance. In: Marshall G, Walters D (eds) Molecular biology in crop protection. Chapman & Hall, London, pp 177–201

Gatehouse AMR, Boulter D, Hilder VA (1992) Potential of plant derived genes in the genetic manipulation of crops for insect resistance. In: Gatehouse AR, Hilder VA, Boulter D (eds) Plant genetic manipulation for crop protection. CAB International, Wallingford, pp 155–181

Girijashankar V, Sharma HC, Sharma KK, Swathisree V, Prasad LS, Bhat BV, Royer M, Secundo BS, Narasu ML, Altosaar I, Seetharama N (2005) Development of transgenic sorghum for insect resistance against the spotted stem borer (Chilo partellus). Plant Cell Rep 24(9):513–522. Epub 2005 Sep 20. PMID: 16172896. https://doi.org/10.1007/s00299-005-0947-7

Goldberg RJ, Tjaden G (1990) Are b.t.t. plants really safe to eat? Biotechnology 8:1011–1014

Goldson SL, Bourdôt GW, Brockerhoff EG, Byrom AE, Clout MN, McGlone MS, Nelson WA, Popay AJ, Suckling DM, Templeton MD (2015) New Zealand pest management: current and future challenges. J R Soc N Z 45(1):31–58. https://doi.org/10.1080/03036758.2014.1000343

Gonzales LA (2002) Likely transcendental effects of Agribiotechnology: the case of Bt hybrid corn in the Philippines. In: Paper presented during the symposium on Bt technology: facts and issues, May 5, Los Baños, Laguna, Philippines

Gouse M (2013) Socioeconomic and farm-level effects of genetically modified crops: the case of Bt crops in South Africa. In: Falck-Zepeda J, Gruère G, Sithole-Niang I (eds) Genetically modified crops in Africa: economic and policy lessons from countries south of the Sahara. IFPRI, Washington, DC, pp 25–59. https://doi.org/10.2499/9780896297951

Grainnet (2007) Monsanto and Dow Agrosciences launch "SmartStax", industry's first-ever eight-gene stgraacked combination in corn. Grainnet. http://www.grainnet.com

Hajek AE, Eilenberg J (2018) Natural enemies: an introduction to biological control. Cambridge University Press, Cambridge

ISAAA (2020) Global status of commercialized biotech/GM crops: ISAAA brief no. 54. In: Executive summary. International Service for the Acquisition of Agri-biotech applications (ISAAA), Ithaca, NY, USA. http://www.isaaa.org/resources/publications/briefs/32/executivesummary

Isman MB (2019) Challenges of pest management in the twenty first century: new tools and strategies to combat old and new foes alike. Front Agron 1:2–4. https://doi.org/10.3389/fagro.2019.00002

Jackson TA (2007) Biological control: a global perspective. In: Vincent C, Goettel MS, Lazarovitts G (eds) A novel bacterium for control of grass grubs. CABI, Wallingford, pp 160–168

James C (2002) Global status of commercialized transgenic crops: 2002. ISAAA briefs no. 26. ISAAA, Ithaca, NY

James C (2011) Global status of commercialized biotech/GM crops: 2011. In: ISAAA Brief No. 43. Executive summary. International Service for the Acquisition of Agri-biotech Applications (ISAAA), Ithaca, NY, USA. http://www.isaaa.org/resources/publications/briefs/43/executivesummary/pdf/Brief%2043%20-%20Executive%20Summary%20-%20English.pdf. Accessed Aug 2020

Junne G (1991) The impacts of biotechnology on international trade. In: Sasson A, Costarini V (eds) Biotechnology in perspective: socioeconomic implications for developing countries. United Nations Educational, Scientific and Cultural Organization (UNESCO), Paris

Kalode MB, Sharma NC (1995) Host plant resistance to insects: progress, problems and future needs. In: Sharma HC, Rao MV (eds) Pests and pest management in India: the changing scenario. Plant protection Association of India, Hyderabad, pp 229–243

Kennedy GG (2008) Integration of insect-resistant genetically modified crops within IPM programs. In: Romeis J et al (eds) Integration of insect-resistant genetically modified crops within IPM programs. Springer, Berlin, pp 1–26

Koch MS, Ward JM, Levine SL, Baum GA, Vicini JL, Hammond BG (2015) The food and environmental safety of Bt crops. Front Plant Sci 6:283–336

Koul O, Dhaliwal GS, Cuperus GW (eds) (2004) Integrated pest management: potential, constraints and challenges. CABI, Wallingford, p 336. ISBN 0-85199-6868

Koul B, Srivastava S, Sanyal I, Tripathi B, Sharma V, Amla DV (2014) Transgenic tomato line expressing modified *Bacillus thuringiensis* cry1Ab gene showing complete resistance to two lepidopteran pests. SpringerPlus 3:84. https://doi.org/10.1186/2193-1801-3-84

Kumar H (2004) Tomato expressing Cry1A(b) insecticidal protein from *Bacillus thuringiensis* protected against tomato fruit borer, Helicoverpa armigera (Hübner) (Lepidoptera: Noctuidae) damage in the laboratory, greenhouse and field. Crop Prot 23(2):135–139

Kumar H, Kumar V (2004) Tomato expressing Cry1A(b) insecticidal protein from *Bacillus thuringiensis* protected against tomato fruit borer, Helicoverpa armigera (Hübner) (Lepidoptera: Noctuidae) damage in the laboratory, greenhouse and field. Crop Prot 23:135–139

Kumar PA, Mandaokar A, Sreenivasu K, Chakrabarti SK, Bisaria S, Sharma SR, Kaur S, Sharma RP (1998) Insect-resistant transgenic brinjal plants. Mol Breed 4:33–37

Kumar P, Gambhir G, Gaur A, Sharma KC, Thakur AK, Srivastava DK (2018) Development of transgenic broccoli with cryIAa gene for resistance against diamondback moth (Plutella xylostella). 3 Biotech 8:299

Laizer HC, Chacha MN, Ndakidemi PA (2019) Farmers' knowledge, perceptions and practices in managing weeds and insect pests of common bean in northern Tanzania. Sustainability 11:4076. https://doi.org/10.3390/su11154076

Larry LM, Richard ES (2002) Lectins and protease inhibitors as plant defenses against insects. J Agric Food Chem 50:6605–6611

Liu SS, De Barro PJ, Xu J, Luan JB, Zang LS, Ruan YM (2007) Asymmetric mating interactions drive widespread invasion and displacement in a whitefly. Science 318(5857):1769–1772

Lu C (2010) The first approved transgenic rice in China. GM Crops 1(3):113–115

Mabubu JI, Nawaz M, Hua H (2016) Advances of transgenic Bt-crops in insect pest management: an overview. J Entomol Zool Stud 4(3):48–52

Malone LA et al (2008) Beyond Bt: alternative strategies for insectresistant genetically modified crops. In: Romeis J et al (eds) Integration of insect-resistant genetically modified crops within IPM programs. Springer, Berlin, pp 357–417

Mandaokar AD, Goyal RK, Shukla A, Bisaria S, Bhalla R, Reddy VS, Chaurasia A, Sharma RP, Altosaar I, Ananda Kumar P (2000) Transgenic tomato plants resistant to fruit borer (Helicoverpa armigera Hübner). Crop Prot 19:307–312

McBride KE, Svab Z, Schaaf DJ, Hogan PS, Stalker DM, Maliga P (1995) Amplification of a chimeric Bacillus gene in chloroplasts leads to an extraordinary level of an insecticidal protein in tobacco. Biotechnology (NY) 13(4):362–365

Mi X, Ji X, Yang J, Liang L, Si H, Wu J, Zhang N, Wang D (2015) Transgenic potato plants expressing cry3A gene confer resistance to Colorado potato beetle. C R Biol 338(7):443–450

Myers SS, Smith MR, Guth S, Golden CD, Vaitla B, Mueller ND, Dangour AD, Huybers P (2017) Climate change and global food systems: potential impacts on food security and undernutrition. Annu Rev Public Health 38(1):259–277

Naranjo SE (2010) Impacts of Bt transgenic cotton on integrated pest management. J Agric Food Chem 59:5842

Naranjo SE (2011) Empact of Bt transgenic cotton on integrated pest management. J Agric Food Chem 59:5842–5851

Nelson CH (2001) Risk perception, behavior, and consumer response to genetically modified organisms. Am Behav Sci 44(8):1371–1388

Nwanze KF (1982) Relationship between cassava root yields and infestations by the mealybugs, Phenacoccus manihoti. Trop Pest Manag 28:27–32

Oerke EC (2006) Crop losses to pests. J Agric Sci 144:31–43. https://doi.org/10.1017/S0021859605005708

Oerke EC, Dehne HW (2004) Safeguarding production—losses in major crops and the role of crop protection. Crop Prot 23:275–285

Paarlberg R (2002) The real threat to GM crops in poor countries: consumer and policy resistance to GM foods in rich countries. Food Policy 27:247–250. https://doi.org/10.1016/S0306-9192(02)00014-3

Paarlberg R (2008) RABESA regional workshop policy options paper: toward a regional policy on GMO crops among COMESA/ASARECA countries. ACTS Press, Nairobi

Panda N, Khush GS (1995) Host pest resistance to insects. CAB International, Wallingford, Oxon, p 341

Pandey P, Irulappan V, Bagavathiannan MV, Senthil-Kumar M (2017) Impact of combined abiotic and biotic stresses on plant growth and avenues for crop improvement by exploiting physio-morphological traits. Front Plant Sci 8(537):1–15

Romeis J, Van Driesche RG, Barratt BIP, Bigler F (2008) Insect-resistant transgenic crops and biological control. In: Romeis J, Shelton AM, Kennedy GG (eds) Integration of insect-resistant genetically modified crops within IPM programs. Progress in biological control, vol 5. Springer, Dordrecht. https://doi.org/10.1007/978-1-4020-8373-0_4

Saker MM, Salama HS, Salama M, El-Banna A, Abdel Ghany NM (2011) Production of transgenic tomato plants expressing cry 2Ab gene for the control of some lepidopterous insects endemic in Egypt. J Genet Eng Biotechnol 9(2):149–155

Schuler TH, Poppy GM, Kerry BR, Denholm I (1998) Insect-resistant transgenic crops. Trends Biotechnol 16(4):168–174

Schünemann R, Knaak N, Fiuza LM (2014) Mode of action and specificity of *Bacillus thuringiensis* toxins in the control of caterpillars and stink bugs in soybean culture. Int Sch Res Notices 2014: 135675. https://doi.org/10.1155/2014/135675

Shade RE, Schroeder HE, Pueyo JJ, Tabe LM, Murdock LL, Higgins TJV, Chrispeels MJ (1994) Transgenic pea seeds, expressing the α-amylase inhibitor of the common bean are resistant to bruchid beetles. Biotechnology 12:793–797

Sharma HC (1993) Host plant resistance to insects in sorghum and its role in integrated pest management. Crop Prot 12:11–34

Sharma HC (2005) Heliothis/Helicoverpa management—emerging trends and strategies for future research. Oxford & IBH Publishing, New Delhi

Shelton AM, Olmstead DL, Burkness EC, Hutchison WD, Dively G, Welty C (2013) Multi-state trials of Bt sweet corn varieties for control of the corn earworm. J Econ Entomol 106:2151–2159

Shelton AM, Hossain MJ, Paranjape V, Azad AK, Raman ML, Khan ASMM et al (2018) Bt eggplant project in Bangladesh: history, present status and future direction. Front Bioeng Biotechnol 6:106. https://doi.org/10.3389/fbioe.2018.00106

Siddiqui HA, Asif M, Asad S, Naqvi RZ, Ajaz S, Umer N, Anjum N, Rauf I, Sarwar M, Arshad M, Amin I, Saeed M, Mukhtar Z, Bashir A, Mansoor S (2019) Development and evaluation of double gene transgenic cotton lines expressing cry toxins for protection against chewing insect pests. Sci Rep 9:11774. https://doi.org/10.1038/s41598-019-48188-z

Smith CM (ed) (2005) Plant resistance to arthropods: molecular and conventional approaches. Springer, Berlin

Stewart LMD, Hirst M, Ferber ML, Merryweather AT, Cayley PJ, Possee RD (1991) Construction of an improved baculovirus insecticide containing an insectspecific toxin gene. Nature 352:85–88

Sun H, Zhi-hong L, Wei L, Jie Z, Kang-lai HE, Li Z, Min L, Da-fang H (2015) Developing transgenic maize (*Zea mays* L.) with insect resistance and glyphosate tolerance by fusion gene transformation. J Integr Agric 14(2):305–313

Surendra KD (2019) The new integrated pest management paradigm for the modern age. J Integr Pest Manag 10(1):12. https://doi.org/10.1093/jipm/pmz010

Tabashnik BE, Carrière Y (2009) Insect resistance to genetically modified crops. In: Ferry N, Gatehouse AMR (eds) Environmental impact of genetically modified crops. CAB International, Wallingford, pp 74–100

Tomalski MD, Miller LK (1991) Insect paralysis by baculovirus-mediated expression of a mite neurotoxin gene. Nature 352:82–85

Tyczewska A, Woźniak E, Gracz J, Kuczyński J, Twardowski T (2018) Towards food security: current state and future prospects of agrobiotechnology. Trends Biotechnol 36(12):1219–1229

USDA-ARS (United States Department of Agriculture-Agricultural Research Service) (2018) A national road map for integrated pest management. https://www.ars.usda.gov/ARSUserFiles/OPMP/IPM%20Road%20Map%20FINAL.pdf

Verheijen FG, Jones RJ, Rickson RJ, Smith CJ (2009) Tolerable versus actual soil erosion rates in Europe. Earth Sci Rev 94(1–4):23–38

Wang WZ, Yang BP, Feng XY, Cao ZY, Feng CL, Wang JG, Xiong GR, Shen LB, Zeng J, Zhao TT, Zhang SZ (2017) Development and characterization of transgenic sugarcane with insect resistance and herbicide tolerance. Front Plant Sci 8:1535. https://doi.org/10.3389/fpls.2017.01535

West PC, Gerber JS, Engstrom PM, Mueller ND, Brauman KA, Carlson KM, Cassidy ES, Johnston M, MacDonald GK, Ray DK, Siebert S (2014) Leverage points for improving global food security and the environment. Science 345(6194):325–328

Wheeler T, Braum VJ (2013) Climate change impact on global food security. Science 341(6145):508–513

Wu KM, Guo Y (2003) Influences of *Bacillus thuringiensis* Berliner cotton planting on population dynamics of the cotton aphid, *Aphis gossypii* glover, in northern China. Environ Entomol 32:312–318

Xue J, Liang G, Crickmore N, Li H, He K, Song F, Feng X, Huang D, Zhang J (2008) Cloning and characterization of a novel Cry1A toxin from *Bacillus thuringiensis* with high toxicity to the Asian corn borer and other lepidopteran insects. FEMS Microbiol Lett 280:95–101

Zalom FG, Bolda MP, Dara SK, Joseph S (2018) UC IPM pest management guidelines: strawberry (insects and mites). University of California Statewide IPM Program, Oakland, CA, publication number 3468

Chapter 9
Improving Insect Control Using Genetically Modified Entomopathogens

R Dhanapal, Achanta Sravika, S Sekar, S Ramesh Babu, and M Gajalakshmi

Abstract Entomopathogens, employed as biological control, were formulated as eco-friendly substitutes for chemical insecticide. Nevertheless, microbial-based insecticides were implicated to be commercially inadequate based on their indiscriminate efficacy and low virulence. The resistance, pathogenicity, and virulence of entomopathogens to adverse conditions were greatly enhanced by genetic engineering. Improvement of virulence was achieved by modifying entomopathogens to express insecticidal proteins/peptides. The continued use of synthetic chemical insecticides remains an apprehension by the public as well as their approval of genetically modified organisms including inventive biological insecticides which provide diverse beneficial and environmentally sustainable options to control insect pests.

Keywords Entomopathogens · Genetic engineering · Entomopathogenic fungi · Entomopathogenic Bacteria · And Entomopathogenic virus

R. Dhanapal (✉)
Department of Entomology, Adhiparasakthi Horticultural College, Tamil Nadu Agricultural University, Kalavai, Ranipet, Tamil Nadu, India

Department of Entomology and Agricultural Zoology, Institute of Agricultural Sciences, Banaras Hindu University, Varanasi, Uttar Pradesh, India
e-mail: dhanapal.agri@gmail.com

A. Sravika
Department of Entomology, College of Agriculture, GKVK, University of Agricultural Sciences, Bengaluru, Karnataka, India

S. Sekar · M. Gajalakshmi
Department of Entomology, Agricultural College & Research Institute, Tamil Nadu Agricultural University, Coimbatore, Tamil Nadu, India

S. Ramesh Babu
Department of Entomology and Agricultural Zoology, Institute of Agricultural Sciences, Banaras Hindu University, Varanasi, Uttar Pradesh, India

S. De Mandal et al. (eds.), *New and Future Development in Biopesticide Research: Biotechnological Exploration*, https://doi.org/10.1007/978-981-16-3989-0_9

9.1 Introduction

Biological control is an earth-friendly, fiscal, and effective management technique of insect pests. The number of biocontrol products like entomopathogens has increased in pest management, but they represent about 1% of the applicable control measures. In recent years, several small and large entrepreneurs have started the viable production of biocontrol agents. However, the major drawbacks are short shelf life, insensitivity in the pest-targeted areas considering their inconsistent capability (Fang and St Leger 2012), and extreme temperatures like too cold and too hot conditions (Junaid et al. 2013).

Bio-pesticides are important constituents of integrated programs for insect pest management and could significantly undermine overall insecticidal load on food, feed, or fiber crops (Dhanapal et al. 2020a). Several insect pathogens, including fungi, bacteria, nematodes, and viruses, could be adopted for regulating target pest populations (Singh et al. 2019). Microbes are usually the most preferred insecticides as they are mostly environmentally safe; their formulations are finally degraded and leave no harmful residues in the environment. These can also be combined into organic agriculture protocols and can be mass-produced for large scale production. They are highly specific in action and do not cause toxicity or infection in alternative organisms. For instance, *Bacillus thuringiensis* (Bt), binds to the peripheral midgut epithelial cells on specific receptors, and organisms that lack specific receptors in their gut cannot be affected by it, and henceforth it is safe for beneficial arthropod insect species such as insect pollinators. The efficiency of these bio-pesticides can be more operative than chemical insecticides on a long-term basis (Gressel 2001). The crucial success of biocontrol is governed by how adequate the search and screening progression is done, which depends on the cropping system, crop and its target pathogen (Dhanapal et al. 2019). Suitable isolates are evaluated for their efficacy under field conditions. The isolates that fail to perform well in the field condition should be subjected to the in vitro evaluation to ascertain the cause of their failure, and necessary traits like virulence and pathogenicity should be increased through genetic improvement. Many of the entomopathogenic strains have low ultraviolet tolerance and deprived host-finding capacity, signifying the prospect of enlightening these characters through molecular tactics.

Comprehension of biocontrol agents and their mechanism of action could enhance its control either by improving the mechanism or using the entomopathogens under predictable circumstances (Sala et al. 2019). Several technologies were developed for the improvement of effectiveness of entomopathogens (Srinivasan et al. 2019). The performance of numerous strains selected artificially was increased under various field environments integrating hybridization of different strains and other rDNA tools. The applied aspects are associated with strain improvements such as extreme abiotic stress and pesticide tolerances, host finding, and infectivity to targeted pests, shelf-life, and formulation. The augmentation of the effectiveness of fungi, bacteria, and viruses has only been possible after the development of recombinant DNA methods. Genetically enhanced entomopathogens led to the next level of improvement and act as biological weaponry to control populations of insect pests in agrarian systems (Routray et al. 2016).

9.2 Entomopathogenic Fungi

Major biocontrol agents such as entomopathogenic fungi were intensively investigated over 100 years. They reproduce both sexually and asexually and generate infectious propagules (Hajek 2004) that attach their spores to insect cuticles and germinate, penetrate the cuticle, and distribute throughout the insect. The infectious disease process produces several secondary metabolites that benefit the suppression of the insect immune system (Subbanna et al. 2019). Eco factors seemingly determine the efficiency of fungal entomopathogens at field level (Dhanapal et al. 2020b). Several entomopathogenic fungi species with a diverse host range of (Lomer et al. 2001) blood-sucking insect pests, for instance, mosquitoes as they need not be ingested by its host. As an alternative, such fungi invade the cuticle by means of direct penetration and proliferates the insect hemocoel. Hyphae re-emerge to cover the cadaver after the death of its insect host and produce a large number of conidia that would infect new hosts.

9.2.1 Improving Virulence

Pathogens have evolved, based on their inherent lack of efficacy, to maintain an evolutionary balance with their host to avoid rapid death even at high doses. The insertion of specific genes into fungal cells comprises productive biocontrol (Gressel 2001). Significant initial candidate genes could presumably be the cuticle-degrading enzymes and toxins which are determined by single genes. Improved virulence was achieved by inserting genes into the genome of *M. anisopliae* that encode controlled cuticle-degrading protease Pr1 and were overexpressed significantly (Table 9.1). The resulting strain showed a mean survival time (LT_{50}) towards *Manduca sexta* was 25% reduced (St. Leger et al. 1996). Several arachnid and scorpion toxins were proclaimed to be the key resources for transgenic plants and biopesticide delivery systems. Improved efficacy of the *M. anisopliae* strain ARSEF 549 strain could be ensued by the incorporation of scorpion toxin (AaIT) (Zlotkin et al. 2000). Enhancement of virulence by genetic engineering with both reduced killing time and lethal dosage, thus improving the infection rates that allow for control of the pest using low concentration of pesticide. It also increases the operative perseverance of the biopesticides with a possibility of decreased inoculum threshold while employing genetically engineered fungus. Similar mortality rates were attained by the modified fungus as in *Manduca sexta* at 22 lesser spore doses than wild type, while the survival times were almost 40% reduced at specific doses (Wang and St. Leger 2007). Similar results were obtained from studies in mosquitoes (nine-fold reduction in LC_{50}) and Broca (coffee berry borer beetle; 16-fold reduction in LC_{50}) (Pava-Ripoll et al. 2008). Synergistic virulence effect was not attained by coexpression of Pr1 and AaIT in *B. bassiana* as the Pr1 produced in the hemolymph digested the expressed AaIT (Lu et al. 2008).

Table 9.1 Entomopathogenic fungal protein encoding genes

Fungus	Gene	Encoding enzymes	References
Beauveria bassiana	*Prt1*	Protease Bassianin I	Joshi et al. (1995) Kim et al. (1999)
	Prt 1—Like	Serine endoprotease	Fang et al. (2005)
	Chit *Bbchit* *1, Bbchit 2*	Chitinase Endonuclease Chitinase	Fang et al. (2005) Yokoyama et al. (2002) Baratto et al. (2003)
	buv1	UV repair	Chelico et al. (2006)
B. brongniartii	*buv1*	UV repair	Chelico et al. (2006)
Metarhizium anisopliae	*Sod*	Superoxide dismutase	Schrank et al. (1993)
	Pr1B	Subtilisin like protease	Joshi et al. (1997)
	Pr1 (A-K)	Protease	Bagga et al. (2004)
	CRR1	DNA binding protein	Screen et al. (1997)
	Nrr1	Nitrogen response regulator	Screen et al. (1998)
	Chit1	Chitinase Chitinase Chitin synthase	Bogo et al. (1998) Kang et al. (1998) Nam et al. (1998)
	Chit2	Chitinase	Baratto et al. (2003, 2006), Screen et al. (2001)
	MeCPAA	Zinc carboxypeptidase	Joshi and St Leger (1999)
	ssgA	Hydrophobin Trehalase Peptide synthase	Bidochka et al. (2001) Zhao et al. (2006) Bailey et al. (1996)
	Trp1	Tryptophan synthase	Staats et al. (2004)
	Pr 1, Pr 2, Pr 4	Subtilisin, trypsin and cysteine proteases	Bagga et al. (2004)
	Mad 1, Mad 2	Adhesin- like proteins	Wang and St. Leger (2007)
	Hyd 1, Hyd 2, Hyd 3	Hydrophobins	Zhang et al. (2011)
	SsgA	Hydrophobin-like protein	St. Leger et al. (1996)

RNA interference (RNAi) is considered a potential technique for insect pest control. But RNAi technology was not yet being used in the field. The high cost involved in practical application, unbalanced efficiency, and complications arising in transporting exogenous ds/siRNA to the target pests were presented as the most possible explanations. Entomopathogenic fungi-based delivery of ds/siRNA is an effective method to deliver the ds/siRNA in insects. Whitefly, especially nymph, is mostly affected by *Isaria fumosorosea*. *I. fumosorosea* acts as a host to generate ds/siRNA, and psTLR7, a dsRNA expression plasmid, created for the silent vector pSilent-1, by hosting Toll-like receptor 7 (TLR7) gene of its B-biotype (Chen et al. 2015). The protoplast of the *I. fumosorosea* strain IfB01 received plasmid psTLR7, while its recombinations expressed the detailed dsRNA of whitefly's *TLR7* gene. The *TLR7* gene was silenced by an altered strain, IfB01-TLR7, as well as recovers its virulence against whitefly (Hu and Wu 2016).

The cuticle degrading protease was constitutively overexpressed with enhanced virulence by the prime recombinant organism, constructed to establish whether fungal dissemination within the insect body would be promoted by cuticle disruption and also accelerate eradication. Constitutive expression of Pr1 by *M. robertsii* in *M. sexta* seemingly triggered a host protease and reduced the survival time (LT_{50}) by 25% owing to massive melanization in the body cavity, instead of fungal dissemination (St. Leger et al. 1996). *B. bassiana* chitinase CHIT1 overproduction condensed the survival time of hosts by 23% (Fang et al. 2005).

Virulence improvement of other entomopathogenic fungi like *M. robertsii* and *B. bassiana*, was achieved by pathogenicity-related gene expression, for instance, *Pr1A* (Gongora 2004). In some cases, virulence was synergistically improved by cuticle degrading enzyme combinations. Accordingly, overexpression of both Pr1 protease BbCDPE1 (*B. bassiana* Pr1 like protease) and Bbchit1 chitinase by engineered *B. bassiana* strain can reduce the time to kill as well as the lethal dose when compared to their individual expression (Fan et al. 2007). Virulence of *B. bassiana* compared to wildtype and those that overexpress native protease alone was explicitly increased, and the fusion of *B. morichitin*-binding domain to chitin-bound Pr1A protease BbCDEP1 released the superior number of peptides from the cuticles of insect (Lovett and St. Leger 2018).

9.2.2 Improve Heat Tolerance

A major limitation to commercialize and substantial application of mycoinsecticides was the changeability in field performance (de Faria and Wraight 2007). A prevalent issue that arises due to solar UV radiation is the reduced fungal viability in the field (Ignoffo and Garcia 1992; Jaronski 2010). Various UV protectants were assessed for the formulation of mycoinsecticides (Jackson et al. 2010). Genetic modification can enhance the tolerance of insect-pathogenic fungi towards stress. UV tolerance and heat shock protein 25 in *M. robertsii* were increased by overexpression of photolyase (Fang and St Leger 2012), as well as improvement in their tolerance to heat, osmotic pressure, and similar stresses (Liao et al. 2014; Lovett and St Leger 2014). Under natural conditions, melanin is not produced by *Metarhizium* spp. However, metabolically engineered *M. anisopliae* can express dihydroxy naphthalene-melanin biosynthesis genes and was effective against Diamond Back Moth (DBM) (Rangel et al. 2008; Lovett et al. 2019).

Stress conditions, including desiccation, cold, heat, and UV radiation, could be withstood by the transformed strain when compared to wildtype (Cheng et al. 2011). A single tyrosinase gene could not be solely correlated with fungal UV tolerance, and these UV resistant mechanisms were regulated by genes involved in DNA repair and photoreactivation (Thomas et al. 2003). Accordingly, the transformation of entomopathogenic fungi with functionally verified gene classes that could improve their UV resistance.

9.2.3 *Synthetic Genes*

Diverse functional genes which are hybrids of different activities, obtained by a vast diversity of protein and peptide sequences already present in nature. Interestingly, chitin-binding domains were seemingly lacking in the chitinase enzyme from *B. bassiana*. Fan et al. (2007) produced many hybrid chitinase enzymes of *B. bassiana* which were amalgamated into the chitin-binding domain which derived from diverse sources such as bacterial, plant and insects. Synthetic chitinase enzyme derived from mulberry silkworm, chitinase of *B. mori* fused with chitinase of *B. bassiana* revealed strong binding with the chitin when compared with another hybrid chitinase. The time to death was 23% reduced by synthetic chitinase gene expression of white muscardine fungus, *B. bassiana* based on its enhanced virulence when compared to wild fungi.

9.3 Entomopathogenic Bacteria

Bacterial entomopathogens and/or their toxins must be ingested by insects and multiply or are activated to initiate disease after they enter their alimentary tract. The midgut cells were targeted by the released bacterial toxins and other virulence factors (enzymes) and penetrate through the main body cavity by disrupting the epithelial barrier. Septicemia occurs due to bacterial proliferation in the hemocoel, which kills the infected host. Insect hosts were killed by both Gram-positive and Gram-negative bacteria, yet most microbial-based insect control is dependent on Gram-positive, spore-forming bacteria of the genus *Bacillus*. For this group, microbial insect pest control was dominated by the most successful microbial pesticide till date, *Bacillus thuringiensis* (Bt). Transgenic Bt crops developed by cloning and transformation of the crystal (Cry) and vegetative insecticidal protein (Vip) toxin genes from Bt, which revolutionized pest control and the agricultural landscape. Constitutive production of Bt toxins by the Bt crops, protect them from insect attack. Commercially developed alternative entomopathogenic bacteria include gram-positive *B. sphaericus, Paenibacillus popilliae*, and gram-negative bacteria of the genus *Serratia*. Current research goals towards bacterial pesticides include the search for novel pathogens and toxins and the development of modified toxins to increase efficacy and extend the activity range of these microbial insect control technologies.

The main aim of genetic engineering is to increase their rapidity of a kill, chiefly by the combination of genes that encode arthropod or bacterially derivative insect-selective toxins, insect hormones, and enzymes.

9.3.1 Improved Virulence

Recombinant DNA technology could introduce several *Cry* genes with desired insecticidal activities to construct novel Bt strains (Azizoglu et al. 2020). Recombinant or resident plasmids could be used to maintain cloned *Cry* genes in *Bt*, or it can be stably integrated into the chromosome by homologous recombination in vivo. Obukowicz et al. (1987), transferred *Cry* gene from *B.t.* subsp. *kurstaki* HD-1 to corn root-colonizing *Pseudomonas fluorescens* strain to develop pesticide efficiency similar to that of *B.t.* subsp. *kurstaki* HD-1 against black cutworm (*Agrotis ipsilon*) using transposon Tn5.

Bt majorly requires genetic development with reverence for strain improvement before expressing into the effective biological pesticide. The *Bt* strains can be combined with the *Cry* toxins, which target a variety of pests and also escalate toxicity and by binding to different sites, the onset of pest-resistance with different modes of action could be delayed. Strain development is mainly tried to incorporate different Cry genes or increase their copy number because total *Bt* strain bioactivity depends upon additive or synergistic interactions of *Cry* proteins present in their comparative amounts (Caccia et al. 2020). New *Bt* strains have been produced through conjugation with additional *Cry* genes over the native strains (Wiwat et al. 1995; Hu et al. 2004). The trans-conjugant gene cry3A is highly expressed in *B.t.* subsp. *kurstaki* HD119, which is a native of *B.t.* subsp. *tenebrionis*. It is expressed effectively without affecting the native gene expression of *B.t.* subsp. *kurstaki* HD119. Sub-cloning the nucleotide sequences alone is required for vector replication in *B.t.* and can achieve compatibility and stability of the vector with native *B.t.* plasmids (Gamel and Piot 1992). The combination of *Cry* genes into the chromosome of the desired recipient strain is the method for strain improvement apart from conjugation. Through phage CP-54 Ber-mediated transduction, a cry1Aa gene was transferred into selected *Bt* strains. A copy of the *cryIA*(*a*) δ-endotoxin gene borne by both shuttle vector pHT3101 and its derivative pHT408, transferred into several subspecies of *Bacillus thuringiensis* through phage CP-54 Ber-mediated transduction which showed greater activity against DBM, *Plutella xylostella*. The dual-specificity displayed by transductants were constructed using novel isolates LM63 and LM79 as recipients, with larvicidal activity against insects of the order Coleoptera. The cloned toxin gene *cryIIIA* was expressed in high levels by pHT7911 in transformants indicating electro transformation was successful. Different recipients established that pHT3101 is a very good expression vector for cloned δ-endotoxin genes. The introduction of cloned crystal protein genes into various *B. thuringiensis* receivers was by CP-54Ber-mediated transduction, thereby creating strains with new mixtures of the gene (Lecad et al. 1992). Through the transducing phage, the *Cry* genes can be chromosomally integrated with the wide range of the insecticidal spectrum. By electroporation method, the cry1C gene was integrated into the chromosome of *B.t.* subsp. *kurstaki* HD73 from *aizawai* (Kalman et al. 1995). Using protein engineering, the expression of *Cry* proteins insecticidal toxicity can be increased in the *Bt* strains to counteract resistance to pests. The expression of the

Cry gene can be enhanced by the alternation of the regulatory elements in the gene. Moreover, the manipulation of the controlling elements such as the promoter can improve the yield of *Cry* protein in naturally occurring strain. Dual cyt1Aa promoters along with a STAB-SD sequence stabilized cry3A transcript-ribosome complex, resulted in a many-fold expression increase of *cry3Aa* gene (Park et al. 1998). Delivery of RNA bio-pesticides could necessarily be provided by alternative ways of using this technology in the field. Genes would be silenced by non-transgenic RNAi-based products without any hereditary changes introduced, and not regulated as GM products, with their release and approval probably within a shorter timeframe (Cagliari et al. 2018). When Lepidopteran cotton bollworm (*Helicoverpa armigera*) larvae were presented with an artificial diet coated with engineered bacteria for 5 days, drastic reductions in body weight, body length, and pupation rate were observed, manifested by their high mortality and inhibition of target gene expression (Ai et al. 2018).

RNAi interference (RNAi) is used to regulate genes encoding vital functions by silencing them. In the noctuid moth *Spodoptera littoralis,* encapsulation and nodulation responses were controlled by the target gene, *Sl 102,* by combined oral administration of ds RNA-Bac artificial diet and *Bt*-based biopesticide (Xentari™). These new *Bt* spray formulations comprising kill ds RNA-Bac, overpowered the insect's immune reaction by synergizing *Bt* toxins. For numerous organisms, a revolution in genome engineering was ignited by CRISPR/Cas9 tools. A vast potential to modify organisms as desired was suggested by CRISPR/Cas9 technology which intended gene editing in a precise manner. It is a potential molecular technique for future implementation towards most aspects of filamentous fungi (Wang and Coleman 2019).

9.3.2 Genetic Regulation of Cry and Cyt Protein Synthesis

Major genetic factors controlling endotoxin synthesis in Bt are promoters, 5′ mRNA stabilizing and 3′ transcriptional termination sequences. Another factor that limits synthesis is the comparative stability of each Bt endotoxin. There are two strong sporulation-dependent promoters *viz*., *Bt* I and *Bt* II which control the Bt endotoxin production. *Bt* I and *Bt* II transcription are strongly regulated by sigma −35 and sigma −28 complexed with RNA polymerase. Although, *Bt* promoters characteristically synthesize Cry 3A, under a weak promoter and active during the vegetative growth stage. mRNA stabilizing 9 nucleotide sequences called STAB-SD, in 5′ cry3A transcript helps in the moderate level synthesis of Cry3A. Expression constructs for several proteins obtained by splicing and influenced by *Bt* sporulation dependent promoters, can increase endotoxin synthesis by ten times. The half-life of the transcript is increased, resulting in greater synthesis of endotoxins than would take place in the absence of these terminators. During or after translation, the synthesis of *Bt* endotoxins was increased by several other factors. For instance, the net synthesis of many endotoxins was increased by a 20-kDa protein that was encoded by the cry11A operons third ORF, which evidently serves as a chaperone. The cry2Aa operon encodes a 29 kDa protein that promotes crystallization and consequently the yield of Cry 2A. Finally, the stability of various endotoxin proteins is different in that some are far more stable (Cry3A) than others (Cry4A). In general, a more stable protein presents a higher yield when synthesized using expression vectors at high levels.

9.3.3 Safety Concerns About the Wild Type and Recombinant Bacterial Insecticides

Previous experiments largely focused on the evaluation of chemical insecticides in the assessments of the kinds of tests conducted to determine the safety of bacterial insecticides. The assessments, however, have been developed over decades and intended to determine the risks associated with *Bt*, especially bacterial infectivity and protein toxicology were used as active ingredients. Three tiers (I–III) of tests were conducted. Tier I evaluates if isolation of Bt subspecies as an unformulated substance, in a series of tests, poses a danger to various groups of non-target organisms when used at high levels, usually 100 times the recommended amount for field use. Acute oral, pulmonary (inhalation) and intraperitoneal content assessments are key tests of the substance against various species of vertebrates, that last from 1 week and can extend more than a month, which depends on the organism. In most crucial experiments, Bt cells either in a vegetative or sporulated state are fed, injected or induced to be inhaled by mammals. The experiments involving invertebrates are mainly studies of feeding and touch. The organisms such as mice, rats, rabbits, guinea pigs, numerous species of birds, fish, predatory and parasitic insects, beneficial insects such as the honeybee, aquatic and marine invertebrates, and plants are representative of non-target vertebrates and invertebrates. If any of these tests demonstrate strong infectivity or acute toxicity, the bacterium will be rejected. Tier-II checks must be performed if any uncertainty persists, though similar to Tier I tests, require several consecutive exposures, particularly species with evidence of toxicity or infectivity as determined by Tier I tests, along with those to assess whether and if bacteria have been extracted from non-target tissues. Tier III tests must be performed if any infectivity, toxicity, mutagenicity, or teratogenicity are observed. These include experiments such as 2-year feeding trials, additional teratogenicity, and mutagenicity monitoring.

9.4 Entomopathogenic Viruses

Baculoviruses are large, occluded, double-stranded DNA viruses isolated largely from Lepidoptera. For insect pest management purposes, baculoviruses have several beneficial characteristics such as host specificity and environmental safety and can be applied without harming the non-target organisms, a key detriment to the current use of classical chemical insecticides. However, the major limitation of its application is that it endures unstability under sunlight and longer time to eradicate the hosts. According to the host-virus combination, baculovirus exerts between a few days to several weeks to succumb to the infection. These limitations has been successfully tackled with genetic engineering by using one or more of three different strategies: (1) removal of a baculovirus gene, (2) insertion of genes that encode a toxin, hormone or enzyme, (3) supplement gut-active toxins into the obstruction body.

Baculoviruses (family Baculoviridae) have two phenotypes. Baculoviruses are found in fields as occlusion bodies (OB) designated as polyhedra and granules for the genera Nucleopolyhedrovirus (NPV) and Granulovirus (GV), respectively. The polyhedrin or granulin template of the OB dissolves in the gut with its alkaline medium upon ingestion by a suitable insect host and releases Occlusion Derived Virus (ODV). ODV infiltrates the peritrophic membrane that lines the gut and infects the midgut epithelial cells. ODV is responsible for a primary infection called the first phenotype. The early progress of replications in the infected cell produces the second phenotype called Budded Virus (BV).

9.4.1 Improving Insect Control

9.4.1.1 Gene Deletion

The baculovirus gene, *egt,* which encodes *ecdysteroid UDP-glycosyltransferase* (EGT), catalysis the conjunction of sugar particles to 20-hydroxy ecdysone (moulting hormone) renders ecdysteroids inactive. The molting of the infected host is inhibited by *egt* expression. The *egt* functions to extend the larval feeding and resulted from an increase in weight after the infection. This *egt* gene is classified as an auxiliary gene, i.e., a gene that is not essential for baculovirus repetition but provides some competitive advantage. The *egt* gene deletion in baculovirus increases the viral pesticide effectiveness significantly. Reduced feeding and earlier mortality was showed by *egt* deleted mutant of AcMNPV infected larvae than the wildtype AcMNPV infected larvae (Bianchi et al. 2000). The same approach also decreased the survival time of larvae infected with *Lymantria dispar* MNPV and *H. armigera* NPV. However, the impact of *egt* deletion on survival time was not consistent between hosts and varied according to the stage used for bioassay.

9.4.1.2 Gene Insertion

The most common approach to increase the speed of killing baculovirus insecticides is to insert genes encoding toxin, hormone, or enzyme into the baculovirus genome. During viral replication in insect host, the toxic agent is expressed that negatively affects the insect cell. An extensive range of genes that encode insect-specific toxins extracted from various sources such as scorpions, spiders, parasitic wasps, and sea anemones were inserted into the baculovirus genomes (Harrison and Bonning 2000). These toxins act on the major ion channels including Na^+, K^+, Ca^{2+}, and Cl^- channels leading to rapid paralysis of host. The excitatory and depressant toxins of the yellow Israeli scorpions of the genus, *Leiurus* have received particular attention. Field testing of recombinant baculoviruses expressing these toxins conducted across the world, and some are competitive with the synthetic pyrethroid insecticides. A recombinant baculovirus insecticide was constructed expressing cathepsin-L, a protease targeting basement membrane and kills insects significantly faster than the wildtype virus.

9.4.2 Improving Virulence

The establishment of improved cell lines is a major area for the recognition and advancement of new strains of baculoviruses. From lepidopteran species, it has been proved that the majority of cell lines act as a precious medium for baculovirus propagation in vitro. The success of Baculoviruses in cell culture depends on virus-host interactions, including pathogenicity, host range, virulence, and latency (Granados et al. 1987; Battu et al. 2002). Attempts for replication of Granulovirus (GV's) in established cell lines or primary organ cultures were minimal in certain cases with no success before 1984. The reason behind this minimal success is based on lack of appropriate viral receptors or vanished host enzymes necessary for virus propagation. In Germany, first in vitro replication of Granulovirus (CpGV) in primary cell lines from fruitfly, *Cydia pomonella* was reported by Miltenburger et al. (1984). Establishment of several new semi-loopers, *T. ni* cell lines, susceptible to TnGV was a major advancement (Granados et al. 1987).

In recombinant technology, baculovirus transfer vector was initially incorporated with the foreign gene. The gene-inserted plasmid propagates in *Escherichia coli* and expression is driven by the selected promoter with the plasmid incorporated into a cloning site downstream. The transfer vector is then mixed with DNA from wildtype baculovirus. Homologous recombination events within the nucleus of cultured insect cells were used to incorporate the engineered DNA into the virus. Precise insertion of foreign DNA without disrupting other genes was allowed by the baculovirus system and randomly incorporates new DNA into its genome.

Through recombinant DNA technology, it is possible to overcome various defects of naturally occurring baculoviruses while maintaining or enhancing their desirable

Table 9.2 Recombinant baculovirus source of toxin and host range

Toxin	Source of toxin	Virus name	Host tested	References
TXP1	Straw itch mite—*Pyemotes tritici*	Vsp-Tox34 Vp6.9tox34 HzEGTGA-26tox34	*Trichoplusia ni* *T. ni* *H. zea*	Tomalski and Miller (1991) Tomalski and Miller (1992) Popham et al. (1997)
AaIT	North African scorpion—*Androctonus australis*	Acst 3 AcAaIT BmAaIT	*T. ni* *Heliothis virescens*	Stewart et al. (1991) Martens et al. (1995) Maeda et al. (1991)
LqhIT1 LqhIT2 LqhIT3	Scorpion—*Leiurus quinquestriatus*	AcLIT1.p10 AcLIT2.p10 AcLqhIT3	*O. nubilalis* *H. armigera* *B. mori*	Harrison and Bonning (2000) Regev et al. (2003) Imai et al. (2000)
As II	Sea anemone—*Anemonia sulcata*	vSAt2p+	*T. ni*	Popham et al. (1998)
μ-Aga-rv	*Agelenopsis sulcata*	vMAgap+	*S. frugiperda* *T. ni*	Popham et al. (1998)
DTX9.2	Primitive weaving spider—*Diguetia canites*	vAcDTX9.2	*T. ni* *H. virescens*	Hughes et al. (1997)

pest-specific characteristics. Genetic manipulation of baculoviruses can be achieved by the exchange of genetic material between different baculoviruses or the addition of alien genes into the baculovirus genome. This genetic alteration of entomopathogenic viruses can help to increase the speed of killing. *Helicoverpa zea* nucleopolyhedrovirus, commercially produced as a pesticide (Elcar), enhanced the speed to kill when it was genetically improved. A greater reduction in time-lapse for a virus to kill larvae was achieved by the insertion of a potent insect-selective neurotoxin gene derived from mite, *tox 34* into viral gene encoding ecdysteroid UDP-glucosyltransferase resulted in 50% mortality or paralysis within 40 h after virus treatment. Genetic capability to improve baculovirus properties as a pesticide promotes the production of a commercially feasible product for the control of insect pests (Holly et al. 1997). Under field conditions, genetically modified baculovirus kills resulted in minimized crop damage compared to wild type. Successful field trials of a genetically engineered nucleopolyhedrovirus of alfalfa looper, *Autographa californica* (AcNPV) expressing an insect-specific toxin (AaIT) derived from the venom of the North African scorpion *Androctonus australis* have been reported (Cory et al. 1994; Kroemer et al. 2015) (Table 9.2). AaIT neurotoxin disrupts the ion conductance across neuron axonal membranes. Kontogiannatos et al. (2013) established that a recombinant BmNPV baculovirus, expressing juvenile hormone esterase-specific hairpin, induced gene-specific knockdown and associated morphologic effects in the Mediterranean corn borer, *Sesamia nonagrioides*.

9.5 Entomopathogenic Nematodes

Entomopathogenic nematodes belonging to Steinernematidae and Heterorhabditidae are intensely used against insects that inhabit soil as biological alternatives to chemicals. The entomopathogenic nematode-bacterium complex *Heterorhabditis bacteriophora* (Poinar)-*Photorhabdus luminescens* (Thomas and Poinar) targets insect pests in the soil and is a safe biological control agent. The nematode-bacterium symbiotic complex (*S. feltiae* and *Xenorhabdus bovienii*) is mass-produced in industrial-scale bioreactors. Short shelf life of approximately 6 weeks causes limits its use in larger-scale plant protection markets, despite all its advantages. Strain improvement programs successfully employed classical genetics techniques including mutagenesis, hybridization and artificial selection. In classical genetics, the routinely used and well-established technique is crossbreeding. The tests were carried out by injecting the infective juveniles of *Steinernema* into greater wax moth, *Galleria melonella* (Poinar 1967). In case of *Heterorhabditis*, second generation amphimictic virgin females are selected and transferred along with males into cadavers of *G. melonella*. In areas of molecular diagnosis and phylogeny utilize molecular genetics, though not widely applied. The whole genome of *Caenorhabditis elegans* is fully sequenced and can be used as a model system in understanding the basic and applied aspects of molecular genetics.

9.5.1 *Improvement of Shelf Life*

The shelf life of entomopathogenic nematode (EPN), *Steinernema feltiae* is relatively inadequate in field exposed condition. During storage, rate of EPN metabolism must be cut down for shelf-life enhancement by decreasing temperature or exposure of third stage Dauer Juveniles (DJ's) to desiccation. Desiccation tolerance limited to *S. feltiae* and can be augmented by adapting to desiccation conditions moderately. Cross strain, HYB01 added effectively at low temperatures from parental strains and other hybrids with AT_{50} and AT_{10} of 0.52 °C and 0.09 °C, respectively. Nimkingrat et al. (2013) studied genetically selected strain for their virulence and reproductive ability and was tested with *C. pomonella* cocooned larvae by cross-hybridization and selective breeding. Genetic engineering seemingly strained the genetic enhancement of entomopathogenic nematodes. Transmission of a heat shock protein generating gene from *C. elegans* into *H. bacteriophora* (HP88) and its subsequent survival increased from 3 into 90% after heat treatment for 1 h at 40 °C. One of the most important attempts to create transgenic EPNs was the insertion of heat shock gene, *hsp70A* of *C. elegans* in to *H. bacteriophora*. However, first reported genetic transformation of *H. bacteriophora* and insertion of this gene by microinjection contributes to a considerable enhancement in heat stress resistance (Hashmi et al. 1995). Superior field efficiency of *H. bacteriophora* was mostly based on the conservation of 19 genes of insulin/IGF-1 signaling pathway and capable genetic employment candidates to augment the durability and/or stress resistance (Bai et al. 2013).

9.5.2 *Improvement of Virulence*

The G-protein-coupled-receptor (GPCR) gene family is essential for the advancement of *H. bacteriophora* with at least 82 predicted GPCRs in its genome. A few sensory receptors involving olfaction and host-seeking behavior due to the field efficacy of GPCRs. Putative olfactory receptors in EPNs are capable of increasing seeking or adjusting the specificity of the host. EPN host-seeking behavior utilizes GPCR, which is crucial for its enhancement or alteration and determines its field efficacy (Bai et al. 2013).

Protease and protease inhibitors are imperative in attack and killing of host insects by *Steinernema*. A crucial role was played by proteases in destroying immunity of insect host as well as deprivation of their tissue (Dillman et al. 2015). The steinernematid genome contains a fascinating gene family encoding the fatty acid and retinol-binding (FAR) proteins. An addition to suppression and evasion of the immune system is the FAR proteins with a pivotal role in parasitism through sequestration of retinoids of host insects (Dillman et al. 2015).

Trait improvement in EPNs have two major advantages, namely, artificial selection and genetic engineering. The genetic mechanism's original selected traits were not necessarily considered for artificial selection. Genomic tools can now be used to appreciate the genetic changes obtained for the chosen traits by domestication and artificial selection. Genetic engineering and acquiesced knowledge could achieve improved traits or improved groupings in several biological systems (Kole et al. 2015). Several traits and genomic tools assist in enhancing the effectiveness of EPNs in categorizing and modifying genes encoding required characteristics, reducing EPN generation time both in vitro and in vivo with refined capability and diverse assembly of species and strains to select from a rich EPN genetic diversity.

9.6 Future Prospects

In the existent scenario of crop production, bio-control agents like entomopathogens are of utmost importance, but the major drawbacks concerning its potential under extreme environmental conditions should be further studied. Commercially produced entomopathogens have not been used efficiently due to their sensitivity to field conditions. Most bioagents are well implemented under laboratory conditions but do not accomplish that efficaciousness when applied to soil or target pests. The versatility of bioagents seems to be restricted by physiological and ecological restrictions. Numerous methods that can underwrite the escalation of the competence of bioagents comprise mutation or polyethylene glycol-promoted protoplasm fusion.

To confound this problem, the selection and characterization of biocontrol agents could be improved through genetic engineering and supplementary molecular tools. Stabilization of choice improvement will be the chief challenge to enable profitable manipulation of genetic enhancement.

References

Ai X, Wei Y, Huang L et al (2018) Developmental control of *Helicoverpa armigera* by ingestion of bacteria expressing dsRNA targeting an arginine kinase gene. Biocontrol Sci Tech 28:253–267

Azizoglu U, Jouzani GS, Yilmaz N et al (2020) Genetically modified entomopathogenic bacteria, recent developments, benefits and impacts: a review. Sci Total Environ 24:98–120

Bagga S, Hu G, Screen SE, St Leger RJ (2004) Reconstructing the diversification of subtilisins in the pathogenic fungus *Metarhizium anisopliae*. Gene 324:159–169. https://doi.org/10.1016/j.gene.2003.09.031. PMID: 14693381

Bai X, Adams BJ, Ciche TA et al (2013) A lover and a fighter: the genome sequence of an entomopathogenic nematode *Heterorhabditis bacteriophora*. PLoS One 8:e69618

Bailey AM, Kershaw MJ, Hunt BA, Paterson IC, Charnley AK, Reynolds SE, Clarkson JM (1996) Cloning and sequence analysis of an intron-containing domain from a peptide synthetase encoding gene of the entomopathogenic fungus *Metarhizium anisopliae*. Gene 173:195–197. https://doi.org/10.1016/0378-1119(96)00212-0

Baratto CM, Silva MV, da Santi L, Passaglia L, Schrank IS, Vainstein MH, Schrank A (2003) Expression and characterization of the 42 kDa chitinase of the biocontrol fungus *Metarhizium anisopliae* in *Escherichia coli*. Can J Microbiol 49:723–726. https://doi.org/10.1139/w03-085. PMID: 14735222

Baratto CM, Dutra V, Boldo JT, Leiria LB, Vainstein MH, Schrank A (2006) Isolation, characterization, and transcriptional analysis of the chitinase chi2 gene (DQ011663) from the biocontrol fungus *Metarhizium anisopliae* var. *anisopliae*. Curr Microbiol 53:217–221. https://doi.org/10.1007/s00284-006-0078-6. PMID: 16874542

Battu GS, Arora R, Dhaliwal GS (2002) Prospects of baculoviruses in integrated pest management. In: Koul O, Dhaliwal GS (eds) Microbial biopesticides. Taylor and Francis, London, pp 215–238

Bianchi FJ, Snoeijing I, van der Werf W et al (2000) Biological activity of SeMNPV, AcMNPV and three AcMNPV deletion mutants against *Spodoptera exigua* larvae (Lepidoptera: Noctuidae). J Invertebr Pathol 75:28–35

Bidochka MJ, Kamp AM, Lavender TM, Dekoning J, De Croos JNA (2001) Habitat association in two genetic groups of the insect pathogenic fungus *Metarhizium anisopliae*: uncovering cryptic species? Appl Environ Microbiol 67:1335–1342. https://doi.org/10.1128/AEM.67.3.1335-1342.2001. PMID: 11229929. PMCID: 92732

Bogo MR, Rota CA, Pinto H Jr, Ocampos M, Correa CT, Vainstein MH, Schrank A (1998) A chitinase encoding gene (chit1 gene) from the entomopathogen *Metarhizium anisopliae*: isolation and characterization of genomic and full-length cDNA. Curr Microbiol 37:221–225. https://doi.org/10.1007/s002849900368. PMID: 9732526

Caccia S, Astarita F, Barra E et al (2020) Enhancement of *Bacillus thuringiensis* toxicity by feeding *Spodoptera littoralis* larvae with bacteria expressing immune suppressive dsRNA. J Pest Sci 93:303–314

Cagliari R, Kremer FS, da Silva PL (2018) Bauhinia lectins: biochemical properties and biotechnological applications. Int J Biol Macromol 119:811–820

Chelico L, Haughian JL, Khachatourians GG (2006) Nucleotide excision repair and photoreactivation in the entomopathogenic fungi *Beauveria bassiana*, *B. brongniartii*, *B. nivea*,

Metarhizium anisopliae, Paecilomyces farinosus, and *Verticillium lecanii*. J Appl Microbiol 100:954–972. https://doi.org/10.1111/j.1365-2672.2006.02844.x. PMID: 16629997
Chen X, Li L, Hu Q et al (2015) Expression of dsRNA in recombinant *Isaria fumosorosea* strain targets the TLR7 gene in *Bemisia tabaci*. BMC Biotechnol 15:1
Cheng LV, Huang B, Qiao M, Wei J, Ding B (2011) Entomopathogenic Fungi on *Hemiberlesia pitysophila*. PLoS One 6(8):e23649. https://doi.org/10.1371/journal.pone.0023649
Cory JS, Hirst ML, Williams T et al (1994) Field trial of a genetically improved baculovirus insecticide. Nature 370:138–140
de Faria MR, Wraight SP (2007) Mycoinsecticides and Mycoacaricides: a comprehensive list with worldwide coverage and international classification of formulation types. Biol Control 43:237–256
Dhanapal R, Kumar DVSR, Lakshmipathy R, Sandhya Rani C, Manoj Kumar R (2019) Pathogenicity testing of indigenous isolates of entomopathogenic fungus, *Lecanicillium lecanii* against tobacco caterpillar, *Spodoptera litura*. J Exp Zool India 22(2):753–756
Dhanapal R, Kumar DVSR, Lakshmipathy R, Sandhya Rani C, Manoj Kumar R (2020a) Exploration of indigenous strains of the green muscardine fungus from soils and their pathogenicity against the tobacco caterpillar, *Spodoptera litura* (Fabricius) (Lepidoptera: Noctuidae). Egypt J Biol Pest Control 30:34. https://doi.org/10.1186/s41938-020-00239-8
Dhanapal R, Kumar DVSR, Lakshmipathy R, Sandhya Rani C, Manoj Kumar R (2020b) Isolation of indigenous strains of the white halo fungus as a biological control agent against 3rd instar larvae of tobacco caterpillar, *Spodoptera litura* (Fabricius) (Lepidoptera: Noctuidae). Egypt J Biol Pest Control 30:89. https://doi.org/10.1186/s41938-020-00286-1
Dillman AR, Macchietto M, Porter CF et al (2015) Comparative genomics of *Steinernema* reveals deeply conserved gene regulatory networks. Genome Biol 16(1):200
Fan Y, Fang W, Guo S, Pei X, Zhang Y, Xiao Y, Li D, Jin K, Bidochka MJ, Pei Y (2007) Increased insect virulence in *Beauveria bassiana* strains overexpressing an engineered chitinase. Appl Environ Microbiol 73(1):295–302. https://doi.org/10.1128/AEM.0197406. PMID: 17085713. PMCID: 1797141
Fang WG, St Leger RJ (2012) Enhanced UV resistance and improved killing of malaria mosquitoes by photolyase transgenic entomopathogenic fungi. PLoS One 7(8):e43069
Fang W, Leng B, Xiao Y, Jin K, Ma J, Fan Y, Feng J, Yang X, Zhang Y, Pei Y (2005) Cloning of *Beauveria bassiana* chitinase gene *Bbchit1* and its application to improve fungal strain virulence. Appl Environ Microbiol 71:363–370. https://doi.org/10.1128/AEM.71.1
Gamel PH, Piot JC (1992) Characterization and properties of a novel plasmid vector for *Bacillus thuringiensis* displaying compatibility with host plasmids. Gene 120:17–26
Gongora CE (2004) Transformacion de *Beauveria bassiana* cepa Bb9112 con les genes de la proteina verde fluorescente y la protease pr1A de *M. anisopliae*. Rev Colomb Entomol 30:1–5
Granados RR, Dwyer KG, Derksen ACG (1987) Production of viral agents in invertebrate cell cultures. In: Maramorosch K (ed) Biotechnology in invertebrate pathology and cell culture. Academic, San Diego, pp 167–181
Gressel J (2001) Potential failsafe mechanisms against the spread and introgression of transgenic hypervirulent biocontrol fungi. Trends Biotechnol 19:149–154
Hajek AE (2004) Natural enemies: an introduction to biological control. Cambridge University Press, Cambridge
Harrison RL, Bonning BC (2000) Use of scorpion neurotoxins to improve the insecticidal activity of *Rachiplusia ou* multicapsid nucleopolyhedrovirus. Biol Control 17:191–201. https://doi.org/10.1006/bcon.1999.0792
Hashmi S, Hashmi G, Gaugler R (1995) Genetic transformation of an entomopathogenic nematode by microinjection. J Invertebr Pathol 66:293–296
Holly JR, Li PY, Lois K et al (1997) Genetic improvement of *Helicoverpa zea* nuclear polyhedrosis virus as a biopesticide. Biol Control 10(2):83–91
Hu Q, Wu W (2016) Recombinant fungal entomopathogen RNAi target insect gene. Bioengineered 7(6):504–507

Hu X, Hansen BM, Eilenberg J et al (2004) Conjugative transfer, stability and expression of a plasmid encoding a cry1Ac gene in *Bacillus cereus* group strains. FEMS Microbiol Lett 231:45–52

Hughes PR, Wood HA, Breen JP, Simpson SF, Duggan AJ, Dybas JA (1997) Enhanced bioactivity of recombinant baculoviruses expressing insect-specific spider toxins in lepidopteran crop pests. J Invertebr Pathol 69:112–118. https://doi.org/10.1006/jipa.1996.4643

Ignoffo CM, Garcia C (1992) Influence of conidial colour on inactivation of several entomogenous fungi by simulated sunlight. Environ Entomol 21:913–917

Imai N, Ali SE, El-Singaby NR, Iwanaga M, Matsumoto S, Iwabuchi K, Maeda S (2000) Insecticidal effects of a recombinant baculovirus expressing scorpion toxin LqhIT2. J Seric Sci Jpn 69:197–205

Jackson MA, Dunlap CA, Jaronski ST (2010) Ecological considerations in producing and formulating fungal entomopathogens for us in insect control. BioControl 55:129–145

Jaronski ST (2010) Ecological factors in the inundative use of fungal entomopathogens. BioControl 55:159–185

Joshi L, St Leger RJ (1999) Cloning, expression, and substrate specificity of MeCPA, a zinc carboxypeptidase that is secreted into infected tissues by the fungal entomopathogen *Metarhizium anisopliae*. J Biol Chem 274:9803–9811. https://doi.org/10.1074/jbc.274.14.9803. PMID: 10092670

Joshi L, St. Leger RJ, Bidochka MJ (1995) Cloning of a cuticledegrading protease from the entomopathogenic fungus, *Beauveria bassiana*. FEMS Microbiol Lett 125:211–217. https://doi.org/10.1111/j.1574-6968.1995.tb07360.x. PMID: 7875568

Joshi L, St Leger RJ, Roberts DW (1997) Isolation of a cDNA encoding a novel subtilisin-like protease (Pr1B) from the entomopathogenic fungus, *Metarhizium anisopliae* using differential display-RT-PCR. Gene 197:1–8. https://doi.org/10.1016/S0378-1119(97)00132-7

Junaid JM, Dar NA, Bhat TA et al (2013) Commercial biocontrol agents and their mechanism of action in the management of plant pathogens. Int J Mod Plant Anim Sci 1(2):39–57

Kalman S, Kiehne KL, Cooper N et al (1995) Enhanced production of insecticidal proteins in *Bacillus thuringiensis* strains carrying an additional crystal protein gene in their chromosomes. Appl Environ Microbiol 61:3063–3068

Kang SC, Park S, Lee DG (1998) Isolation and characterization of a chitinase cDNA from the entomopathogenic fungus, *Metarhizium anisopliae*. FEMS Microbiol Lett 165:267–271. https://doi.org/10.1111/j.1574-6968

Kim HK, Hoe HS, Suh DS, Kang SC, Hwang C, Kwon ST (1999) Gene structure and expression of the gene from *Beauveria bassiana* encoding bassiasin I, an insect cuticle-degrading serine protease. Biotechnol Lett 21:777–783. https://doi.org/10.1023/A:1005519323748

Kole C, Muthamilarasan M, Henry R et al (2015) Application of genomics-assisted breeding for generation of climate resilient crops: progress and prospects. Front Plant Sci 6:563

Kontogiannatos D, Swevers L, Maenaka K et al (2013) Functional characterization of a juvenile hormone esterase related gene in the moth *Sesamia nonagrioides* through RNA interference. PLoS One 8:e73834

Kroemer JA, Bonning BC, Harrison RL (2015) Expression, delivery and function of insecticidal proteins expressed by recombinant baculoviruses. Viruses 7:422–455

Lecad MM, Chaufaux J, Ribier J et al (1992) Construction of novel *Bacillus thuringiensis* strains with different insecticidal specificities by transduction and by transformation. Appl Environ Microbiol 58:840–849

Liao X, Lu HL, Fang W et al (2014) Over expression of a *Metarhizium robertsii* HSP25 gene increases thermotolerance and survival in soil. Appl Microbiol Biotechnol 98(2):777–783

Lomer CJ, Bateman RP, Johnson DL et al (2001) Biological control of locusts and grasshoppers. Annu Rev Entomol 46:667–702

Lovett B, St Leger RJ (2014) Stress is the rule rather than the exception for *Metarhizium*. Curr Genet 61:253–261

Lovett B, St. Leger RJ (2018) Genetically engineering better fungal biopesticides. Pest Manag Sci 74(4):781–789

Lovett B, Bilgo E, Diabate A, St. Leger RJ (2019) A review of progress towards field application of transgenic mosquitocidal entomopathogenic fungi. Pest Manag Sci 75:2316–2324

Lu DD, Pava-Ripoll M, Li ZZ et al (2008) Insecticidal evaluation of *Beauveria bassiana* engineered to express a scorpion neurotoxin and a cuticle degrading protease. Appl Microbiol Biotechnol 81:515–522

Maeda S, Volrath SL, Hanzlik TN, Harper SA, Majima K, Maddox DW, Hammock BD, Fowler E (1991) Insecticidal effects of an insect-specific neurotoxin expressed by a recombinant baculovirus. Virology 184:777–780. https://doi.org/10.1016/0042-6822(91)90451-G

Martens JW, Knoester M, Weijts F, Groffen SJ, Hu Z, Bosch D, Vlak JM (1995) Characterization of baculovirus insecticides expressing tailored *Bacillus thuringiensis* CryIA(B) crystal proteins. J Invert Path 66:249–257. https://doi.org/10.1006/jipa.1995.1097

Miltenburger HG, Naser WL, Harvey JP (1984) The cellular substrate: a very important requirement for baculovirus *in vitro* replication. Z Naturforsch Biosci 39:993–1002

Nam JS, Lee DH, Lee KH, Park HM, Bae KS (1998) Cloning and phylogenic analysis of chitin synthase genes from the insect pathogenic fungus, *Metarhizium anisopliae* var. *anisopliae*. FEMS Microbiol Lett 159:77–84. https://doi.org/10.1111/j.1574-6968.1998.tb12844.x. PMID: 9485597

Nimkingrat P, Khanam S, Strauch O et al (2013) Hybridisation and selective breeding for improvement of low temperature activity of the entomopathogenic nematode *Steinernema feltiae*. BioControl 58:417–426

Obukowicz MG, Perlak FJ, Bolten SL et al (1987) IS*50L* as a non-self-transposable vector used to integrate the *Bacillus thuringiensis* delta-endotoxin gene into the chromosome of root-colonizing pseudomonads. Gene 51(1):91–96

Park HW, Ge B, Bauer LS et al (1998) Optimization of cry3A yields in *Bacillus thuringiensis* by use of sporulation dependent promoters in combination with the STAB-SDS mRNA sequence. Appl Environ Microbiol 64:3932–3938

Pava-Ripoll M, Posada FJ, Momen B et al (2008) Increased pathogenicity against coffee berry borer, *Hypothenemus hampei* (Coleoptera: Curculionidae) by *Metarhizium anisopliae* expressing the scorpion toxin (AaIT) gene. J Invertebr Pathol 99:220–226

Poinar GO (1967) Description and taxonomic position of the DD-136 nematode (Steinernematidae, Rhabditoidea) and its relationship to *Neoaplectana carpocapsae* Weiser. Proc Entomol Soc Wash 34:199–209

Popham HJR, Li Y, Miller LK (1997) Genetic improvement of *Helicoverpa zea* nuclear polyhedrosis virus as a biopesticide. Biol Control 10:83–91. https://doi.org/10.1006/bcon.1997.0552

Popham HJR, Pellock BJ, Robson M, Dierks PM, Miller LK (1998) Characterization of a variant of *Autographa californica* nuclear polyhedrosis virus with a nonfunctional ORF 603. Biol Control 12:223–230. https://doi.org/10.1006/bcon.1998.0627

Rangel DEN, Anderson AJ, Roberts DW (2008) Evaluating physical and nutritional stress during mycelial growth as inducers of tolerance to heat and UV-B radiation in *Metarhizium anisopliae* conidia. Mycol Res 112:1362–1372

Regev A, Rivkin H, Inceoglu B, Gershburg E, Hammock BD, Gurevitz M, Chejanovsky N (2003) Further enhancement of baculovirus insecticidal efficacy with scorpion toxins that interact cooperatively. FEBS Lett 537:106–110. https://doi.org/10.1016/S0014-5793(03)00104-2

Routray S, Dey D, Baral S et al (2016) Genetic improvement of natural enemies: a review. Agric Rev 37(4):325–332

Sala A, Barrena R, Artola A et al (2019) Current developments in the production of fungal biological control agents by solid-state fermentation using organic solid waste. Crit Rev Environ Sci Technol 49:655–694

Schrank A, Bassanesi MC, Pinto JH, Costa SV, Bogo MR, Silva MSN (1993) Superoxide dismutases in the entomopathogenic fungus *Metarhizium anisopliae*. Ciência e Cultura 45:200–205

Screen S, Bailey A, Charnley K, Cooper R, Clarkson J (1997) Carbon regulation of the cuticle-degrading enzyme PR1 from *Metarhizium anisopliae* may involve a trans-acting DNA-binding protein CRR1, a functional equivalent of the *Aspergillus nidulans* CREA protein. Curr Genet 31:511–518. https://doi.org/10.1007/s002940050238. PMID: 9211795

Screen S, Bailey A, Charnley K, Cooper R, Clarkson J (1998) Isolation of a nitrogen response regulator gene (nrr1) from *Metarhizium anisopliae*. Gene 221:17–24. https://doi.org/10.1016/S0378-1119(98)00430-2

Screen SE, Hu G, St Leger RJ (2001) Transformants of *Metarhizium anisopliae* sf. *anisopliae* overexpressing chitinase from *Metarhizium anisopliae* sf. *acridum* show early induction of native chitinase but are not altered in pathogenicity to *Manduca sexta*. J Invert Pathol 78:260–266. https://doi.org/10.1006/jipa.2001.5067. PMID: 12009808

Singh A, Bhardwaj R, Singh IK (2019) Biocontrol agents: potential of biopesticides for integrated pest management. In: Giri B, Prasad R, Wu QS, Varma A (eds) Biofertilizers for sustainable agriculture and environment. Soil biology, vol 55. Springer, Cham

Srinivasan R, Sevgan S, Ekesi S et al (2019) Biopesticide based sustainable pest management for safer production of vegetable legumes and brassicas in Asia and Africa. Pest Manag Sci 75:2446–2454

St. Leger RJ, Joshi L, Bidochka MJ et al (1996) Construction of an improved mycoinsecticide overexpressing a toxic protease. Proc Natl Acad Sci U S A 93(13):6349–6354. https://doi.org/10.1073/pnas.93.13.6349

Staats CC, Silva MS, Pinto PM, Vainstein MH, Schrank A (2004) The *Metarhizium anisopliae* trp1 gene: cloning and regulatory analysis. Curr Microbiol 49:66–70. https://doi.org/10.1007/s00284-004-4256-0

Stewart LM, Hirst M, Lopez Ferber M, Merryweather AT, Cayley PJ, Possee RD (1991) Construction of an improved baculovirus insecticide containing an insect-specific toxin gene. Nature 352:85–88. https://doi.org/10.1038/352085a0

Subbanna ARNS, Stanley J, Rajasekhara H et al (2019) Perspectives of microbial metabolites as pesticides in agricultural pest management. In: Merillon JM, Ramawat KG (eds) Co-evolution of secondary metabolites. Springer, Cham, pp 1–28

Thomas MB, Watson EL, Valverde-Garcia P (2003) Mixed infections and insect-pathogen interactions. Ecol Lett 6:183–188

Tomalski MD, Miller LK (1991) Insect paralysis by baculovirus-mediated expression of a mite neurotoxin gene. Nature 352:82–85. https://doi.org/10.1038/352082a0

Tomalski MD, Miller LK (1992) Expression of a paralytic neurotoxin gene to improve insect baculoviruses as biopesticides. Biotechnology 10:545–549. https://doi.org/10.1038/nbt0592-545

Wang Q, Coleman JJ (2019) Progress and challenges: development and implementation of CRISPR/Cas9 technology in filamentous fungi. Comput Struct Biotechnol J 17:761–769

Wang CS, St. Leger RJ (2007) A scorpion neurotoxin increases the potency of a fungal insecticide. Nat Biotechnol 25:1455–1456

Wiwat C, Panbangred W, Mongkolsuk S et al (1995) Inhibition of a conjugation-like gene transfer process in *Bacillus thuringiensis* subsp. *israelensis* by the anti-S-layer protein antibody. Curr Microbiol 30:69–75

Yokoyama E, Yamagishi K, Hara A (2002) Group-I intron containing a putative homing endonuclease gene in the small subunit ribosomal DNA of *Beauveria bassiana* IFO 31676. Mol Biol Evol 19:2022–2025. https://doi.org/10.1093/oxfordjournals.molbev.a004025. PMID: 12411610

Zhang S, Xia YX, Kim B, Keyhani NO (2011) Two hydrophobins are involved in fungal spore coat rodlet layer assembly and each play distinct roles in surface interactions, development and pathogenesis in the entomopathogenic fungus, *Beauveria bassiana*. Mol Micriobiol 80(3):811–826. https://doi.org/10.1111/j.1365-2958.2011.07613.x

Zhao H, Charnley AK, Wang ZK, Yin YP, Li ZL, Li YL, Xia YX (2006) Identification of an extracellular acid trehalase and its gene involved in fungal pathogenesis of *Metarizium anisopliae*. J Biochem 140(3):319e327

Zlotkin E, Fishman Y, Elazar M (2000) AaIT: from neurotoxin to insecticide. Biochimie 82:869–881

Chapter 10
Importance of Metabolic Enzymes and Their Role in Insecticide Resistance

Muthusamy Ranganathan, Mathiyazhagan Narayanan, and Suresh Kumarasamy

Abstract Insects are a vital component in the world as they do harmful and harmless effects on human beings. The medically and agriculturally essential insects occupy more space for their habitats and better surveillance. Consequently, insects' population increased and reduced agricultural products' productivity and served as a vector for many threatening diseases. The use of chemical insecticides to combat pests has resulted in the creation of resistance in many insect species. This may respond to either resistance to other chemicals with the same action mode and sometimes produce multiple resistance and cross to different insecticide classes. However, insects develop resistance to various chemical groups; the mechanism and mode method of insecticide resistance action are similar. Insects become intoxicated at four different stages of pharmacological interactions: behavioral alteration, increased enzymatic metabolism, altered target site response, ingestion of the toxicant or decreased penetration. Metabolic resistance, which is regulated by advanced enzymes and results in transforming more complex toxic molecules into less toxic compounds, is a more general resistance process. The resistance mediated by metabolic mechanisms results from enhanced production of enzymes and the increased rate and expression levels of some related metabolic enzymes. Studying insecticide resistance among insects will help us understand its response to particular chemical compounds and the resistance mechanism.

Keywords Metabolic resistance · Detoxification · Cytochrome P450 · Esterase · Glutathione –S transferase

M. Ranganathan (✉) · M. Narayanan · S. Kumarasamy
PG and Research Center in Biotechnology, MGR College , Hosur, Tamil Nadu, India
e-mail: rmuthusamy.dr@gmail.com

S. De Mandal et al. (eds.), *New and Future Development in Biopesticide Research: Biotechnological Exploration*, https://doi.org/10.1007/978-981-16-3989-0_10

10.1 Introduction

Insects/pests are dangerous to crops and forests and are directly associated with food availability. These pests interfere with agriculture productivity and storage, processing, marketing, transport, etc. According to the recent report, an estimated 7–50% of crop loss has occurred annually (Oliveira et al. 2014). Apart from the direct damage and losses caused by the insect, indirectly, they served as a vector for pathogens like viruses and bacteria, thereby threatening the public and environment. Hence it is essential to control such pest's population to protect the economy.

There are different methods available for pest control, such as:

(a) Cultural control
(b) Mechanical/Physical control
(c) Biological control and
(d) Chemical control.

A cultural pest control method involves a modified farm process to avoid insect pests or make them unsuitable to their habituating environment. Mechanical pest control methods practice a manual hand collection and killing of the larval caterpillar to reduce its populations. Biological control uses natural enemies of insect pests. These natural enemies are categorized into predators, parasites, and pathogens. Chemical insecticides and their use are one of the most effective pest control techniques. These insecticides are chemical substances that can be used to destruct and control the pest, and every year a billion kilogram of insecticides are being used (Alavanja 2009). Pesticide overuse harms agriculture and human health. The extensive and discriminative uses of pesticides create resistance mechanisms in insects. Functionally resistance can be defined as an organism's ability to survive a dose of toxicants that is lethal to the susceptible one. There have been 500 different insect pests' species that target major crops such as tobacco, peanuts, cotton etc., has developed resistance to the novel insecticides. Moreover, the constant spread of this resistance in the future population poses a serious challenge towards controlling these pests (Connor et al. 2011; Stratonovitch et al. 2014).

10.2 Insecticide Resistance and Its Evolution in Insects

Insecticide tolerance can be evolved by four stages of pharmacological interactions in which insects become intoxicated: improved enzymatic metabolism, altered target site insensitivity, behavioral alteration, decreased penetration, or ingestion of the toxicant. Pest organisms can evolve more than one of these mechanisms simultaneously, or the mechanism can operate on more than one category of insecticides (e.g. oxidative metabolism), resulting in cross resistance.

The first case of resistance to insecticide in scale insects was reported by Melander (1914). The evolution of DDT an organic insecticide developed resistance

Table 10.1 Insecticide resistance cases in different insect species and its mechanism

Insect species	Insecticides and class	Resistance mechanism
Bemisia tabaci	Deltamethrin (type II pyrethroid) and Monocrotophos (organophosphate)	Metabolic resistance (Ahmad et al. 2002)
Aedes aegypti	Temephos (organophosphate) and Deltamethrin (type II pyrethroid)	Metabolic resistance & *kdr* Knock down resistance (Marcombe et al. 2009)
Spodoptera littoralis	Methoxyfenozide (Diacylhydrazine)	Metabolic resistance (Mosallanejad and Smagghe 2009)
Plutella xylostella	Cypermethrin (type II Pyrethroid)	Metabolic resistance (Baek et al. 2010)
Spodoptera litura	Chlorantraniliprole (Anthranilic diamide)	Metabolic resistance (Muthusamy et al. 2014)
Aedes aegypti	Temephos (organophosphate)	Metabolic resistance & AChE insensitivity (Muthusamy and Shivakumar 2015a, b, c, d)
Aedes aegypti	Permethrin (type I Pyrethroid)	Metabolic resistance & *kdr* Knock down resistance (Muthusamy and Shivakumar 2015a)
Cimex lectularius and *Cimex hemipterus*	Dichloro-diphenyl-trichloroethane (organochlorine) & Imidacloprid (neonicotinoid)	Metabolic resistance, AChE insensitivity, Knock down resistance, GABA receptor insensitivity and altered nAChRs (Dang et al. 2017)
Helicoverpa armigera	Indoxacarb (organochlorine)	Metabolic resistance (Cui et al. 2018)
Anopheles stephensi	Permethrin (type I Pyrethroid) Deltamethrin (type II pyrethroid) and malathion (organophosphate)	Metabolic resistance, AChE & *kdr* insensitivity (Safi et al. 2017)
Odontotermes brunneus	Cypermethrin (type II Pyrethroid)	Metabolic resistance (Mamatha et al. 2020)
Amsacta albistriga	Cypermethrin (type II Pyrethroid)	Metabolic resistance (Mathiyazhagan et al. 2020)
Drosophila melanogaster	Propoxur (carbamate)	AChE insensitivity (You et al. 2020)

was an issue of the past. Unfortunately, by 1947 housefly (*Musca domestica*) resistance to DDT was documented. Resistance to insecticides has increased dramatically in recent decades, owing to introducing new insecticide classes such as carbamates, cyclodienes, pyrethroids, formamidines, organophosphates and microbial biological pest control agents. *Bacillus thuringiensis* (www.irac-nline.org 2010). Examples of insecticide resistance cases in different insect species and its mechanism are given in Table 10.1.

Insecticide resistance affects many species and affects all large insecticide types. Today an estimation of 447 cases of arthropod resistance species exists in the world. Several insects have developed resistance to newer insecticide chemistry with a different mode of action. Over the past few decades, 90% of the arthropod resistance cases reported in different species populations are either Hemiptera (in the broad sense, 14%), Lepidoptera (15%), Diptera (35%), mites (14%) or Coleoptera (14%). Few studies relatively involve the stable tracing of Arthropod resistance. One such classical study was carried out by Sukhoruchenko and Dolzhenko (2008) on agricultural insect pests in Russia. They reported that 36 arthropod species had developed resistance to regularly used plant conservation products. They also report the development of the group, cross, and multiple resistances in economically essential pests.

10.3 Metabolic Resistance Mechanism

Metabolic resistance, which is regulated by advanced enzymes and results in transforming more complex toxic molecules into a less toxic compound, is a more general resistance process. Three main enzyme mechanisms, carboxylesterases, cytochrome P450 regulated monooxygenases, and glutathione S-transferases, are involved in the metabolic tolerance pathway and are responsible for various insecticide metabolism. Increased metabolism can modify enzymes in the available form and make the insecticide more degradable (Siegfried and Scharf 2001).

The metabolic detoxification of insecticide involves three phases. The first phase includes CYPs reducing substrate toxicity. Using GSTs and carboxylesterases (COEs), hydrophobic toxic compounds are converted to hydrophilic materials in phase II, allowing for easier excretion. ATP binding cassette (ABC) and main membrane transporters, which can pump conjugated xenobiotics out of the cell, are involved in phase III. Insects use various strategies to shield themselves from harmful substances, including evasion, sequestration, excretion, target site mutation, susceptibility alteration, overexpression, and the development of various isoforms of detoxifying enzymes (Chapman 2003; Silva et al. 2001). CYP- or COE43-mediated reactions result in toxin reduction or oxidation, which is the most common biochemical pathway for metabolic detoxification of toxic chemicals. GSTs then use glutathione conjugation to convert the detoxified molecule into a more water-soluble form, which aids in eliminating the cell (Enayati et al. 2005). This can be achieved by either overexpression (Silva et al. 2001) or duplicated isoforms of these enzymes are expressed. Alternatively, modifying the target site (mutation an amino acid residue) could cause insects to become insensitive to toxic chemicals or react to them. Sequestration is concerned with the selective transport and preservation of toxic compounds and avoiding their interaction with natural physiological processes (You et al. 2013).

Among resistance mechanisms, metabolic enzyme-mediated resistance poses a significant challenge to pest control. Resistant individuals possessing this mechanism can render more toxic substances to less toxic to escape from its effect.

10.3.1 Carboxylesterases

Carboxylesterase is an important enzyme which metabolizes the exogenous and endogenous chemical compounds. This large enzyme family can be characterized by substrate specificities and inhibitors or electrophoretic mobilities (Dauterman 1985; Soderlund 1997). Insect carboxylesterase plays a significant part in the biotransformation and detoxification of exogenous xenobiotics through hydrolysis. Synthetic pesticides, like pyrethroids, are an important class of xenobiotics metabolized by this enzyme (Crow et al. 2012). The non-insecticidal 1-napthyl acetate, an artificial substrate, is often used to detect the carboxylesterase activity in a colorimetric biochemical assay (Fig. 10.1). The quantitative alterations in the esterase coding region, like mutational substitution, may change the esterase specificity to its naphthyl acetate substrate; by changing enzymatic nature could cause resistance to insecticide (Claudianos et al. 2006). The great substitution (Trp224Ser) in the OP resistance *Culex* esterase gene revealed a modified enzymatic nature of esterase that decreased the carboxylesterase activity in resistance mosquito and other insect species (Cui et al. 2011).

In many insect species, the higher activity of the esterase enzyme has been correlated with insecticide resistance (Latif and Subrahmanyam 2010; Muthusamy and Shivakumar 2015b).

Since most chemical insecticides contain an ester moiety in their composition, improved detoxification and sequestration by carboxylesterase confers tolerance to organophosphate, pyrethriod, and carbamates in insects (Hemingway et al. 2004; Oakeshott et al. 2005; Li et al. 2007). In many studies of pyrethroid metabolic resistance, an exalted esterase activity or synergism by esterase enzyme inhibitors revealed the contribution of esterase to the resistance in insects (Oakeshott et al. 2010). Most notably, non-denaturing PAGE studies revealed that the staining intensity of one or more esterase bands with various electrophoretic mobilities could be involved (Farnsworth et al. 2010). In addition to these pathways in insects, higher esterase activities can result in gene amplification, which can lead to insecticide tolerance. In some cases, over-expression of carboxylesterase with higher fold amplification was found in some insect species (Small and Hemingway 2000; Cui et al. 2007; Muthusamy and Shivakumar 2015a, b, c, d).

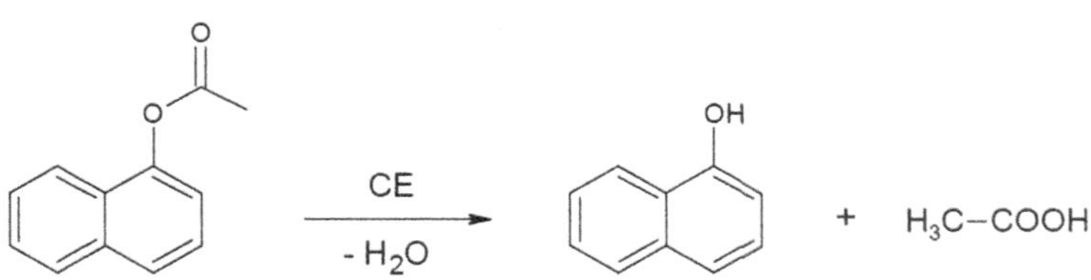

Fig. 10.1 Carboxylesterases catalyze the hydrolysis of 1-naphthyl acetate (artificial substrate). (Modified from Konanz (2009))

10.3.2 Cytochrome P450 Dependent Monooxygenases

Cytochrome P450-dependent monooxygenases are ubiquitous enzymes occurring in microbes, plants, mammals, and insects involved in the digestion of xenobiotics such as pesticides and plant toxins (Nelson 2011). These are hemoprotein-related microsomal oxidases named after their reduced form showed a typical absorbance peak at 450 nm when complexed with carbon monoxide. P450 is responsible for the metabolism of a wide range of xenobiotic compounds in insects and is also involved in their growth, development, and reproduction. P450 also plays a key role in converting herbicide molecules in plants through oxidation and peroxidation reactions (Feyereisen 2005; Hlavica and Lehnerer 2010; Li et al. 2012; Muthusamy and Shivakumar 2015b). Monooxygenases can be present in various tissues of insects, including the fat body, Malpighian tubules, and the midgut (Hodgson 1985; Scott 1999). P450 system activation was found in microsomes (endoplasmic reticulum-bound) and mitochondria in the insect subcellular distribution (Hodgson 1985). Many model substrates, such as p-nitroanisole, methoxyresorufin, NADPH cytochrome c reductase, TMBZ peroxidation, p-Nitroanisole O-Demethylase, and ethoxyresorufin, were commonly used for the biochemical identification of monooxygenase activity in insects. The oxidation of Tetramethylbenzidine (TMBZ) by peroxidase is used to measure the resistant in insects (Kranthi 2005) (Fig. 10.2).

In insect P450s are grouped into four major clades based on their evolutionary relationship: the mitochondrial P450s, CYP2, CYP3, and CYP9). Among them, the CYP3 clade, CYP6, and CYP9 P450 families, Insecticide detoxification and metabolism are critical in various insect species (Poupardin et al. 2010; Musasia et al. 2013). Overexpression of cytochrome P450 genes from various families has been shown to impart insecticide resistance in various insect species. Deltamethrin resistance in *Tribolium castaneum* was also documented when the expression of CYP6BQ9 was knocked down (Zhu and Snodgrass 2003). The overexpression of *Cyp12a4 I* associated with the lufenuron *resistance in Drosophila melanogaster* (Bogwitz et al. 2005). Similarly, the detoxification ability and expression level of four novel P450s were studied in honey bees of *Apis cerana cerana* (Zhang et al. 2019). The P450s also help in the detoxification of the toxic phytochemical, including aflatoxin B1 present in the diet of honey bees (Mao et al. 2009; Niu et al. 2011; Zhang et al. 2019). The *CYP4G11* gene has been reported to protect honeybees from the damage caused by insecticides (Shi et al. 2013). Also, the CYP9Q family of

$$TMB_{red} + H_2O_2 \xrightarrow{\text{Horseradish peroxidase}} TMB_{ox} + 2\,H_2O_2$$

Fig. 10.2 Microsomal proteins catalyze peroxidation of Tetramethylbenzidine with hydrogen peroxide as co-substrate. (Modified from Kranthi (2005))

bumblebees P450 plays a significant role in studying the insecticide sensitivity to different classes (Manjon et al. 2018).

In many cases, regulatory changes of insects are responsible for the metabolic resistance mechanism. Up-regulation of metabolic enzymes through mutations in trans and cis-acting regulatory loci or gene amplification encoding the enzyme is typically the mechanism for increased development (Hemingway and Karunaratne 1998); monooxygenases confer resistance to a wide range of insecticides, including organophosphates, carbamates, pyrethroids, and inhibitors of chitin biosynthesis (Li et al. 2007; Atoyebi et al. 2020; Mamatha et al. 2020). It has been reported that Neonicotinoid resistance is linked with CYP6A1, CYP6D1, and CYP6D3 genes are overexpressed in *Musca domestica* where the CYP6D1 and CYP6D3 males are overexpressed and female resistant housefly, respectively (Markussen and Kristensen 2010). Similarly, the DDT resistance by *Drosophila melanogaster* is associated with the two resistance loci of p-450 gene subunits, i.e., CG10737 and Cyp6w1 (Schmidt et al. 2017). It has also been reported that the overexpression of 3 cytochrome P450 genes, *CYP6CY14, CYP6CY22*, and *CYP6UN1,* are responsible for the dinotefuran (the third-generation neonicotinoid) resistance in *Aphis gossypii* Glover in China (Chen et al. 2020). Similarly, the resistant *Anopheles mosquitoes* showed overexpressed P450 enzymes, *CYP4G16* and *CYP4G17* (Ingham et al. 2014).

10.3.3 Glutathione S-Transferase

Glutathione S-transferases (GST) are a multifunctional intracellular enzyme present in most aerobic microorganisms, plants, and animals, including insects, and play an important role in intracellular transportation, hormone biosynthesis, and oxidative stress protection (Ketterman et al. 2011; Listowsky et al. 1998; Enayati et al. 2005). GST proteins are also recognized as MAPEG proteins and belong to the superfamily of mitochondrial, cytosolic, and microsomal proteins. Subclasses of the cytosolic superfamilies are included in the detoxification process and include Delta, Epsilon, Omega, Sigma, Theta, Mu, and Zeta (Che-Mendoza et al. 2009). In insects, GSTs are categorized as microsomal and cytosolic. The number of cytosolic GSTs is much higher than the number of microsomal GSTs divided into six classes. The subclasses (Delta) and (Epsilon) are insect-specific, while the Omega, Sigma, Theta, and Zeta are present in a variety of species types (Low et al. 2007). GST can detoxify various chemical compounds by glutathione conjugation and plays a key role in the resistance production of various insecticide groups, including organophosphates and pyrethroids, due to the availability of a wide variety of substrates for individual enzymes (Yamamoto et al. 2009). Furthermore, they aid in the removal of harmful oxygen-free radicals produced by pesticides (Fig. 10.3).

Many endogenous, hydrophobic and foreign compounds form water-soluble conjugates with GSH, making detoxification easier. In many vertebrate and non-vertebrates systems, GSTs are responsible for detoxifying chemical substances;

Fig. 10.3 The conjugation of the artificial substrate 1-hcloro-2, 4-dinitrobenzene with GSH was induced by glutathione S-transferase. (Modified from Konanz (2009))

protect them against oxidative damage, and transporting numerous endogenous metabolites and intracellular hormones (Sanil et al. 2014). Insecticides may also be metabolized by promoting reductive dehydrochlorination or eliminating oxygen free radicals generated by pesticides (Hayes et al. 2005). Increased GST enzyme production using gene amplification or overexpression is also associated with GST-based insecticide resistance (Vontas et al. 2002).

The high activity GSTs in organophosphate and DDT resistance has been studied in *Musca domestica* (Motoyama and Dauterman 1980). The number of cytosolic GSTs is much higher than the number of microsomal GSTs divided into six classes. The subclasses (Delta) and (Epsilon) are insect-specific, while the Omega, Sigma, Theta, and Zeta are present in a variety of species types (Low et al. 2007). Studies have shown that GSTs were responsible for many detoxifying classes of chemical insecticides such as organophosphate (OPs), synthetic pyrethroids (SPs), and chlorine (Ketterman et al. 2011; Mamatha et al. 2020). Increased activity and expression level of one or more GST genes was described to cause insecticide resistance in many insects (Hemingway 2000; Ranson et al. 2001). In many studies, the GST was associated with resistance and other enzymes (Pavlidi et al. 2018; Mathiyazhagan et al. 2020).

10.4 Behavioral Resistance

Behavioral resistance mechanism necessitates alteration in the insect behavior by which they can avoid insecticides. The ability of an insect's resistance to behavioral and penetration response is the least mechanism. Insects' behavioral resistance can be (1) stimulus-dependent following direct contact or without contact (2) stimulus-independent like zoophily or exophily (Chareonviriyaphap et al. 2013). Stimulus-dependent habits necessitate the insect's sensory restoration in order to reveal a toxin-nursed surface before receiving a lethal dose, causing a delayed response. In insecticide resistant mosquito vectors, stimulus-independent activity has been observed, accompanied by extensive insecticide use (Meyers et al. 2016; Moiroux et al. 2012). Similar behavior resistance has been reported in numerous insect pests. In such a study, Sarfraz et al. (2005) observed that the laboratory developed *P. xylostella* laid more eggs near the soil instead of laying eggs on the stem and leaves of the host plant exposed to the insecticide.

Behavioral resistance to insecticides on simple repellency or avoidance has been observed in German cockroach gel bait with sucrose, maltose and fructose, which are commonly feedants to sensitive laboratory strains of *B. germanica* (Wang et al. 2004). Whereas in the resistant strain of *B. germanica* excluded all of these semiochemicals from their diet. Despite studies documenting insecticide resistance to insect behavior, the gene responsible for toxic chemical metabolism is unknown (Mamidala et al. 2011).

10.5 Penetration Resistance

This type of resistance involves the modifications in the cuticle leading to the slowdown in the penetration of insecticide inside the insects' body. Cuticle thickening and cuticle structure change are two distinct pathways for resistance to penetration. The event of resistance to penetration in insects through physiological changes is basic reason. However, in some instances, reduced insecticide penetration through the insect cuticle has been identified as an alternative resistance mechanism. Only a few studies have reported the association of insecticide penetration or cuticular thickness with resistance (Strycharz et al. 2013).

Many insect species overcome insecticides' effects through reduced cuticular penetration (Pan et al. 2009; Wood et al. 2010). This style of resistance is often linked to other types of opposition (Zhu et al. 2013; Dang et al. 2017). Balabanidou et al. (2016) made significant strides in understanding cuticular tolerance, identifying the basic changes observed in resistant mosquito cuticles, and, more importantly, expanding on previous research to include further evidence of a role for the CYP4G subfamily of P450s in this process. The thickening of cuticle in the *Triatoma infestans* vector has been associated with pyrethroid resistance and is deduced from the transcriptional gene analysis in *An. stephensi* (Vontas et al. 2007). Another study from West Africa revealed pyrethroids and DTT resistance in *An. gambiae*, the overexpression of CPLCG3 and CPRs have been linked to a thicker procuticle in the femur leg segment and a phenotype (Yahouédo et al. 2017). It has been reported that the femur cuticle was thicker in the resistant strain of *Culex pipiens* compared to the susceptible one. The CPLCG5 gene is silenced, resulting in a thinner cuticle and greater insecticide resistance. This demonstrates CPLCG5's function in insect resistance (Huang et al. 2018). Pedrini et al. (2009) revealed that the decreased penetration rates across the cuticle are associated with the lower insecticide inoculation in the internal organs, leading to metabolically-mediated detoxification.

10.6 Resistance by Target-Site Insensitivity

In insects, exposure to altered target site insensitivity is a critical mechanism. A genetically-based modification is made to the target-site where the insecticide normally binds, such as a single-nucleotide polymorphism that causes difference in the amino acid sequence within the target protein's binding region (Liu 2015). The resistance can also be thought of as a preadaptive phenotype. A small number of individuals have one or more resistance alleles that allow them to survive exposure to the stressor. As a result, effective insecticide resistance testing and a thorough understanding of the factors that contribute to and the processes that regulate resistance growth are critical to the effectiveness of pest management and vector-borne disease control (Butler 2011).

10.6.1 Altered Acetylcholinesterases

Acetylcholinesterase is a crucial enzyme that hydrolyzes acetylcholine in cholinergic synapses rapidly (Rosenberry 1975). Organophosphate (OP) insecticides mainly attack AChE, which phosphorylates the serine residues in its active site and blocks the hydrolysis of acetylcholine, causing the insect to die (Menozzi et al. 2004). AChE is used in two ways in insects. A globular disulfide-linked dimeric protein (ca. 150 kDa) is one of the most common forms, with a glycolipid anchor connecting it to the membrane. The AChE active site is divided into two subsites: the esteratic catalytic site, which has a distinct catalytic triad of amino acid residues (serine, glutamic acid, and histidine), and the anionic choline-binding site (Fournier et al. 1992; Fournier and Mutero 1994). In several pest species, the insensitive AChE has become an important tool for insecticide resistance (Chen et al. 2001; Weill et al. 2002; Muthusamy et al. 2013). According to molecular studies, in AChE encoding genes, point mutations associated with target-site insensitivity confer structural modifications (Kozaki et al. 2001). According to the findings, in *Drosophila melanogaster* (Brochier et al. 2001), decreased AChE insensitivity was found to be a typical resistance mechanism to OP/carbamates in other insect species (Lee et al. 2006; Seong et al. 2012).

10.6.2 Altered GABA Receptors

The GABA receptor belongs to a family of ligand-gated ion channels that act as fast inhibitory neurotransmission in insects. According to molecular studies, point mutations in genes encoding insecticide targets have been linked to insensitivity to the target site (Bloomquist 2001). A single common point mutation (alanine to serine at position 302) in the dieldrin resistance (Rdl) subunit is well defined in many insect organisms (Soderlund 1997; Ozoe and Akamatsu 2001; Wondji et al. 2011; Heong et al. 2013). In fly (*D. melanogaster*) and other insects, the dieldrin (Rdl) gene, which primarily functions in encoding GABA receptors composed of five subunits arranged around a central gated ion channel, showed insecticide resistance (cyclodiene) (Remnant et al. 2014). There is only one Rdl gene in most pests, but certain insects/pests have Rdl genes in various allelic variants. The natural function of (Rdl) is affected by a single nucleotide polymorphism (SNP). A single mutation of alanine to serine at position 302 in the second transmembrane region of the Rdl subunit causes the resistant phenotype. *Myzus persicae* has four types of alleles, with the wild form (called allele A) encoding 'Ala302,' while the other three alleles encoding 'Gly302,' also known as allele 'G,' TCG codon encoding 'Ser302,' and 'AGT' codon encoding 'Ser302' (known as allele 'S'). The central causes of resistance are alanine and glycine, while resistance is not caused by the other loci of two serine-containing "s" alleles (Assel et al. 2014). Resistance is caused by the S/S locus, while A/G has a resistance function to GABA receptors and is resistant to dieldrin (Bass et al. 2014).

10.6.3 Altered Sodium Channel Proteins: Nerve Insensitivity

Voltage-gated sodium channels (vgSChs) are transmembrane proteins responsible for electrical conductivity in the nervous system by inducing action potentials in the neuronal membranes of most excitable cells. When these channels open, Na+ current is produced in the insect nervous system, which causes the membrane potential to depolarize. Many insecticides, such as synthetic pyrethroids, DDT, and oxadiazines, as well as a few synthetic and natural toxins, target insect sodium channels (Narahashi 2000; Vais et al. 2001; Dong 2003). The voltage-gated Na+ channel in a cell membrane has four homologous domains from I to IV, each with six hydrophobic segments (S1 to S6). The S4 and S6 segments are voltage sensors that create a pore in the channel when combined with the S5 segment, connecting the P-loops (Martins and Valle 2012).

Resistance has developed in many species due to the widespread use of pyrethroids and DDT in insect control. Reduced target-site vulnerability, also known as knockdown resistance or kdr, is an essential mechanism that confers resistance to all insecticides in insects (Zlotkin 2001). The housefly was the first species to be tested for this kind of tolerance (*Musca domestica*). Pyrethroid insecticide tolerance in insects was investigated by comparing the coding sequences of para orthologous sodium channel genes in susceptible and resistant animals (Whalon et al. 2008, 2010). Kdr resistance in insects was discovered to be caused by a mutation(s) in the sodium channel gene, according to molecular studies. Several mutations linked to kdr or super-kdr resistance in the housefly have been discovered in recent years (Williamson et al. 1993) and some other essential pest species (Soderlund 2010; Dong 2007; Davies et al. 2007; Thiaw et al. 2018; Kushwah et al. 2020).

10.7 Future Prospective/Conclusion

Insecticide resistance has become an increasing problem in the world today. However, the pest control program mainly relies on synthetic insecticides. In general, insect resistance has developed mainly by increasing pesticide quantity or replacing the older one with a modern, more significant compound. It is also essential to gather information about the resistance mechanism underlying the insecticide in the population before deciding on alternative insecticides or increasing the doses. However, insects' resistance can be delayed either by applying insecticide with different chemical groups, and the addition of synergist, plant growth regulators and biological pesticides derived from natural products can also be made possible.

References

Ahmad M, Arif MI, Ahmad Z, Denholm I (2002) Cotton whitefly (*Bemisia tabaci*) resistance to organophosphate and pyrethroid insecticides in Pakistan. Pest Manag Sci 58:203–208

Alavanja M (2009) Pesticides use and exposure extensive worldwide. Rev Environ Health 24:303–309

Assel M, Wolf C, Noack S, Williams H, Ilg T (2014) The novel isoxazoline ectoparasiticide fluralaner: selective inhibition of arthropod γ-aminobutyric acid and *L*-glutamate-gated chloride channels and insecticidal/acaricidal activity. Insect Biochem Mol Biol 45:111

Atoyebi SM, Tchigossou GM, Akoton R, Riveron JM, Irving H, Weedall G, Tossou E, Djegbe I, Oyewole IO, Bakare AA, Wondji CS, Djouaka R (2020) Investigating the molecular basis of multiple insecticide resistance in a major malaria vector *Anopheles funestus* (*sensu stricto*) from Akaka-Remo, Ogun State, Nigeria. Parasit Vectors 13:423

Baek JH, Clark JM, Lee SH (2010) Cross-strain comparison of cypermethrin-induced cytochrome P450 transcription under different induction conditions in diamondback moth. Pestic Biochem Physicl 96:43–50

Balabanidou V, Kampouraki A, MacLean M, Blomquist GJ, Tittiger C, Juárez MP, Mijailovsky SJ, Chalepakis G, Anthousi A, Lynd A, Antoine S, Hemingway J, Ranson H, Lycett GJ, Vontas J (2016) Cytochrome P450 associated with insecticide resistance catalyzes cuticular hydrocarbon production in Anopheles gambiae. Proc Natl Acad Sci U S A 113:9268–9273

Bass C, Puinean AM, Zimmer CT, Denholm I, Field LM, Foster SP, Gutbrod O, Nauen R, Slater R, Williamson MS (2014) The evolution of insecticide resistance in the peach potato aphid, *Myzus persicae*. Insect Biochem Mol Biol 51:41

Bloomquist JR (2001) GABA and glutamate receptors as biochemical sites for insecticide action. In: Ishaaya I (ed) Biochemical sites of insecticide action and resistance. Springer, Berlin, pp 17–41

Bogwitz MR, Chung H, Magoc L, Rigby S, Wong W, O'Keefe M, Daborn PJ (2005) Cyp12a4 confers lufenuron resistance in a natural population of Drosophila melanogaster. Proc Natl Acad Sci U S A 102(36):12807–12812

Brochier L, Pontie Y, Willison M, Estrada-Mondaca S, Czaplicki J, Klaebe A, Fournier D (2001) Involvement of deacylation in activation if substrate hydrolysis by *Drosophila* acetylcholinesterase. J Biol Chem 276:18296–18302

Butler D (2011) Mosquitoes score in chemical war. Nature 475:19

Chapman RF (2003) Contact chemoreception in feeding by phytophagous insects. Annu Rev Entomol 48:455–484

Chareonviriyaphap T, Bangs MJ, Suwonkerd W, Kongmee M, Corbel V, Ngoen-Klan R (2013) Review of insecticide resistance and behavioral avoidance of vectors of human diseases in Thailand. Parasit Vectors 6:280

Che-Mendoza A, Penilla RP, Rodríguez DA (2009) Insecticide resistance and glutathione S-transferases in mosquitoes: a review. Afr J Biotechnol 8(8)

Chen Z, Newcomb R, Forbes E, Mckenzie J, Batterham P (2001) The acetylcholinesterase gene and organophosphorus resistance in the Australian sheep blowfly, *Lucilia cuprina*. Insect Biochem Mol Biol 31:805–816

Chen A, Zhang H, Shan T, Shi X, Gao X (2020) The overexpression of three cytochrome P450 genes CYP6CY14, CYP6CY22 and CYP6UN1 contributed to metabolic resistance to Dinotefuran in melon/cotton aphid, Aphis gossypii glover. Pestic Biochem Physiol 167:104601

Claudianos C, Ranson H, Johnson RM, Biswas S, Schuler MA, Berenbaum MR, Feyereisen R, Oakeshott JG (2006) A deficit of detoxification enzymes: pesticide sensitivity and environmental response in the honeybee. Insect Mol Biol 15:615–636

Connor DJ, Loomis RS, Cassman KG (2011) Crop ecology: productivity and management in agricultural systems. Cambridge University Press, Cambridge

Crow JA, Bittles V, Herring KL, Borazjani A, Potter PM, Ross MK (2012) Inhibition of recombinant human carboxylesterase 1 and 2 and monoacylglycerol lipase by chlorpyrifos oxon, paraoxon and methyl paraoxon. Toxicol Appl Pharmacol 258:145–150

Cui F, Weill M, Berthomieu A, Raymond M, Qiao CL (2007) Characterization of novel esterases in insecticide-resistant mosquitoes. Insect Biochem Mol Biol 37:1131–1137

Cui F, Wang H, Liu S, Chang HRG, Qiao C, Raymond M, Kang L (2011) Two single mutations commonly cause qualitative change of nonspecific carboxylesterases in insects. Insect Biochem Mol Biol 41:1–8

Cui L, Wang QQ, Qi HL, Wang QY, Yuan HZ, Rui CH (2018) Resistance selection of indoxacarb in *Helicoverpa armigera* (Hübner) (Lepidoptera: Noctuidae): cross-resistance, biochemical mechanisms and associated fitness costs. Pest Manag Sci 74:2636–2644

Dang K, Doggett SL, Singham GV, Lee CY (2017) Insecticide resistance and resistance mechanisms in bed bugs, *Cimex* spp. (Hemiptera: Cimicidae). Parasite Vectors 10:318

Dauterman WC (1985) Insect metabolism: extramicrosomal. In: Kerkut GA, Gilbert LI (eds) Comprehensive insect physiology, biochemistry and pharmacology. Pergamon, Oxford, pp 713–730

Davies TG, Field LM, Usherwood PN, Williamson MS (2007) A comparative study of voltage-gated sodium channels in the Insecta: implications for pyrethroid resistance in Anopheline and other Neopteran species. Insect Mol Biol 16(3):361–375

Dong K (2003) Voltage-gated sodium channels as insecticide targets. In: Voss G, Ramos G (eds) Chemistry of crop protection. Progress and prospects in science and regulation. Wiley-VCH, Weinheim, pp 167–176

Dong K (2007) Insect sodium channels and insecticide resistance. Invertebr Neurosci 7:17–30

Enayati AA, Ranson H, Hemingway J (2005) Insect glutathione transferases and insecticide resistance. Insect Mol Biol 14:3–8

Farnsworth CA, Teese MG, Yuan G, Li Y, Scott C, Zhang X, Wu Y, Russell RJ, Oakeshott JG (2010) Esterase-based metabolic resistance to insecticides in heliothine and spodopteran pests. J Pestic Sci 35:275–289

Feyereisen R (2005) Insect cytochrome P450. In: Gilbert LI, Iatrou K, Gill SS (eds) Comprehensive molecular insect science, vol 4. Elsevier, pp 1–77

Fournier D, Mutero A (1994) Modification of acetylcholinesterase as a mechanism of resistance to insecticides. Comp Biochem Physiol C Toxicol Pharmacol 108:19–31

Fournier D, Bride JM, Hoffmann F, Karch F (1992) Acetylcholinesterase. Two types of modifications confer resistance to insecticide. J Biol Chem 267:14270–14274

Hayes JD, Flanagan JU, Jowsey IR (2005) Glutathione transferases. Annu Rev Pharmacol Toxicol 45:51–88

Hemingway J (2000) The molecular basis of two contrasting metabolic mechanisms of insecticide resistance. Insect Biochem Mol Biol 30:1009–1015

Hemingway J, Karunaratne SH (1998) Mosquito carboxylesterases: a review of the molecular biology and biochemistry of a major insecticide resistance mechanism. J Med Vet Entomol 12 (1):1–12

Hemingway J, Hawkes NJ, McCarroll L, Ranson H (2004) The molecular basis of insecticide resistance in mosquitoes. Insect Biochem Mol Biol 34:653–665
Heong KL, Tan KH, Garcia CPF, Liu Z, Lu Z (2013) Research methods in toxicology and insecticide resistance monitoring of rice planthoppers. IRRI Books, International Rice Research Institute (IRRI), number 164475
Hlavica P, Lehnerer M (2010) Oxidative biotransformation of fatty acids by cytochromes P450: predicted key structural elements orchestrating substrate specificity, regioselectivity and catalytic efficiency. Curr Drug Metab 11:85–104
Hodgson E (1985) Microsomal monooxygenases. In: Kerkut GA, Gilbert LI (eds) Comprehensive insect physiology, biochemistry and pharmacology. Pergamon, Oxford, pp 225–331
Huang Y, Guo Q, Sun X, Zhang C, Xu N, Xu Y, Zhou D, Sun Y, Ma L, Zhu C, Shen B (2018) Culex pipiens pallens cuticular protein CPLCG5 participates in pyrethroid resistance by forming a rigid matrix. Parasit Vectors 11:6
Ingham VA, Jones CM, Pignatelli P, Balabanidou V, Vontas J, Wagstaff SC et al (2014) Dissecting the organ specificity of insecticide resistance candidate genes in Anopheles gambiae: known and novel candidate genes. BMC Genomics 15(1):1–9
Ketterman AJ, Saisawang C, Wongsantichon J (2011) Insect glutathione transferases. Drug Metab Rev 43(2):253–265
Konanz S (2009) Characterization of mechanisms of resistance to common insecticides in noctuid pest species and resistance risk assessment for the new lepidopteran specific compound flubendiamide. Ph.D thesis, p 5
Kozaki TT, Shono T, Tomita Y (2001) Fenitroxon insensitive acetylcholinesterase of the housefly, *Musca domestica* associated with point mutations. Insect Biochem Mol Biol 31:991–997
Kranthi KR (2005) Insecticide resistance—monitoring, mechanisms and management manual. CICR, Nagpur
Kushwah RBS, Kaur T, Dykes CL, Kumar RH, Kapoor N, Singh OMP (2020) A new knockdown resistance (*kdr*) mutation, F1534L, in the voltage-gated sodium channel of *Aedes aegypti*, co-occurring with F1534C, S989P and V1016G. Parasite Vectors 13:327
Latif AO, Subrahmanyam B (2010) Pyrethroid synergists suppress esterase-mediated resistance in Indian strains of the cotton bollworm, *Helicoverpa armigera* (Hübner). Pestic Biochem Physiol 97:279–288
Lee DW, Kim SS, Shin SW, Kim WT, Boo KS (2006) Molecular characterization of two acetylcholinesterase genes from the oriental tobacco budworm, *Helicoverpa assulta* (Guenée). Biochem Biophys Acta 1760:125–133
Li X, Schuler MA, Berenbaum MR (2007) Molecular mechanisms of metabolic resistance to synthetic and natural xenobiotics. Annu Rev Entomol 52:231–253
Li H, Melø TB, Arellano JB, Razi Naqvi K (2012) Temporal profile of the singlet oxygen emission endogenously produced by photosystem II reaction centre in an aqueous buffer. Photosynth Res 112:75–79
Listowsky IM, Abramovitz H, Niitsu HY (1998) Intracellular binding and transport of hormones and xenobiotics by glutathione S-transferase. Drug Metab Rev 19:305–318
Liu N (2015) Insecticide resistance in mosquitoes: impact, mechanisms, and research directions. Annu Rev Entomol 60:537–559
Low WY, Ng HL, Morton CJ, Parker MW, Batterham P, Robin C (2007) Molecular evolution of glutathione S-transferases in the genus *Drosophila*. Genetics 177(3):1363–1375
Mamatha V, Muthusamy R, Murugan JM, Kweka EJ (2020) Effect of cypermethrin on worker and soldier termites of subterranean termites *Odontotermes brunneus* (Hagen) (Termitidae: Isoptera). Proc Zool Soc 73:40–45
Mamidala P, Jones SC, Mittapalli O (2011) Metabolic resistance in bed bugs. Insects 2:36–48
Manjon C, Troczka BJ, Zaworra M, Beadle K, Randall E, Hertlein G, Singh KS, Zimmer CT, Homem RA, Lueke B, Reid R, Kor L, Kohler M, Benting J, Williamson MS, Davies TGE, Field LM, Bass C, Nauen R (2018) Unravelling the molecular determinants of bee sensitivity to neonicotinoid insecticides. Curr Biol 28(7):1137–1143

Mao W, Rupasinghe SG, Johnson RM, Zangerl AR, Schuler MA, Berenbaum MR (2009) Quercetin-metabolizing CYP6AS enzymes of the pollinator Apis mellifera (Hymenoptera: Apidae). Comp Biochem Physiol B: Biochem Mol Biol 154(4):427–434

Marcombe S, Poupardin R, Darriet F, Reynaud S, Bonnet J, Strode C, Brengues C, Yébakima A, Ranson H, Corbel V, David JP (2009) Exploring the molecular basis of insecticide resistance in the dengue vector *Aedes aegypti*: a case study in Martinique Island (French West Indies). BMC Genomics 10:494

Markussen MD, Kristensen M (2010) Cytochrome P450 monooxygenase-mediated neonicotinoid resistance in the house fly *Musca domestica* L. Pestic Biochem Physiol 98(1):50–58

Martins AJ, Valle DS (2012) The pyrethroid knockdown resistance. In: Soloneski S (ed) Insecticides—basic and other applications. InTech. ISBN: 978-953-51-0007-2. http://www.intechopen.com/books/insecticides-basic-and-other-applications/the-pyrethroid-knockdownresistance

Mathiyazhagan N, Muthusamy R, Shivakumar MS, Suresh K, Sabariswaran K (2020) Toxicity of cypermethrin and enzyme inhibitor synergists in red hairy caterpillar *Amsacta albistriga* (Lepidoptera: Arctiidae). J Basic Appl Zool 81:45

Melander AL (1914) Can insects become resistant to sprays. J Econ Entomol 7:167–173

Menozzi P, Shi MA, Lougarre A, Tang ZH, Fournier D (2004) Mutations of acetylcholinesterase which confer insecticide resistance in *Drosophila melanogaster* populations. BMC Evol Biol 4:4

Meyers JI, Pathikonda S, Popkin-Hall ZR, Medeiros MC, Fuseini G, Matias A, Garcia G, Overgaard HJ, Kulkarni V, Reddy VP, Schwabe C, Lines J, Kleinschmidt I, Slotman MA (2016) Increasing outdoor host-seeking in *Anopheles gambiae* over 6 years of vector control on Bioko Island. Malar J 15(1):239

Moiroux N, Gomez MB, Pennetier C, Elanga E, Djènontin A, Chandre F, Djègbé I, Guis H, Corbel V (2012) Changes in *Anopheles funestus* biting behavior following universal coverage of long-lasting insecticidal nets in Benin. J Infect Dis 206:1622–1629

Mosallanejad H, Smagghe G (2009) Biochemical mechanisms of methoxyfenozide resistance in the cotton leafworm *Spodoptera littoralis*. Pest Manag Sci 65(1):732–736

Motoyama N, Dauterman WC (1980) Glutathione S-transferases: their role in the metabolism of organophosphorus insecticides. Rev Biochem Toxicol 2:49–69

Musasia FK, Isaac AO, Masiga DK, Omedo IA, Mwakubambanya R, Ochieng R, Mireji PO (2013) Sex-specific induction of CYP6 cytochrome P450 genes in cadmium and lead tolerant *Anopheles gambiae*. Malar J 12:97

Muthusamy R, Shivakumar MS (2015a) Effect of lambda cyhalothrin and temephos on detoxification enzyme systems in *Culex quinquefasciatus* (Diptera: Culicidae). J Environ Biol 36 (1):235–239

Muthusamy R, Shivakumar MS (2015b) Resistance selection and molecular mechanisms of cypermethrin resistance in red hairy caterpillar (*Amsacta albistriga* Walker). J Pestic Biochem Physiol 117:54–61

Muthusamy R, Shivakumar MS (2015c) Susceptibility status of *Aedes aegypti* to temephos from three districts of Tamil Nadu, India. J Vector Dis 52:159–165

Muthusamy R, Shivakumar MS (2015d) Involvement of metabolic resistance and F1534C kdr mutation in the pyrethroid resistance mechanisms of *Aedes aegypti* in India. Acta Trop 148:137–141

Muthusamy R, Suganya R, Gowri M, Shivakumar MS (2013) Biochemical mechanisms of organophosphate and pyrethroid resistance in red hairy caterpillar *Amsacta albistriga* (Lepidoptera: Arctiidae). Saudi J Agric Sci 12:47–52

Muthusamy R, Vishnupriya M, Shivakumar MS (2014) Biochemical mechanism of chlorantraniliprole resistance in *Spodoptera litura* (Fab) (Lepidoptera: Noctuidae). J Asia Pac Entomol 17:865–869

Narahashi T (2000) Neuroreceptors and ion channels as the basis for drug action: past, present, and future. J Pharmacol Exp Ther 294:1–26

Nelson DR (2011) Progress in tracing the evolutionary paths of cytochrome P450. Biochem Biophys Acta 1814:14–18

Niu G, Johnson RM, Berenbaum MR (2011) Toxicity of mycotoxins to honeybees and its amelioration by propolis. Apidologie 42(1):79–87

Oakeshott JG, Claudianos C, Campbell PM, Newcomb RD, Russell RJ (2005) Biochemical genetics and genomics of insect esterases. In: Gilbert LI, Latrou K, Gill SS (eds) Comprehensive molecular insect science—pharmacology, vol 5. Elsevier, Oxford, pp 309–381

Oakeshott JG, Johnson RM, Berenbaum MR, Ranson H, Cristino AS, Claudianos C (2010) Metabolic enzymes associated with xenobiotic and chemosensory responses in *Nasonia vitripennis*. Insect Mol Biol 19(1):147–163

Oliveira C, Auad A, Mendes S, Frizzas M (2014) Crop losses and the economic impact of insect pests on Brazilian agriculture. Crop Prot 56:50–54

Ozoe Y, Akamatsu M (2001) Non-competitive GABA antagonists: probing the mechanisms of their selectivity for insect *versus* mammalian receptors. Pest Manag Sci 57:923–931

Pan C, Zhou Y, Mo J (2009) The clone of lactase gene and its potential function in cuticular penetration resistance of *Culex pipiens pallens* to fenvalerate. Pestic Biochem Physiol 93:105–111

Pavlidi N, Vontas J, Van Leeuwen T (2018) The role of glutathione S-transferases (GSTs) in insecticide resistance in crop pests and disease vectors. Curr Opin Insect Sci 27:97–102

Pedrini N, Mijailovsky SJ, Girotti JR, Stariolo R, Cardozo RM, Gentile A, Juarez MP (2009) Control of pyrethroid-resistant Chagas disease vectors with entomopathogenic fungi. PLoS Negl Trop Dis 3:e434

Poupardin R, Riaz MA, Vontas J, David JP, Reynaud S (2010) Transcription profiling of eleven cytochrome P450s potentially involved in xenobiotic metabolism in the mosquito *Aedes aegypti*. Insect Mol Biol 19:185–193

Ranson H, Rossiter L, Ortelli F, Jensen B, Wang X, Roth CW, Collins FH, Hemingway J (2001) Identification of a novel class of insect glutathione S-transferases involved in resistance to DDT in the malaria vector *Anopheles gambiae*. Biochem J 359:295–304

Remnant EJ, Morton CJ, Daborn PJ, Lumb C, Yang YT, Ng HL, Parker MW, Batterham P (2014) The role of Rdl in resistance to phenylpyrazoles in *Drosophila melanogaster*. Insect Biochem Mol Biol 54:11

Rosenberry TL (1975) Acetylcholinesterase. Adv Enzymol 43:103–218

Safi NH, Ahmadi AA, Nahzat S, Ziapour SP, Nikookar SH, Fazeli-Dinan M, Enayati A, Hemingway J (2017) Evidence of metabolic mechanisms playing a role in multiple insecticides resistance in *Anopheles stephensi* populations from Afghanistan. Malar J 16:100

Sanil D, Shetty V, Shetty N (2014) Differential expression of glutathione s-transferase enzyme in different life stages of various insecticide-resistant strains of *Anopheles Stephensi*: a malaria vector. J Vect Borne Dis 51(2):95–105

Sarfraz M, Keddie AB, Dosdall LM (2005) Evidence for behavioural resistance by the diamondback moth, *Plutella xylostella* (L.). J Appl Entomol 129:340–341

Schmidt JM, Battlay P, Gledhill-Smith RS, Good RT, Lumb C, Fournier-Level A, Robin C (2017) Insights into DDT resistance from the Drosophila melanogaster genetic reference panel. Genetics 207(3):1181–1193

Scott JG (1999) Cytochromes P450 and insecticide resistance. Insect Biochem Mol Biol 29:757–777

Seong KM, Kim YH, Kwon DH, Lee SH (2012) Identification and characterization of three cholinesterases from the common bed bug, *Cimex lectularius*. Insect Mol Biol 21:149–159

Shi W, Sun J, Xu B, Li H (2013) Molecular characterization and oxidative stress response of a cytochrome P450 gene (CYP4G11) from Apis cerana cerana. Zeitschrift für Naturforschung C 68(11–12):509–521

Siegfried BD, Scharf ME (2001) Mechanisms of organophosphate resistance in insects. In: Ishaaya I (ed) Biochemical sites of insecticide action and resistance. Springer, Berlin, pp 269–321

Silva CP, Terra WR, de Sa MFG, Samuels RI, Isejima EM, Bifano TD, Almeida JS (2001) Induction of digestive alpha amylases in larvae of *Zabrotes subfasciatus* (Coleoptera: Bruchidae) in response to ingestion of common bean alpha-amylase inhibitor 1. J Insect Physiol 47:1283–1290

Small GJ, Hemingway J (2000) Molecular characterization of the amplified carboxylesterase gene associated with organophosphorus insecticide resistance in the brown plant hopper *Nilaparvata lugens*. Insect Mol Biol 9:647–653

Soderlund DM (1997) Molecular mechanisms of insecticide resistance. In: Sjut V (ed) Molecular mechanisms of resistance to agrochemicals, vol 13. Springer, Heidelberg, pp 21–56

Soderlund DM (2010) State-dependent modification of voltage-gated sodium channels by pyrethroids. Pestic Biochem Physiol 97(2):78–86

Stratonovitch P, Elias J, Denholm I, Slater R, Semenov MA (2014) An individual-based model of the evolution of pesticide resistance in heterogeneous environments: control of *Meligethes aeneus* population in oilseed rape crops. PLoS One 9(12):e115631

Strycharz JP, Lao A, Li H, Qiu X, Lee SH, Sun W, Yoon KS, Doherty JJ, Pittendrigh BR, Clark JM (2013) Resistance in the highly DDT-resistant 91-R strain of *Drosophila melanogaster* involves decreased penetration, increased metabolism, and direct excretion. Pestic Biochem Physiol 107 (2):207–217

Sukhoruchenko GI, Dolzhenko VI (2008) Problems of resistance development in arthropod pests of agricultural crops in Russia. OEPP/EPPO Bull 38:119–126

Thiaw O, Doucouré S, Sougoufara S, Bouganali C, Konaté L, Diagne N, Faye O, Sokhna C (2018) Investigating insecticide resistance and knock-down resistance (*kdr*) mutation in Dielmo, Senegal, an area under long lasting insecticidal-treated nets universal coverage for 10 years. Malar J 17:123

Vais H, Williamson MS, Devonshire AL, Usherwood PNR (2001) The molecular interactions of pyrethroid insecticides with insect and mammalian sodium channels. Pest Manag Sci 57:877–888

Vontas JG, Small GJ, Nikou DC, Ranson H, Hemingway J (2002) Purification, molecular cloning and heterologous expression of a glutathione S-transferase involved in insecticide resistance from the rice brown planthopper, Nilaparvata lugens. Biochem J 362(2):329–337

Vontas J, David JP, Nikou D, Hemingway J, Christophides GK, Louis C, Ranson H (2007) Transcriptional analysis of insecticide resistance in *Anopheles Stephensi* using cross-species microarray hybridization. Insect Mol Biol 16:315–324

Wang C, Scharf ME, Bennett GW (2004) Behavioral and physiological resistance of the German cockroach to gel baits (Blattodea: Blattellidae). J Econ Entomol 97:2067–2072

Weill M, Fort P, Berthomieu A (2002) A novel acetylcholinesterase gene in mosquitoes codes for the insecticide target and is non-homologous to the ace gene in *Drosophila*. Proc R Soc Lond Ser B 269:2007–2016

Whalon ME, Sanchez MD, Hollingworth RM (2008) Global pesticide resistance in arthropods. CAB International, Wallingford. ISBN 978-1-84593-353-1

Whalon ME, Sanchez MD, Hollingworth RM, Duynslager L (2010) Arthropod pesticide resistance database. www.pesticideresistance.com

Williamson MS, Denholm I, Bell CA, Devonshire AL (1993) Knockdown resistance (*kdr*) to DDT and pyrethroid insecticides maps to a sodium channel gene locus in the housefly (*Musca domestica*). Mol Gen Genet 240:17–22

Wondji CS, Dabire RK, Tukur Z, Irving H, Djouaka R, Morgan JC (2011) Identification and distribution of a GABA receptor mutation conferring dieldrin resistance in the malaria vector *Anopheles funestus* in Africa. Insect Biochem Mol Biol 41:484–491

Wood OR, Hanrahan S, Coetzee M, Koekemoer LL, Brooke BD (2010) Cuticle thickening associated with pyrethroid resistance in the major malaria vector *Anopheles funestus*. Vectors 3:67

Yahouédo GA, Chandre F, Rossignol M, Ginibre C, Balabanidou V, Mendez NGA, Pigeon O, Vontas J, Cornelie S (2017) Contributions of cuticle permeability and enzyme detoxification to pyrethroid resistance in the major malaria vector *Anopheles gambiae*. Sci Rep 7(1):11091

Yamamoto KY, Shigeoka Y, Aso Y, Banno M, Kimura T (2009) Molecular and biochemical characterization of a zeta-class glutathione S-transferase of the silk moth. Pesticide Biochem Physiol 95:125–128

You MS, Yue Z, He WY, Yang XH, Yang G, Xie M, Zhan DL, Baxter SW, Vasseur L, Gurr GM, Douglas CJ, Bai JL, Wang P, Cui K, Huang SG, Li XC, Zhou Q, Wu ZY, Chen QL, Liu CH, Wang B, Li XJ, Xu XF, Lu CX, Hu M, Davey JW, Smith SM, Chen MS, Xia XF, Tang WQ, Ke FS, Zheng DD, Hu YL, Song FQ, You YC, Ma XL, Peng L, Zheng YK, Liang Y, Chen YQ, Yu LY, Zhang YN, Liu YY, Li GQ, Fang L, Li JX, Zhou X, Luo YD, Gou CY, Wang JY, Wang J, Yang HM, Wang J (2013) A heterozygous moth genome provides insights into herbivory and detoxification. Nat Genet 45:220–225

You C, Shan C, Xin J, Li J, Ma Z, Zhang Y, Zeng X, Gao X (2020) Propoxur resistance associated with insensitivity of acetylcholinesterase (AChE) in the housefly, *Musca domestica* (Diptera: Muscidae). Sci Rep 10:8400

Zhang W, Yao Y, Wang H, Liu Z, Ma L, Wang Y, Xu B (2019) The roles of four novel P450 genes in pesticides resistance in *Apis cerana cerana* Fabricius: expression levels and detoxification efficiency. Front Genet 15(10):1000

Zhu YC, Snodgrass GL (2003) Cytochrome P450 CYP6X1 cDNAs and mRNA expression levels in three strains of the tarnished plant bug Lygus lineolaris (Heteroptera: Miridae) having different susceptibilities to pyrethroid insecticide. Insect Mol Biol 12(1):39–49

Zhu F, Gujar H, Gordon JR, Haynes KF, Potter MF, Palli SR (2013) Bed bugs evolved unique adaptive strategy to resist pyrethroid insecticides. Sci Rep 3:1456

Zlotkin E (2001) Insecticides affecting voltage-gated ion channel. In: Ishaaya I (ed) Biochemical sites of insecticide action and resistance. Springer, Berlin, pp 43–76

Chapter 11
Insect Microbiota and Host Immunity: An Emerging Target for Pest Control

Muhammad Shakeel, Abrar Muhammad, Shuzhong Li, Surajit De Mandal, Xiaoxia Xu, and Fengliang Jin

Abstract The study of insects and their associated microbial communities is an important field in agriculture, primarily due to the role of insects as pests. Recent advances in next-generation sequencing technology have aided in improving our understanding of the microbial communities associated with insects, revealing wide diversity in their taxonomy and function. The resident microorganisms can contribute to insect fitness by providing certain amino acids, vitamin B, and sterols for fungal partners. Some organisms protect their insect host by either making toxins or modifying the immune system of insects. Though, *Drosophila melanogaster* has been used as a model framework system to generate new insights into gut immunity and physiology; however, to date, little is known about the microbiota of other most damaging agricultural pests worldwide. In this chapter, we have discussed the omics approaches applied to study the insect gut microbiome and the advances made in this field. We have then reviewed the role of gut microbiota and the current understanding of intestinal epithelial immune responses of insects.

Keywords Insect · Microbial community · Pest · Immune response

11.1 Introduction

Insects are considered the most diverse and abundant animals on Earth, present in a wide range of ecological habitats. They have evolved to harbor diverse gut microbiota that helps them in physiological and ecological advantages. Insect gut microbiota provide protection against predators, pathogens, and parasites; helps in

M. Shakeel · S. Li · S. De Mandal · X. Xu · F. Jin (✉)
Key Laboratory of Bio-Pesticide Innovation and Application of Guangdong Province, College of Plant Protection, South China Agricultural University, Guangzhou, China
e-mail: jflbang@scau.edu.cn

A. Muhammad
Max Planck Partner Group, Institute of Sericulture and Apiculture, College of Animal Sciences, Zhejiang University, Hangzhou, China

S. De Mandal et al. (eds.), *New and Future Development in Biopesticide Research: Biotechnological Exploration*, https://doi.org/10.1007/978-981-16-3989-0_11

the digestion, pheromone production, reproduction; contribute to inter-and intraspecific communication and plays a significant role in the overall growth and development of the host (Gil et al. 2004; Jang and Kikuchi 2020; Tang et al. 2012). Ceja-Navarro et al. (2019) have shown how the wood feeding beetle and its microbiome have co-evolved to support their functions. They revealed that the gut contains specialized compartments, each with a unique microbiome that works together to turn the wood into energy, food for its young ones, and nutrients for forest growth. They also found that specialized gut compartments allow performing reactions to happen in the presence or absence of oxygen. Coevolution of insects and their gut microbes help the host digest plant polymers such as lignin and cellulose and produce by-products such as acetate and biofuels like hydrogen, ethanol, and methane (Ceja-Navarro et al. 2019). Shelomi et al. (2016) found that stick insects can make microbial enzymes themselves to digest their foods. They showed that specific gene codes for the essential enzymes jumped into the insect from the ancestral gut microbe using a "horizontal gene transfer" mechanism (Shelomi et al. 2016). It was also found that larval growth rate can be influenced by the gut microbial composition or joint effect of the microbiota and the host plant species. This supported the hypothesis that diet can affect the gut microbiota, and the composition of the gut microbiota affects the larval growth rate. It was also reported that bacterial communities are often entered into the insect through their foods and surrounding insects. Hammer et al. (2017) reported that microbes in the gut of the caterpillar species are extremely low in abundance and mainly derived from the leaf and host-specific resident symbionts are largely absent (Hammer et al. 2017).

Several synthetic pesticides are being used to control insect pests of crops around the globe. However, increasing pest resistance towards these chemical pesticides is a global concern, and scientific exploration is highly needed to understand the mechanisms that develop such resistance. Most of the widely studied resistance mechanisms are attributed to the evolutionary alternations in the insect genomes, including the up-regulation of the degrading enzymes, alteration of the drug target sites, and enhancement of the drug excretion (Kikuchi et al. 2012). However, the gut microbiota of insects often increases the insecticide resistance in insects, and thus, understanding the role of gut microbiota in insecticide resistance is highly important to fight against destructive pests. Various symbiotic bacteria have been identified, which help in detoxification of the insecticide, leading to the emergence of insecticide resistance among the pests (Boush and Matsumura 1967; Kikuchi et al. 2011, 2012). A recent study by Mason et al. (2019) demonstrates the importance of the gut microbiome for the development of insect-resistant crops. Gut microbiota is either directly involved in the biodegradation of pesticides or by inducing the immune response of the host (Mason et al. 2019). Although gut microbiota is associated with the beneficial interaction with the host, they were also reported to cause harmful effects in some hosts. It was found that *Bacillus thuringiensis* (*Bt*) is unable to cause infection in the hemolymph of the live insect; however, this entomopathogenic bacteria cause fatal septicemia after interaction with other microorganisms of the larval midgut (Broderick et al. 2006, 2009). Understanding this mechanism underlying the possible association between Bt and gut bacterial community and their insect-killing abilities would potentially lead to the development of novel biopesticides against pests. Host deregulated immune genotype often changes the gut microbial community composition, leading to host mortality (Broderick et al. 2006; Hammer et al. 2017; Ryu et al. 2008).

Besides, it has been shown that diet, life stage, and the surrounding environment played a vital role in shaping the gut microbial community in insects (Colman et al. 2012; Malacrinò et al. 2018). Studies across a diverse group of insects have shown that insect gut is associated with a very simple microbial community despite their important functions. This is due to the specific environmental conditions in the insect gut and the host regulatory mechanism that restrict the development of only specific gut microbial communities (Broderick and Lemaitre 2012; Cariveau et al. 2014; Corby-Harris et al. 2014; Hu et al. 2014; Huang et al. 2013; Jung et al. 2014; Kim et al. 2013; Köhler et al. 2012). Furthermore, it has also been observed that the composition of the gut microbial community is different between the field-collected and laboratory-reared insects (Belda et al. 2011). The present chapter aims to illustrate the significance of gut microbiota in regulating host immunity. Understanding the role of symbiotic bacteria in host immunity in important insect pests may provide novel leads towards the development of novel pest management techniques.

11.2 Omics Approaches Applied to Study the Insect Gut Microbiome

Several studies conducted on insect gut microbiota are based on culture-dependent approaches that often produce biased results. However, with the introduction of omics-based approaches, it now possible to obtain a detailed understanding of the composition of the insect gut microbial communities and their role in the host. The introduction of next-generation sequencing drastically decreased the cost of sequencing that led to the generation of a huge amount of sequencing data, mainly 16S rRNA gene sequences, which is universally present in bacteria. This gene contains both conserved and variable regions suitable for PCR amplification as well as sequencing (Dave et al. 2012). Different NGS platforms such as Illumina, 454 pyrosequencing, Ion Torrent have been used to study the microbiota based on the 16S rRNA. Amplicon sequencing is used to decipher taxonomic profiling using marker genes and contains common steps such as denoising, Chimera detection, OTU clustering, Taxonomic classification, and Statistical analysis (De Mandal et al. 2015). This technique is used in several insect species to identify gut microbial communities (Yun et al. 2014). On the other hand, shotgun sequencing is used to study the microbial community's taxonomic and functional aspects. Common steps involve assembly, binning, functional annotation, and assignment to the predicted protein-coding sequences (De Mandal et al. 2015). Metagenomics sequencing is used to characterize the mixed populations of microorganisms. For example, the study of the fungus garden microbiome of leafcutter ants identifies diverse fungal and bacterial member. Metagenomics study has shown that these bacteria might involve in the symbiotic degradation of plant biomass. Therefore, this tool provides important opportunities for understanding the community functions (Suen et al. 2010). Moreover, the introduction of several bioinformatics tools such as Tax4Fun, PICRUSt has also been used to predict the functional potential of the microorganism based on the 16S rRNA gene sequences (Langille et al. 2013).

11.3 Gut Microbiota of Red Palm Weevil

Rhynchophorus ferrugineus (Olivier) (Coleoptera: Dryophthoridae), commonly known as red palm weevil (RPW) is arguably one of the deadliest pests of the palm cultivations worldwide (Al-Dosary et al. 2016; Ju et al. 2011; Peng et al. 2016). RPW is a concealed tissue borer feeding on the pith that renders the fronds yellow and dry (Faleiro 2006; Muhammad et al. 2017). The pest is native to South Asia and Melanesia but has invaded all major palm growing countries in Asia, Africa, Australia, and Europe (Murphy and Briscoe 1999; Wan and Yang 2016). Several features, including its concealed nature, high reproductive rate, enhanced sperm storage capability, broad host range, adaptability to a wide range of climatic conditions, strong flight ability, presence of several overlapping generations, and strong immunity, have contributed to the invasion success of this pest (Ajlan and Abdulsalam 2000; El-Mergawy and Al-Ajlan 2011). One often neglected yet crucial attribute to this species' evolutionary success is its symbiotic association with diverse microbiota residing in the gut (Muhammad et al. 2017). The gut microbiota of insects plays many key roles in host physiology and influence many aspects, including nutrition metabolism (Janssen and Kersten 2015; Muhammad et al. 2017; Wong et al. 2014), growth and development (Blatch et al. 2010; Wong et al. 2014), innate immunity (Kim et al. 2015; Muhammad et al. 2019a, b), mating behavior (Ami et al. 2010), and homeostasis (Douglas 2015; Engel and Moran 2013).

The gut microbiota of RPW has been deciphered using the culture-dependent and sequencing-based culture-independent methods (Jia et al. 2013; Montagna et al. 2015; Muhammad et al. 2017; Tagliavia et al. 2014). Mainly, RPW gut microbiota is represented by Proteobacteria, Firmicutes, Bacteroidetes, and Actinobacteria in the families; Enterobacteriaceae, Acetobacteriaceae, Lactobacillaceae, Enterococcaceae, Entomoplasmataceae, and Streptococcaceae (Jia et al. 2013; Montagna et al. 2015; Muhammad et al. 2017; Tagliavia et al. 2014). These microbes have improved host fitness and immunity through nutrient metabolism and immune system regulation (Habineza et al. 2019; Muhammad et al. 2017; Muhammad et al. 2019a, b) (Fig. 11.1). Germ-free (GF) RPW larvae have been generated with the aim to dissect the complex interplay between the host and its gut microbiota (Muhammad et al. 2019a, b). Subsequent studies revealed that the RPW gut microbiota positively impacts the host's biology through energy metabolism and digestion of dietary carbohydrates (Habineza et al. 2019; Muhammad et al. 2019a, b). Several

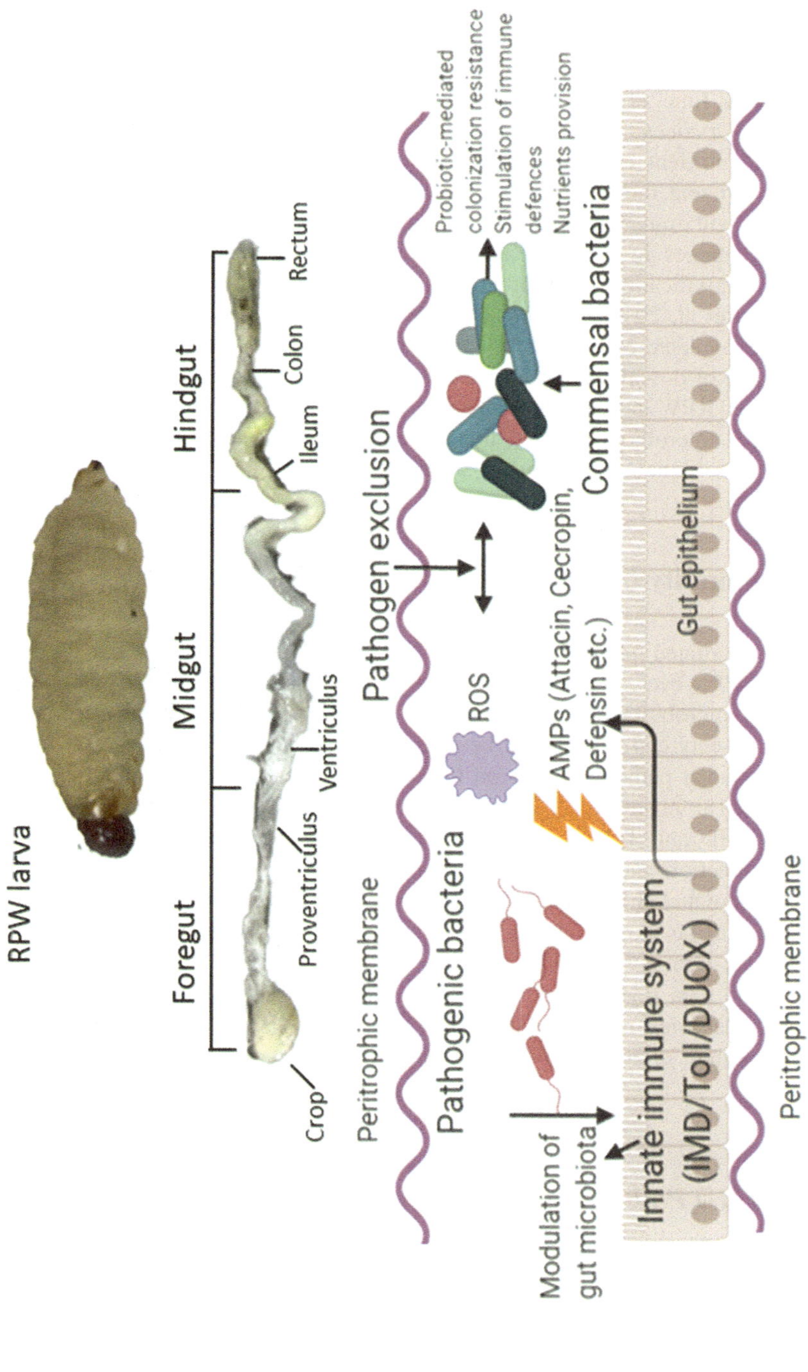

Fig. 11.1 Intestinal immune system regulation and promoting effects of RPW gut microbiota

cellulose-degrading bacterial symbionts, including *Salmonella enterica, Enterobacter cloacae, Citrobacter koseri, Raoultella, Klebsiella pneumonia, K. variicola,* and *K. oxytoca* have been isolated from RPW guts (Muhammad et al. 2017). RPW devoid of gut microbiota exhibited decreased body mass and prolonged developmental time, affirming their promoting effects (Habineza et al. 2019). Furthermore, *Lactococcus lactis* and *E. cloacae,* when introduced into the GF larvae, restored the nutrient contents (protein, glucose, and triglycerides), which were otherwise deficient the bacteria lacking counterparts (Habineza et al. 2019). The comparative gut transcriptome of GF and conventionally reared RPW revealed significant changes in the transcript profile of unigenes involved in immunity, digestion, detoxification, and other metabolic functions of this insect species (Muhammad et al. 2019a, b). Interestingly, the expression of these unigenes was found to be regulated by gut microbiota as the supplementation of symbionts led to normalization of downstream consequences and was closely resemble the conventionally reared RPW individuals. Moreover, the immuno-suppressed state of the host resulted from microbiota deletion underscores the immune stimulatory effects of RPW gut microbiota (Muhammad et al. 2019a, b).

11.4 Immunity of Red Palm Weevil

Symbionts-mediated physiological traits in insects' biology are relatively well established. However, we only have a nascent understanding of the interactions between gut microbiota and immune system regulation in RPW. The innate immune system of insects can distinguish between self and non-self-entities and confers resistance against pathogen infection (Casanova-Torres and Goodrich-Blair 2013; Lin et al. 2018; Zhang et al. 2015). Genetic studies have identified two NF-κB pathways (IMD and Toll) that induce immune effector molecules in response to microbial infection and maintains the homeostasis of gut microbiota (Guo et al. 2014; Lemaitre and Hoffmann 2007). Also, the dual oxidase (DUOX) pathway has been known to be the primary immune response in gut immunity of *Drosophila melanogaster* (Ha et al. 2009a, b; Phoebe Tzou et al. 2000). The RPW gut transcriptome data revealed more than 550 immune-related genes from different functional categories, namely immune recognition, signal modulation, signal transduction, immune effectors, and other immune-related genes (Muhammad et al. 2019a, b).

11.4.1 IMD Pathway

The evolutionary conserved IMD pathway was first revealed by identifying its role in systemic immunity to activate the production of antimicrobial peptides (Lee et al. 2017; Lemaitre and Hoffmann 2007). Later on, it was found that the IMD pathway also regulates the gut bacterial load and antagonizes their excessive proliferation (Guo et al. 2014). The IMD pathway generates a cascade of events in response to indigenous danger include; (1) recognition of pathogen-associated molecular patterns (PAMPs), which is executed by pathogen recognition receptors (PRRs) such as peptidoglycan recognition proteins (PGRPs), C-type lectins, scavenger receptors, β-1,3-GRPs and so on (2) the intracellular signaling cascade that leads to the translocation of *Relish*, a transcription factor homolog of Nuclear Factor-kappa Beta (NF-κB), (3) the synthesis of immune effector molecules such as antimicrobial peptide and lysozymes, and finally (4) activation of negative regulatory mechanisms to maintain the inherent homeostasis (Lee et al. 2017). Bacterial derived peptidoglycan (PGN) from gram-negative or gram-positive bacteria are recognized by a family of long (PGRP-LB and PGRP-LC) or short (PGRP-SC) proteins (Lee et al. 2017). In RPW, a secreted catalytic PGRP called *Rf*PGRP-LB was highly expressed in the gut and fat body, indicating its role in host immune response. Its expression was induced in response to *E. coli,* a gram-negative bacterium but did not induce in response to a gram-positive (*S. aureus*) bacterium (Dawadi et al. 2018). Recombinant protein r*Rf*PGRP-LB has been shown to cause agglutination of *S. aureus* and *E. coli,* suggesting its antibacterial activity. The role of *Rf*PGRP-LB in gut microbiota homeostasis of RPW was confirmed by RNAi (Dawadi et al. 2018). The knockdown of *Rf*PGRP-LB led to an increase in the gut bacterial load and alterations in the community structure and composition of RPW gut microbiota. Moreover, the knockdown increased the transcript abundance of antimicrobial peptide Attacin, inferring its function as a negative regulator of the IMD pathway in the RPW gut (Dawadi et al. 2018). Several pathogen recognition proteins, including PGRP-LB, PGRP-SC1a, PGRP-SC1b, and PGRP-SC2, have been shown to negatively regulate the IMD signals due to their PGN degrading amidase activity (Bischoff et al. 2006; Guo et al. 2014; Paredes et al. 2011). The negative regulation can also be achieved by the activation of regulators such as PIRK that interacts with PGRP-LC, PGRP-LE, and Imd, which subsequently disrupts the formation of the IMD signaling complex and inhibits the overactivation of the pathway (Aggarwal and Silverman 2008). Besides, the negative regulatory proteins contain Dnr1 (defense repressor) and Caspar for Dredd inhibition, Tarbid, CYLD (cylindromatosis), Skp A, Caudal (transcriptional repressor), and the Oct 1homolog Nubbin (transcriptional repressor) (Guntermann et al. 2009; Kim et al. 2006; Lee et al. 2017).

Moreover, the NF-κB transcription factor Relish in RPW (*Rf*Relish) was highly expressed in the immunity-related tissues (gut, fat body, and hemolymph) (Xiao et al. 2019). The *Rf*Relish induced strongly upon bacterial oral and systemic infection, suggesting its involvement in warding off the invaded bacteria. Besides, the knock-down of *Rf*Relish altered the gut bacterial load and relative abundance of the major taxa (Xiao et al. 2019). Together, these observations indicate that *Rf*Relish is critical for immune defense against the bacterial challenge and modulating the homeostasis of gut microbiota of RPW through the regulation of antimicrobial peptides such as cecropin and defensin (Xiao et al. 2019). Recently, a ligand of Toll receptors in RPW (*Rf*Spätzle) has been shown to play a role in host immunity and the homeostasis of gut microbiota (Muhammad et al. 2020). Upon infection (mainly by gram-positive bacteria or fungi), the spätzle ligands bind the Toll receptors and initiate the downstream cascade of antimicrobial peptides secretion. The inhibition of *Rf*Spätzle in RPW led to downregulation of coleoptericin and cecropin that rendered the host vulnerable to microbial infection. Also, the inhibition of *Rf*Spätzle impacted the relative proportion of RPW gut microbiota (Muhammad et al. 2020). Thus, it has become clear that the IMD and Toll signaling pathways in RPW are involved in the immunity and maintaining the homeostasis of gut microbiota (Dawadi et al. 2018; Muhammad et al. 2020; Xiao et al. 2019).

11.4.2 The DUOX (Dual Oxidase) Pathway

For a more robust immune response and efficient elimination of ingested microorganisms, a parallel defense mechanism in the gut epithelium is always working alongside the IMD pathway called the dual oxidase pathway (DUOX) through the production of reactive oxygen species (ROS) (Ha et al. 2005a, b). ROS is a vital bactericidal molecule against the oral infection of non-commensal bacteria in the gut epithelium. Immune regulatory catalases (IRC) are the gut cells that remove ROS because excess ROS is harmful to gut physiology and homeostasis (Ha et al. 2005a, b). Flies with the reduced or knock-downed secretory IRC activity exhibited increased lethality due to an excessive ROS level in the gut lumen that should have otherwise normalized by IRC (Ha et al. 2005a, b). ROS is generated by a member of the family nicotinamide adenine dinucleotide phosphate oxidase (NOX) and acts as the first line of defense in gut immunity (Lee et al. 2017). The NOX/DUOX family shares a catalytic gp91phox domain. In addition, DUOX contains a peroxidase homolog domain (PHD). In the gut epithelium of *D. melanogaster*, the NOX/DUOX homolog produces HOCl (Hypochlorous acid) or an additional protein with a similar activity (Lee et al. 2017; Lemaitre and Hoffmann 2007). Genetic studies in *D. melanogaster* have demonstrated the role of DUOX in gut immunity. For example, DUOX knockout flies showed decreased survival ability upon oral bacterial or yeast infection and negatively impacted the bacterial clearing ability towards gut infected pathogen (Ha et al. 2005a, b).

Interestingly, the DUOX signaling system is activated explicitly with the transient microorganism while tolerating the presence of indigenous gut microbiota (Chen et al. 2016). For instance, oral infection with fly indigenous microbiota such as *Lactobacillus plantarum, Acetobacter pomorum,* and *Commensalibacter intestini* did not induce ROS generation, whereas it was induced by the oral infection of exogenous *Erwinia carotovora carotovora ECC15* (Lee et al. 2013). This suggests that the DUOX signaling system can recognize the bacterial derived substances from transient and indigenous microorganisms distinctively (Lee et al. 2017). Further studies on *Drosophila* system have confirmed that the sole infection of bacterial ligands (PGNs) was unable to induce ROS production. Still, the whole bacteria can activate the ROS synthesis (Ha et al. 2009a, b). High-performance liquid chromatography (HPLC) analysis identified a nucleobase in uridine called uracil which was found to be the ligand responsible for DUOX activation. Later, it was confirmed that uracil is secreted from non-commensal bacteria such as *Vibrio, Klebsiella, ECC15, Serratia, Pseudomonas,* and *Shigella* but not from the commensal bacterium (*C. intestini*) (Lee et al. 2013). Regulation of DUOX activity requires Gαq- and PLCβ-dependent calcium mobilization from the enterocytes (Lee et al. 2017). DUOX expression and its enzymatic activity are regulated by transcription factor ATF2 and its upstream p38 pathway (Lee et al. 2017). Given that, it is obvious that DUOX activation is achieved through the PGN-dependent pathway or uracil-dependent PLCβ pathway (Ha et al. 2009a, b). In the non-infection state, p38 in the presence of commensal microbiota remains inactive, suggesting the presence of a negative regulator (MKP3). Whereas in a state of infection, a massive activation of DUOX occurs that is capable of tolerating a certain level of PGNs from commensal microbiota and destroy the pathogen (Ha et al. 2009a, b). Hedgehog signaling has also been recently known to regulate DUOX activation (Lee et al. 2015). When uracil activates the Hedgehog pathway, it employs Cad99C present in the apical membrane of enterocytes to induce the DUOX-dependent ROS generation (Lee et al. 2015). Flies with MKPs knockdown (Ha et al. 2009a) or reduced Hedgehog signaling activity were more susceptible to enteric infection, confirms its role in ROS production and insect immunity (Lee et al. 2015). The role of the DUOX pathway in the regulation of RPW immunity and gut microbiota population structure is yet to be confirmed.

11.5 Gut Microbiota of Diamondback Moth

The diamondback moth, *Plutella xylostella,* is one of the devastating pests of cruciferous vegetable crops worldwide, which has developed resistance to various insecticides in the field (Furlong et al. 2013; Shakeel et al. 2017a, b). Insect gut microbiota has diverse functions in insect behavior and physiology, such as olfactory behavior and development in *D. melanogaster* (Qiao et al. 2019), nutrition in honey bee (Engel et al. 2012), pathogen invasion in locust (Dillon et al. 2005), and mediate pesticide resistance in *P. xylostella* (Xia et al. 2018). The most dominant gut bacterial member of *P. xylostella* during whole life stages is *Bacillus* sp., and the most abundant phyla in the larval gut is Proteobacteria (Lin et al. 2015). Insecticide-resistant lines of *P. xylostella* had more Lactobacillales and fewer Enterobacteriales compared with the susceptible strain (Xia et al. 2013), which indicated that *P. xylostella* gut microbiota has a possible relationship with insecticide resistance. Although much research work has confirmed that *P. xylostella* gut microbiota mediates insecticide resistance, the underlying mechanism is still unclear. In *Bactrocera dorsalis*, gut microbiota could direct biodegradation of the pesticide

(Cheng et al. 2017); this is one important aspect but not the main mechanism for the diamondback moth (Xia et al. 2018). Another key process is the modulation of the immune response, as an increase in midgut microbiota could induce immune priming in the host insect and caused *Spodoptera exigua* more tolerant to *B. thuringiensis* (Hernandez-Martinez et al. 2010). Besides mediated pesticide resistance, the gut microbiota of *P. xylostella* also plays a key role in food digestion and adaptation.

There exist some specific gut bacteria of *P. xylostella* that participate in the detoxification of plant phenolic, leads host insect to adapt to a specific group of plants (Xia et al. 2017). These gut bacteria secrete a series of enzymes and metabolites which could help the host digest and absorb nutrients and minimize damage from plant defense compounds. While the composition of *P. xylostella* gut microbiota is complex, the functional bacteria species involved in food adaption are few, and composition and diversity of gut microbiota alters upon switching the hosts (Yang et al. 2020). This reminds us that the co-evolution between gut microbiota and host insect cope with the different food sources. In addition to the defensive components of plants in food, insects also have to deal with a variety of pathogenic microorganisms from food and the environment. It has been reported that the extracellular symbiont (*E mundtii*) of *S. littoralis* selectively clears host intestinal lumen pathobionts by secreting a stable antimicrobial peptide, improving normal intestinal growth and reducing the risk of infection in the gut (Shao et al. 2017). *E. mundtii* and *Carnobacterium maltaromaticum* dominate the composition of midgut microbiota in deltamethrin-resistant and deltamethrin-susceptible populations of *P. xylostella* (Li et al. 2017). One strain of *E. mundtii* isolated from larval feces of the *Ephestia kuehniella* can protect *Tribolium castaneum* against Bt infection (Grau et al. 2017). Moreover, for a long time the controversy over the role of gut microbiota in Bt pathogenicity has been in the spotlight. Based on a study conducted in 2006, it was found that the survival of gypsy moth varies according to the midgut microbiota against different units of Bt (Broderick et al. 2006). But recent studies have demonstrated that gut bacteria are not required for insecticidal activity of Bt in *Manduca sexta* (Johnston and Crickmore 2009). Since then, various experimental approaches have been taken to address this key research issue, while until now it is difficult to achieve a consistent conclusion. Bt toxin treatment disturbed the dynamic balance of *P. xylostella* gut microbiota, including changed the bacteria composition, abundance, and distribution, and decreased bacterial diversity. Meanwhile, Bt toxin treatment-induced host midgut immune response, and the gut microbiota could mediate midgut immune response and shifted the basic level of the midgut immune system (Li et al. 2020a, b). This indicated that *P. xylostella* midgut immune response plays important roles during the Bt infection process. To better understand the role of insect gut microbiota in Bt infection, it is necessary to clarify the tripartite interaction relationship between Bt infection, host gut microbiota, and midgut immune response (Li et al. 2020a, b).

11.6 Immune Response of Diamondback Moth

The immune responses of the diamondback moth, *P. xylostella*, are important for defense against pathogen infection, and as one of the major lepidopteran pest, the study of *P. xylostella* immune response is helpful for the development of biological control strategies. Like model insect *D. melanogaster*, the immune response of *P. xylostella* consists of the cellular and humoral immune response (Casanova-Torres and Goodrich-Blair 2013). Cellular immune response mainly relies on the action of hemocytes (Kanost et al. 2004), while humoral immune response includes phenoloxidases mediated melanization and the production of immune effector molecules (Cerenius and Soderhall 2004; Xia et al. 2015). *P. xylostella* midgut is directly exposed to many microorganisms from food or the environment; some of them can be pathogenic and induce local midgut immune response, such as *B. thuringiensis* (Lin et al. 2020). Local midgut immune response is executed by the generation of reactive oxygen species (ROS) and synthesis of various antimicrobial peptides (AMPs) (Buchon et al. 2013). A DUOX protein produces ROS with an extracellular peroxidase domain, which in the presence of chloride can convert H_2O_2 into HOCl and is therefore detoxified in the presence of IRC catalase (Ha et al. 2005a, b). The two main pathways (Toll and IMD) of the immune system produce pathogen-specific AMPs. These pathways regulate the AMPs expression to either Gram-negative bacterial infection or Gram-positive bacterial or fungal infection (Tanji and Ip 2005). In the following part, we will focus on the Toll and IMD pathway and DUOX pathway.

11.6.1 Toll and IMD Pathway

The insect Toll pathway is triggered by fungi, yeast, and Gram-positive bacteria, while the IMD pathway is induced mostly by Gram-negative bacteria (Buchon et al. 2014). These two pathways work synergistically to defend against microbial infection. And the insect becomes more susceptible to many pathogens in the absence of the Toll and IMD pathway (Tzou et al. 2002). The components of the Toll signaling pathway include the tube, MyD88, Pelle, Cactus, Dorsal, and Dif (Sun et al. 2004); these pathway genes were all identified except MyD88 in *P. xylostella* (Xia et al. 2015). The activation of the Toll and IMD pathway needs initial recognition of pathogen-associated molecular patterns (PAMPs), which rely on host pattern recognition receptors (PRRs) (Mogensen 2009). PRRs, including peptidoglycan recognition proteins (PGRPs), β-1,3 Glucan recognition proteins (βGRPs), scavenger receptors, and lectins (Hultmark 2003). It has been reported that PGRP-SA triggers the Toll signaling pathway through recognition of Gram-positive bacteria (Valanne et al. 2011), and the IMD pathway is initiated by PGRP-LC (Gottar et al. 2002). Most of the identified PGRPs expression was suppressed after infected with *Isaria fumosorosea* in *P. xylostella*, the Toll signaling pathway was not induced except MyD88 and spätzle post *I. fumosorosea* infection, and Imd pathway was also not triggered post infected with *I. fumosorosea* at different time points (Xu et al. 2017a, b). This indicated that *I. fumosorosea* can suppress *P. xylostella* humoral immune response. While in another study, Toll and IMD pathways were both activated in response to the virulence factors of fungi and destruxin A (Shakeel et al. 2017a, b).

Inset immune response is tissue-specific among different species. For example, some researchers reported that the Toll pathway might involve in midgut immunity in *Bombyx mori* (Ma et al. 2013; Wu et al. 2010). The IMD pathway plays a crucial role in regulating the expression of AMPs in *D. melanogaster*; however, the Toll pathway is not involved in it directly (Nehme et al. 2007; Vodovar et al. 2005). But in *P. xylostella*, it has been confirmed that both the Toll and IMD pathways participated in midgut immunity (Lin et al. 2018). In contrast, the IMD signaling pathway's function is more conserved in insect midgut immunity, which can induce the expression of AMPs to fight off pathogen infection. During the host immune response process, many components of the *P. xylostella* Toll and IMD pathway were regulated by miRNAs. It has been found that one conserved microRNA, miR-8, block the activation of *P. xylostella* Toll pathway by upregulating Serpin 27 expression (Etebari and Asgari 2013). After that, many research also found that the host *P. xylostella* miRNA expression profiles were shifted after different pathogen infection (Li et al. 2019; Shakeel et al. 2018; Xu et al. 2017a, b), in which many of the differentially expressed miRNAs targeted the Toll and IMD signaling pathway genes, the detail functions of specific miRNAs are needed and deserved further studies.

11.6.2 DUOX Pathway

The production of reactive oxygen species (ROS) is mainly generated by DUOX pathway (Ha et al. 2005a, b), and it is one of the main immune responses in *P. xylostella* midgut. To keep gut redox balance, overproduced ROS is detoxified by the antioxidant enzymes such as peroxidase, and catalase in insect gut (Ha et al. 2005a, b). In *P. xylostella*, several detoxifying enzymes were induced after different microbial infections (Lin et al. 2018), which indicated that the production of ROS was also triggered in the midgut of *P. xylostella*. The ROS and AMPs can work synergistically to regulate the proliferation and growth of invading microorganisms and maintain the gut homeostasis (Yao et al. 2016). In contrast with the activation of insect Toll and IMD pathway, the DUOX pathway only can only be activated by pathogens and not by gut symbionts (Bae et al. 2010; Lee et al. 2013). It has recently been reported that ROS level is also related to *P. xylostella* defense against Bt infection, and reduce ROS level through a knockdown expression of DUOX gene caused significantly increased Btk pathogenicity (Sajjadian and Kim 2020). The

Duox is involved in paracrine signaling in zebrafish epidermal cells to eliminate wound inflammation. In *Anopheles gambiae*, peroxidases and Duox help create a dityrosine network between the intestinal epithelial and peritrophic membranes, leading to reduced intestinal permeability of the immune elicitors. As a result of the disruption in this network, the host triggers pathogen-specific immune responses. In contrast, the function of DUOX pathway genes in *P. xylostella* needs to be explored further in future studies.

11.7 Concluding Remarks and Future Perspectives

This chapter highlights our current knowledge on the omics approaches applied to study the insect gut microbiome. The availability of next-generation sequencing has helped to broaden our knowledge base about the microbiota of insects. In the future, NGS-based studies will enhance our understanding of the bacterial communities' ecology linked to these insects, helping us consider the various factors affecting their composition. In addition, we have elaborated on the interactions between gut microbiota and immune system regulation in insects, particularly in RPW and DBM. The existence of intestinal microbiota may significantly improve the survival rate of RPW challenged by bacteria via upregulating the essential immune genes to enhance host immunocompetence, indicating that intestinal bacteria have strong stimulating effects on the immune system of RPW. Since insects cannot fend off pathogens on their own, innate immunity is the only way these pests will fight off pathogens. However, in *P. xylostella*, when the Cry1Ac toxin activates the immune system, the microbiota plays a significant role in facilitating the killing of *P. xylostella*. Treatment with Cry1Ac induces intestinal microbiota dysbiosis, and the host immune system stimulated to maintain intestinal microbiota homeostasis; in turn, intestinal microbiota dysbiosis, together with Bt toxin, exaggerates the host immune response, resulting in exacerbated harm to the host tissues, eventually leading to host death. The link between Bt toxin and gut immunity and the microbiota might highlight the important role played by these factors in the insecticidal process. These findings could lead to new methods of agricultural biocontrol. Besides, from an applied perspective, these findings have important implications, setting the stage for the development of selective RNAi-based insect pest control strategies to enhance the effectiveness of natural antagonists by reducing the host's immunocompetence.

Acknowledgments This work was supported by grants from the National Natural Science Foundation of China (31972345), Natural Science Foundation of Guangdong, China (2018A030313402, 2019A1515011221) and the Key-Area Research and Development Program of Guangdong Province (2019B020218009).

References

Aggarwal K, Silverman N (2008) Positive and negative regulation of the Drosophila immune response. BMB Rep 41(4):267–277

Ajlan AM, Abdulsalam K (2000) Efficiency of pheromone traps for controlling the red palm weevil Rhynchophorus ferrugineus Olivier (Coleoptera: Curculionidae), under Saudi Arabia conditions. Bull Entomol Soc Egypt (Econ Ser) 27:109–120

Al-Dosary NMN, Al-Dobai S, Faleiro JR (2016) Review on the management of red palm weevil *Rhynchophorus ferrugineus* Olivier in date palm Phoenix dactylifera L. Emirates J Food Agri 28 (1):34

Ami E, Yuval B, Jurkevitch E (2010) Manipulation of the microbiota of mass-reared Mediterranean fruit flies *Ceratitis capitata* (Diptera: Tephritidae) improves sterile male sexual performance. ISME J 4(1):28–37. https://doi.org/10.1038/ismej.2009.82

Bae YS, Choi MK, Lee WJ (2010) Dual oxidase in mucosal immunity and host-microbe homeostasis. Trends Immunol 31(7):278–287. https://doi.org/10.1016/j.it.2010.05.003

Belda E, Pedrola L, Peretó J, Martínez-Blanch JF, Montagud A, Navarro E, Porcar M (2011) Microbial diversity in the midguts of field and lab-reared populations of the European corn borer Ostrinia nubilalis. PLoS One 6(6):e21751

Bischoff V, Vignal C, Duvic B, Boneca IG, Hoffmann JA, Royet J (2006) Downregulation of the Drosophila immune response by peptidoglycan-recognition proteins SC1 and SC2. PLoS Pathog 2(2):e14

Blatch SA, Meyer KW, Harrison JF (2010) Effects of dietary folic acid level and symbiotic folate production on fitness and development in the fruit fly Drosophila melanogaster. Fly (Austin) 4 (4):312–319. https://doi.org/10.4161/fly.4.4.13258

Boush MG, Matsumura F (1967) Insecticidal degradation by Pseudomonas melophthora, the bacterial symbiote of the apple maggot. J Econ Entomol 60(4):918–920

Broderick NA, Lemaitre B (2012) Gut-associated microbes of Drosophila melanogaster. J Gut Microbes 3(4):307–321

Broderick NA, Raffa KF, Handelsman J (2006) Midgut bacteria required for Bacillus thuringiensis insecticidal activity. Proc Natl Acad Sci U S A 103(41):15196–15199. https://doi.org/10.1073/pnas.0604865103

Broderick NA, Robinson CJ, McMahon MD, Holt J, Handelsman J, Raffa KF (2009) Contributions of gut bacteria to Bacillus thuringiensis-induced mortality vary across a range of Lepidoptera. BMC Biol 7(1):11

Buchon N, Broderick NA, Lemaitre B (2013) Gut homeostasis in a microbial world: insights from Drosophila melanogaster. Nat Rev Microbiol 11(9):615–626. https://doi.org/10.1038/nrmicro3074

Buchon N, Silverman N, Cherry S (2014) Immunity in Drosophila melanogaster from microbial recognition to whole-organism physiology. Nat Rev Immunol 14(12):796–810. https://doi.org/10.1038/nri3763

Cariveau DP, Powell JE, Koch H, Winfree R, Moran NA (2014) Variation in gut microbial communities and its association with pathogen infection in wild bumble bees (Bombus). ISME J 8(12):2369–2379

Casanova-Torres AM, Goodrich-Blair H (2013) Immune signaling and antimicrobial peptide expression in Lepidoptera. Insects 4(3):320–338. https://doi.org/10.3390/insects4030320

Ceja-Navarro JA, Karaoz U, Bill M, Hao Z, White RA, Arellano A, Conrad ME (2019) Gut anatomical properties and microbial functional assembly promote lignocellulose deconstruction and colony subsistence of a wood-feeding beetle. Nat Microbiol 4(5):864

Cerenius L, Soderhall K (2004) The prophenoloxidase-activating system in invertebrates. Immunol Rev 198:116–126. https://doi.org/10.1111/j.0105-2896.2004.00116.x

Chen XQ, Lee KA, Ren XT, Ryu JC, Kim G, Ryu JH, Yoon J (2016) Synthesis of a highly HOCl-selective fluorescent probe and its use for imaging HOCl in cells and organisms. Nat Protoc 11 (7):1219–1228. https://doi.org/10.1038/nprot.2016.062

Cheng DF, Guo ZJ, Riegler M, Xi ZY, Liang GW, Xu YJ (2017) Gut symbiont enhances insecticide resistance in a significant pest, the oriental fruit fly Bactrocera dorsalis (Hendel). Microbiome 5:13. https://doi.org/10.1186/s40168-017-0236-z
Colman DR, Toolson EC, Takacs-Vesbach C (2012) Do diet and taxonomy influence insect gut bacterial communities? Mol Ecol 21(20):5124–5137
Corby-Harris V, Maes P, Anderson KE (2014) The bacterial communities associated with honey bee (Apis mellifera) foragers. PLoS One 9(4):e95056
Dave M, Higgins PD, Middha S, Rioux KPJTR (2012) The human gut microbiome: current knowledge, challenges, and future directions. Transl Res 160(4):246–257
Dawadi B, Wang XH, Xiao R, Muhammad A, Hou YM, Shi ZH (2018) PGRP-LB homolog acts as a negative modulator of immunity in maintaining the gut-microbe symbiosis of red palm weevil, Rhynchophorus ferrugineus Olivier. Dev Comp Immunol 86:65–77. https://doi.org/10.1016/j.dci.2018.04.021
De Mandal S, Panda A, Bisht S, Kumar N (2015) Microbial ecology in the era of next generation sequencing 1:2
Dillon RJ, Vennard CT, Buckling A, Charnley AK (2005) Diversity of locust gut bacteria protects against pathogen invasion. Ecol Lett 8(12):1291–1298. https://doi.org/10.1111/j.1461-0248.2005.00828.x
Douglas AE (2015) Multiorganismal insects: diversity and function of resident microorganisms. Annu Rev Entomol 60:17–34. https://doi.org/10.1146/annurev-ento-010814-020822
El-Mergawy R, Al-Ajlan A (2011) Red palm weevil, Rhynchophorus ferrugineus (Olivier): economic importance, biology, biogeography and integrated pest management. J Agric Sci Technol A 1:1–23
Engel P, Moran NA (2013) The gut microbiota of insects—diversity in structure and function. FEMS Microbiol Rev 37(5):699–735. https://doi.org/10.1111/1574-6976.12025
Engel P, Martinson VG, Moran NA (2012) Functional diversity within the simple gut microbiota of the honey bee. Proc Natl Acad Sci U S A 109(27):11002–11007. https://doi.org/10.1073/pnas.1202970109
Etebari K, Asgari S (2013) Conserved microRNA miR-8 blocks activation of the toll pathway by upregulating serpin 27 transcripts. RNA Biol 10(8):1356–1364. https://doi.org/10.4161/rna.25481
Faleiro J (2006) A review of the issues and management of the red palm weevil Rhynchophorus ferrugineus (Coleoptera: Rhynchophoridae) in coconut and date palm during the last one hundred years. Int J Trop Insect Sci 26(03):135–154
Furlong MJ, Wright DJ, Dosdall LM (2013) Diamondback moth ecology and management: problems, progress, and prospects. Annu Rev Entomol 58:517–541. https://doi.org/10.1146/annurev-ento-120811-153605
Gil R, Latorre A, Moya A (2004) Bacterial endosymbionts of insects: insights from comparative genomics. J Environ Microbiol 6(11):1109–1122
Gottar M, Gobert V, Michel T, Belvin M, Duyk G, Hoffmann JA, Royet J (2002) The Drosophila immune response against gram-negative bacteria is mediated by a peptidoglycan recognition protein. Nature 416(6881):640–644. https://doi.org/10.1038/nature734
Grau T, Vilcinskas A, Joop G (2017) Probiotic Enterococcus mundtii isolate protects the model insect Tribolium castaneum against Bacillus thuringiensis. Front Microbiol 8:1261. https://doi.org/10.3389/fmicb.2017.01261
Guntermann S, Primrose DA, Foley E (2009) Dnr1-dependent regulation of the Drosophila immune deficiency signaling pathway. Dev Comp Immunol 33(1):127–134
Guo L, Karpac J, Tran SL, Jasper H (2014) PGRP-SC2 promotes gut immune homeostasis to limit commensal dysbiosis and extend lifespan. Cell 156(1–2):109–122
Ha EM, Oh CT, Bae YS, Lee WJ (2005a) A direct role for dual oxidase in Drosophila gut immunity. Science 310(5749):847–850. https://doi.org/10.1126/science.1117311

Ha EM, Oh CT, Ryu JH, Bae YS, Kang SW, Jang IH, Lee WJ (2005b) An antioxidant system required for host protection against gut infection in Drosophila. Dev Cell 8(1):125–132. https://doi.org/10.1016/j.devcel.2004.11.007

Ha EM, Lee KA, Park SH, Kim SH, Nam HJ, Lee HY, Lee WJ (2009a) Regulation of DUOX by the Gαq-phospholipase Cβ-Ca2+ pathway in Drosophila gut immunity. Dev Cell 16(3):386–397

Ha EM, Lee KA, Seo YY, Kim SH, Lim JH, Oh BH, Lee WJ (2009b) Coordination of multiple dual oxidase–regulatory pathways in responses to commensal and infectious microbes in Drosophila gut. Nat Immunol 10(9):949

Habineza P, Muhammad A, Ji T, Xiao R, Yin X, Hou Y, Shi Z (2019) The promoting effect of gut microbiota on growth and development of red palm weevil, Rhynchophorus ferrugineus (Olivier) (Coleoptera: Dryophthoridae) by modulating its nutritional metabolism. Front Microbiol 10:1212. https://doi.org/10.3389/fmicb.2019.01212

Hammer TJ, Janzen DH, Hallwachs W, Jaffe SP, Fierer N (2017) Caterpillars lack a resident gut microbiome. Proc Natl Acad Sci U S A 114(36):9641–9646

Hernandez-Martinez P, Naseri B, Navarro-Cerrillo G, Escriche B, Ferre J, Herrero S (2010) Increase in midgut microbiota load induces an apparent immune priming and increases tolerance to Bacillus thuringiensis. Environ Microbiol 12(10):2730–2737. https://doi.org/10.1111/j.1462-2920.2010.02241.x

Hu Y, Łukasik P, Moreau CS, Russell JA (2014) Correlates of gut community composition across an ant species (Cephalotes varians) elucidate causes and consequences of symbiotic variability. J Mol Ecol 23(6):1284–1300

Huang XF, Bakker MG, Judd TM, Reardon KF, Vivanco JM (2013) Variations in diversity and richness of gut bacterial communities of termites (Reticulitermes flavipes) fed with grassy and woody plant substrates. J Microb Ecol 65(3):531–536

Hultmark D (2003) Drosophila immunity: paths and patterns. Curr Opin Immunol 15(1):12–19. https://doi.org/10.1016/s0952-7915(02)00005-5

Jang S, Kikuchi Y (2020) Impact of the insect gut microbiota on ecology, evolution, and industry. J Curr Opin Insect Sci 41:33–39

Janssen AW, Kersten S (2015) The role of the gut microbiota in metabolic health. FASEB J 29 (8):3111–3123. https://doi.org/10.1096/fj.14-269514

Jia S, Zhang X, Zhang G, Yin A, Zhang S, Li F, Yu J (2013) Seasonally variable intestinal metagenomes of the red palm weevil (Rhynchophorus ferrugineus). Environ Microbiol 15 (11):3020–3029. https://doi.org/10.1111/1462-2920.12262

Johnston PR, Crickmore N (2009) Gut bacteria are not required for the insecticidal activity of Bacillus thuringiensis toward the tobacco hornworm, Manduca sexta. Appl Environ Microbiol 75(15):5094–5099. https://doi.org/10.1128/AEM.00966-09

Ju RT, Wang F, Wan FH, Li B (2011) Effect of host plants on development and reproduction of Rhynchophorus ferrugineus (Olivier)(Coleoptera: Curculionidae). J Pest Sci 84(1):33–39

Jung J, Heo A, Park YW, Kim YJ, Koh H, Park WJJMB (2014) Gut microbiota of Tenebrio molitor and their response to environmental change. J Microbiol Biotechnol 24(7):888–897

Kanost MR, Jiang H, Yu XQ (2004) Innate immune responses of a lepidopteran insect, Manduca sexta. Immunol Rev 198:97–105. https://doi.org/10.1111/j.0105-2896.2004.0121.x

Kikuchi Y, Hosokawa T, Fukatsu T (2011) An ancient but promiscuous host–symbiont association between Burkholderia gut symbionts and their heteropteran hosts. ISME J 5(3):446

Kikuchi Y, Hayatsu M, Hosokawa T, Nagayama A, Tago K, Fukatsu T (2012) Symbiont-mediated insecticide resistance. Proc Natl Acad Sci U S A 109(22):8618–8622

Kim M, Lee JH, Lee SY, Kim E, Chung J (2006) Caspar, a suppressor of antibacterial immunity in Drosophila. Proc Natl Acad Sci 103(44):16358–16363

Kim JK, Kim NH, Am Jang H, Kikuchi Y, Kim CH, Fukatsu T, Lee BL (2013) Specific midgut region controlling the symbiont population in an insect-microbe gut symbiotic association. J Appl Environ Microbiol 79(23):7229–7233

Kim JK, Lee JB, Huh YR, Jang HA, Kim CH, Yoo JW, Lee BL (2015) Burkholderia gut symbionts enhance the innate immunity of host Riptortus pedestris. Dev Comp Immunol 53(1):265–269. https://doi.org/10.1016/j.dci.2015.07.006

Köhler T, Dietrich C, Scheffrahn RH, Brune A (2012) High-resolution analysis of gut environment and bacterial microbiota reveals functional compartmentation of the gut in wood-feeding higher termites (Nasutitermes spp.). J Appl Environ Microbiol 78(13):4691–4701

Langille MG, Zaneveld J, Caporaso JG, McDonald D, Knights D, Reyes JA, Clemente JC, Burkepile DE, Thurber RL, Knight R, Beiko RG (2013) Predictive functional profiling of microbial communities using 16S rRNA marker gene sequences. Nat Biotechnol 31(9):814–821

Lee KA, Kim SH, Kim EK, Ha EM, You H, Kim B, Lee WJ (2013) Bacterial-derived uracil as a modulator of mucosal immunity and gut-microbe homeostasis in Drosophila. Cell 153 (4):797–811. https://doi.org/10.1016/j.cell.2013.04.009

Lee KA, Kim B, Bhin J, Kim DH, You H, Kim EK, Lee WJ (2015) Bacterial uracil modulates Drosophila DUOX-dependent gut immunity via Hedgehog-induced signaling endosomes. Cell Host Microbe 17(2):191–204

Lee JH, Lee KA, Lee WJ (2017) Microbiota, gut physiology, and insect immunity. In: Advances in insect physiology, vol 52. Elsevier, pp 111–138

Lemaitre B, Hoffmann J (2007) The host defense of Drosophila melanogaster Annu Rev Immunol, 25, 697–743. doi: https://doi.org/10.1146/annurev.immunol.25.022106.141615

Li WH, Jin DC, Shi CH, Li FL (2017) Midgut bacteria in deltamethrin-resistant, deltamethrin-susceptible, and field-caught populations of Plutella xylostella, and phenomics of the predominant midgut bacterium Enterococcus mundtii. Sci Rep 7:1947. https://doi.org/10.1038/s41598-017-02138-9

Li S, Xu X, Zheng Z, Zheng J, Shakeel M, Jin F (2019) MicroRNA expression profiling of Plutella xylostella after challenge with B. thuringiensis. Dev Comp Immunol 93:115–124. https://doi.org/10.1016/j.dci.2018.12.008

Li S, De Mandal S, Xu X, Jin F (2020a) The tripartite interaction of host immunity-Bacillus thuringiensis infection-gut microbiota. Toxins 12(8). https://doi.org/10.3390/toxins12080514

Li S, Xu X, De Mandal S, Shakeel M, Hua Y, Shoukat RF, Jin F (2020b) Gut microbiota mediate Plutella xylostella susceptibility to Bt Cry1Ac protoxin is associated with host immune response. Environ Pollut 271:116271. https://doi.org/10.1016/j.envpol.2020.116271

Lin XL, Pan QJ, Tian HG, Douglas AE, Liu TX (2015) Bacteria abundance and diversity of different life stages of *Plutella xylostella* (Lepidoptera: Plutellidae), revealed by bacteria culture-dependent and PCR-DGGE methods. Insect Sci 22(3):375–385. https://doi.org/10.1111/1744-7917.12079

Lin JH, Xia XF, Yu XQ, Shen JH, Li Y, Lin HL, You MS (2018) Gene expression profiling provides insights into the immune mechanism of *Plutella xylostella* midgut to microbial infection. Gene 647:21–30. https://doi.org/10.1016/j.gene.2018.01.001

Lin J, Yu XQ, Wang Q, Tao X, Li J, Zhang S, You M (2020) Immune responses to Bacillus thuringiensis in the midgut of the diamondback moth, Plutella xylostella. Dev Comp Immunol 107:103661. https://doi.org/10.1016/j.dci.2020.103661

Ma ZG, Li CF, Pan GQ, Li ZH, Han B, Xu JS, Zhou ZY (2013) Genome-wide transcriptional response of silkworm (Bombyx mori) to infection by the microsporidian Nosema bombycis. PLoS One 8(12):e84137. https://doi.org/10.1371/journal.pone.0084137

Malacrinò A, Campolo O, Medina RF, Palmeri V (2018) Instar-and host-associated differentiation of bacterial communities in the Mediterranean fruit fly Ceratitis capitata. PLoS One 13(3): e0194131

Mason CJ, Ray S, Shikano I, Peiffer M, Jones AG, Luthe DS, Felton GW (2019) Plant defenses interact with insect enteric bacteria by initiating a leaky gut syndrome. Proc Natl Acad Sci U S A 201908748

Mogensen TH (2009) Pathogen recognition and inflammatory signaling in innate immune defenses. Clin Microbiol Rev 22(2):240–273, Table of contents. https://doi.org/10.1128/CMR.00046-08

Montagna M, Chouaia B, Mazza G, Prosdocimi EM, Crotti E, Mereghetti V, Daffonchio D (2015) Effects of the diet on the microbiota of the red palm weevil (Coleoptera: Dryophthoridae). PLoS One 10(1):e0117439. https://doi.org/10.1371/journal.pone.0117439

Muhammad A, Fang Y, Hou YM, Shi ZH (2017) The gut entomotype of red palm weevil *Rhynchophorus ferrugineus* Olivier (Coleoptera: Dryophthoridae) and their effect on host nutrition metabolism. Front Microbiol 8:2291. https://doi.org/10.3389/fmicb.2017.02291

Muhammad A, Habineza P, Hou YM, Shi ZH (2019a) Preparation of red palm weevil *Rhynchophorus Ferrugineus* (Olivier) (Coleoptera: Dryophthoridae) germ-free larvae for host-gut microbes interaction studies. Bio-protocol 9(24):e3456. https://doi.org/10.21769/BioProtoc.3456

Muhammad A, Habineza P, Ji T, Hou Y, Shi Z (2019b) Intestinal microbiota confer protection by priming the immune system of red palm weevil Rhynchophorus ferrugineus Olivier (Coleoptera: Dryophthoridae). Front Physiol 10:1303. https://doi.org/10.3389/fphys.2019.01303

Muhammad A, Habineza P, Wang XH, Xiao R, Ji TL, Hou YM, Shi ZH (2020) Spatzle homolog-mediated toll-like pathway regulates innate immune responses to maintain the homeostasis of gut microbiota in the red palm weevil, Rhynchophorus ferrugineus Olivier (Coleoptera: Dryophthoridae). Front Microbiol 11:846. https://doi.org/10.3389/fmicb.2020.00846

Murphy S, Briscoe B (1999) The red palm weevil as an alien invasive: biology and the prospects for biological control as a component of IPM. Biocontrol News Information 20(1):35–46

Nehme NT, Liegeois S, Kele B, Giammarinaro P, Pradel E, Hoffmann JA, Ferrandon D (2007) A model of bacterial intestinal infections in *Drosophila melanogaster*. PLoS Pathog 3 (11):1594–1709. https://doi.org/10.1371/journal.ppat.0030173

Paredes JC, Welchman DP, Poidevin M, Lemaitre B (2011) Negative regulation by amidase PGRPs shapes the Drosophila antibacterial response and protects the fly from innocuous infection. Immunity 35(5):770–779

Peng L, Miao Y, Hou Y (2016) Demographic comparison and population projection of *Rhynchophorus ferrugineus* (Coleoptera: Curculionidae) reared on sugarcane at different temperatures. Sci Rep 6:31659

Qiao H, Keesey IW, Hansson BS, Knaden M (2019) Gut microbiota affects development and olfactory behavior in *Drosophila melanogaster*. J Exp Biol 222(Pt 5):jeb192500. https://doi.org/10.1242/jeb.192500

Ryu JH, Kim SH, Lee HY, Bai JY, Nam YD, Bae JW, Lee WJ (2008) Innate immune homeostasis by the homeobox gene caudal and commensal-gut mutualism in Drosophila. Science 319 (5864):777–782

Sajjadian SM, Kim Y (2020) Dual oxidase-derived reactive oxygen species against Bacillus thuringiensis and its suppression by eicosanoid biosynthesis inhibitors. Front Microbiol 11:528. https://doi.org/10.3389/fmicb.2020.00528

Shakeel M, Farooq M, Nasim W, Akram W, Khan FZA, Jaleel W, Jin F (2017a) Environment polluting conventional chemical control compared to an environmentally friendly IPM approach for control of diamondback moth, Plutella xylostella (L.), in China: a review. Environ Sci Pollut Res Int 24(17):14537–14550. https://doi.org/10.1007/s11356-017-8996-3

Shakeel M, Xu XX, Xu J, Zhu X, Li SZ, Zhou XQ, Jin FL (2017b) Identification of immunity-related genes in Plutella xylostella in response to fungal peptide destruxin A: RNA-Seq and DGE analysis. Sci Rep 7:10966. https://doi.org/10.1038/s41598-017-11298-7

Shakeel M, Xu X, Xu J, Li S, Yu J, Zhou X, Jin F (2018) Genome-wide identification of destruxin A-responsive immunity-related microRNAs in diamondback moth, Plutella xylostella. Front Immunol 9:185. https://doi.org/10.3389/fimmu.2018.00185

Shao YQ, Chen BS, Sun C, Ishida K, Hertweck C, Boland W (2017) Symbiont-derived antimicrobials contribute to the control of the lepidopteran gut microbiota. Cell Chem Biol 24(1):66–75. https://doi.org/10.1016/j.chembiol.2016.11.015

Shelomi M, Danchin EG, Heckel D, Wipfler B, Bradler S, Zhou X, Pauchet Y (2016) Horizontal gene transfer of pectinases from bacteria preceded the diversification of stick and leaf insects. Sci Rep 6:26388

Suen G, Scott JJ, Aylward FO, Adams SM, Tringe SG, Pinto-Tomás AA, Barry KW (2010) An insect herbivore microbiome with high plant biomass-degrading capacity. PLoS Genet 6(9): e1001129

Sun H, Towb P, Chiem DN, Foster BA, Wasserman SA (2004) Regulated assembly of the toll signaling complex drives Drosophila dorsoventral patterning. EMBO J 23(1):100–110. https://doi.org/10.1038/sj.emboj.7600033

Tagliavia M, Messina E, Manachini B, Cappello S, Quatrini P (2014) The gut microbiota of larvae of Rhynchophorus ferrugineus Oliver (Coleoptera: Curculionidae). BMC Microbiol 14:136. https://doi.org/10.1186/1471-2180-14-136

Tang X, Freitak D, Vogel H, Ping L, Shao Y, Cordero EA, Boland W (2012) Complexity and variability of gut commensal microbiota in polyphagous lepidopteran larvae. PLoS One 7(7): e36978

Tanji T, Ip YT (2005) Regulators of the toll and Imd pathways in the Drosophila innate immune response. Trends Immunol 26(4):193–198. https://doi.org/10.1016/j.it.2005.02.006

Tzou P, Ohresser S, Ferrandon D, Capovilla M, Reichhart J-M, Lemaitre B, Imler J-L (2000) Tissue-specific inducible expression of antimicrobial peptide genes in Drosophila surface epithelia. Immunity 13(5):737–748

Tzou P, Reichhart JM, Lemaitre B (2002) Constitutive expression of a single antimicrobial peptide can restore wild-type resistance to infection in immunodeficient Drosophila mutants. Proc Natl Acad Sci U S A 99(4):2152–2157. https://doi.org/10.1073/pnas.042411999

Valanne S, Wang JH, Ramet M (2011) The Drosophila toll signaling pathway. J Immunol 186 (2):649–656. https://doi.org/10.4049/jimmunol.1002302

Vodovar N, Vinals M, Liehl P, Basset A, Degrouard J, Spellman P, Lemaitre B (2005) Drosophila host defense after oral infection by an entomopathogenic Pseudomonas species. Proc Natl Acad Sci U S A 102(32):11414–11419. https://doi.org/10.1073/pnas.0502240102

Wan FH, Yang NW (2016) Invasion and management of agricultural alien insects in China. Annu Rev Entomol 61:77–98

Wong AC, Dobson AJ, Douglas AE (2014) Gut microbiota dictates the metabolic response of Drosophila to diet. J Exp Biol 217(Pt 11):1894–1901. https://doi.org/10.1242/jeb.101725

Wu S, Zhang XF, Chen XM, Cao PS, Beerntsen BT, Ling EJ (2010) BmToll9, an arthropod conservative toll, is likely involved in the local gut immune response in the silkworm, Bombyx mori. Dev Comp Immunol 34(2):93–96. https://doi.org/10.1016/j.dci.2009.08.010

Xia XF, Zheng DD, Zhong HZ, Qin BC, Gurr GM, Vasseur L, You MS (2013) DNA sequencing reveals the midgut microbiota of diamondback moth, Plutella xylostella (L.) and a possible relationship with insecticide resistance. PLoS One 8(7):e68852. https://doi.org/10.1371/journal.pone.0068852

Xia X, Yu L, Xue M, Yu X, Vasseur L, Gurr GM, You M (2015) Genome-wide characterization and expression profiling of immune genes in the diamondback moth, Plutella xylostella (L.). Sci Rep 5:9877. https://doi.org/10.1038/srep09877

Xia XF, Gurr GM, Vasseur L, Zheng DD, Zhong HZ, Qin BC, You MS (2017) Metagenomic sequencing of diamondback moth gut microbiome unveils key holobiont adaptations for herbivory. Front Microbiol 8:663. https://doi.org/10.3389/fmicb.2017.00663

Xia X, Sun B, Gurr GM, Vasseur L, Xue M, You M (2018) Gut microbiota mediate insecticide resistance in the diamondback moth, Plutella xylostella (L.). Front Microbiol 9:25. https://doi.org/10.3389/fmicb.2018.00025

Xiao R, Wang X, Xie E, Ji T, Li X, Muhammad A, Shi Z (2019) An IMD-like pathway mediates the intestinal immunity to modulate the homeostasis of gut microbiota in Rhynchophorus ferrugineus Olivier (Coleoptera: Dryophthoridae). Dev Comp Immunol 97:20–27. https://doi.org/10.1016/j.dci.2019.03.013

Xu J, Xu X, Li S, Wang S, Xu X, Zhou X, Jin F (2017a) Genome-wide profiling of Plutella xylostella immunity-related miRNAs after Isaria fumosorosea infection. Front Physiol 8:1054. https://doi.org/10.3389/fphys.2017.01054

Xu J, Xu XX, Shakeel M, Li SZ, Wang S, Zhou XQ, Jin FL (2017b) The entomopathogenic fungi isaria fumosorosea plays a vital role in suppressing the immune system of Plutella xylostella: RNA-Seq and DGE analysis of immunity-related genes. Front Microbiol 8:1–14. https://doi.org/10.3389/fmicb.2017.01421

Yang FY, Saqib HSA, Chen JH, Ruan QQ, Vasseur L, He WY, You MS (2020) Differential profiles of gut microbiota and metabolites associated with host shift of Plutella xylostella. Int J Mol Sci 21(17):6283. https://doi.org/10.3390/ijms21176283

Yao ZC, Wang AL, Li YS, Cai ZH, Lemaitre B, Zhang HY (2016) The dual oxidase gene BdDuox regulates the intestinal bacterial community homeostasis of Bactrocera dorsalis. ISME J 10 (5):1037–1050. https://doi.org/10.1038/ismej.2015.202

Yun J-H, Roh SW, Whon TW, Jung M-J, Kim M-S, Park D-S, Choi J-H (2014) Insect gut bacterial diversity determined by environmental habitat, diet, developmental stage, and phylogeny of host. J Appl Environ Microbiol 80(17):5254–5264

Zhang L, Wang YW, Lu ZQ (2015) Midgut immune responses induced by bacterial infection in the silkworm, Bombyx mori. J Zhejiang Univ Sci B 16(10):875–882. https://doi.org/10.1631/jzus.B1500060

Chapter 12
Eco-Smart Biorational Approaches in Housefly *Musca domestica* L. 1758 Management

R Senthoorraja, P. Senthamarai Selvan, and S. Basavarajappa

Abstract The houseflies are a mechanical carrier of pathogens, causing enteric and deadly diseases in humans and livestock production. Chemical insecticides used in the control of *M. domestica* have become less acceptable due to their persistence in the environment, toxicity to non-target organisms, and resistance against synthetic chemical pesticides. Due to housefly's nuisance to public health, livestock issues, and economic loss over the globe pondered the scientists to find a solution in their expertise field. In this current review, we penned the effective tools which were employed against housefly control from the decades to till date. The paper focused on the pesticide-free application, which are inbuilt with cutting-edge technologies accompanied by biological materials at the latest hour, especially easy, handy tools, economically affordable, and easily accessible by the end-users. The authors assure that this chapter will provide collective information and create ideas about housefly control for the researchers, farmers, and the learners who are the pioneer and newer on this aspect.

Keywords *Musca domestica* · Botanochemicals · Pheromones · Parasitoids · Nanotechnology · Chemo-ecology

12.1 Introduction

Housefly *Musca domestica* L. 1758 is a cosmopolitan pest and considers as vectoring etiological agents of bacterial, protozoan, and viral infections. This fact turns this insect into a threat to livestock industries, public health concerns, and a major food contaminator across the globe. Houseflies disturb normal human activities and affect

R. Senthoorraja · S. Basavarajappa (✉)
DOS in Zoology, University of Mysore, Mysore, Karnataka, India
e-mail: ornithoraj11@gmail.com

P. Senthamarai Selvan
Department of Zoology, Annamalai University, Annamalainagar, Tamil Nadu, India

S. De Mandal et al. (eds.), *New and Future Development in Biopesticide Research: Biotechnological Exploration*, https://doi.org/10.1007/978-981-16-3989-0_12

human health via., Anthrax, Cholera, Conjunctivitis (epidemic), Dysentery, Food poisoning/gastroenteritis, Leprosy, Poliomyelitis, Trachoma, Tuberculosis, Typhoid fever, Yaws, Myiasis, and economic losses due to health risks. *M. domestica* is also considered a major threat to humans than any other non-biting fly species. India is one of the leading countries across the world for poultry farming, and the fly problem is a serious issue internationally wherever poultry farming is an important economic activity. The poultry farm environment favors housefly breeding since the eggs can thrive in the manure (Mehta et al. 2002). They act as food contaminators at restaurants and food industries where the potential for bacterial contamination by the fly is tremendous, as the fly travels freely between decomposing organic matter found in the restaurant, garbage heaps, exposed kitchen surfaces and foods, dining tables, and even restaurant bathrooms. In addition, flies have been shown to transmit *Campylobacter jejuni* to chickens from sheep under laboratory and field conditions and Turkey Corona Virus on Turkeys. Additional indirect losses can include legal expenses and forced closures of livestock industries. Poultry farms were also unable to reach or fail to implement standard poultry farm hygiene practices due to fly resistance and cost-effectivity as they rely on synthetic chemical pesticides. So, there is a high demand for finding a suitable control measure to manage house flies.

12.2 Housefly Biology

The housefly has four distinct bio-stages in its lifecycle: egg, larva, pupa, and adult. Based on the environmental condition, quality, and quantity of food sources available, their developmental period changes at each stage, and it reaches adult from the egg by 6–42 days. The length of life is usually 2–3 weeks, but it may be as long as 3 months in cold conditions. Eggs are typically laid in masses on organic material such as manure and garbage. Hatching occurs within a few hours (12–38 h). The young larvae usually grow in the breeding materials such as garbage heaps, poultry manures, bio wastages, dumped raw and cooked food materials etc., and usually prefer optimum aerated moisture materials as breeding sources. After the feeding stage is over, the larvae migrate towards the drier place for pupation. Once larvae are found in the drier area, they shrink their body segments and form a capsule-like structure, and the colour changes from pale orange to dark brown. They complete their whole lifecycle within the same place. The pupal stage usually takes 2–10 days; at the end, the well-grown new fly forces from inside and break the pupal wall, followed by spreads its wings, and the body dries and hardens. The adult fly is grey in color, 6–9 mm in length, and has four dark stripes running lengthwise on the dorsal body. The adult is capable of reproducing their young ones within 5 days after emergence under favourable conditions. The adult female rarely lay eggs more than five times, and each time lays eggs in a batch of 120–130 eggs (Keiding 1986).

The housefly is an esculent organism of all kinds of foods, garbage, and excreta, including sweat, animal dung, milk, sugar, syrup, blood, meat broth, and many other materials found in human settlements. Due to the flies' mouthparts structure, they can only intake the food materials in a liquid state or promptly soluble in the salivary gland secretions or the crop. So, the water content is a vital part of a fly's diet, and flies do not survive more than 48 h without access to it. The flies evidently need to feed at least two or three times a day.

12.3 Chemical Insecticides in Housefly Control and Its Effects

The broad-scale use of chemical pesticides (organophosphates, carbamates, pyrethroids and Imidacloprid etc.,) against disease-causing vectors contributed greatly to their control. The persistent use of chemicals at higher doses has led to the deterioration of the environment and health issues to humans and livestock. According to the World Health Organization (WHO), about 20,000 people die annually due to adverse effects of pesticide exposure, while three million are poisoned, and there are nearly 750,000 new cases of chronic pesticide exposure every year (Shah et al. 2016). Intensive use of chemical insecticides to control *M. domestica* have led to the formation of resistant strains. Though chemical insecticides are used as a promising tool in housefly control, they leave high degrees of residues that cause serious problems to humans and all non-target organisms in the globe. Therefore, an alternative, environment friendly biorational approach is required to control the insect vector or pest.

12.4 Biorational Approaches in Housefly Management

The house fly can be controlled by adopting various tools such as botanochemicals, fungal/bacterial pathogens, and parasitoids/ predators and semiochemicals with the aid of cutting-edge technologies (Fig. 12.1).

12.4.1 Plant-Derived Insecticides

Pest control by directly or indirectly by utilization of natural green products, including essential oils, are a promising possible approach in recent days. Aromatic plants and their derivatives are extensively used for various industries such as, flavours and fragrances, condiments or spices, medicines, antimicrobial agents, insecticides, and repellents. These can be used as insecticidal agents at all bio-stages, while some could be repellents, antifeedants, ovipositional deterrents, and insect growth regulators against house fly as well as many other insect pests (Senthamarai Selvan et al. 2021). The lipophilic characteristics of plant essential oils can intervene with the basic metabolic and normal activities of insects.

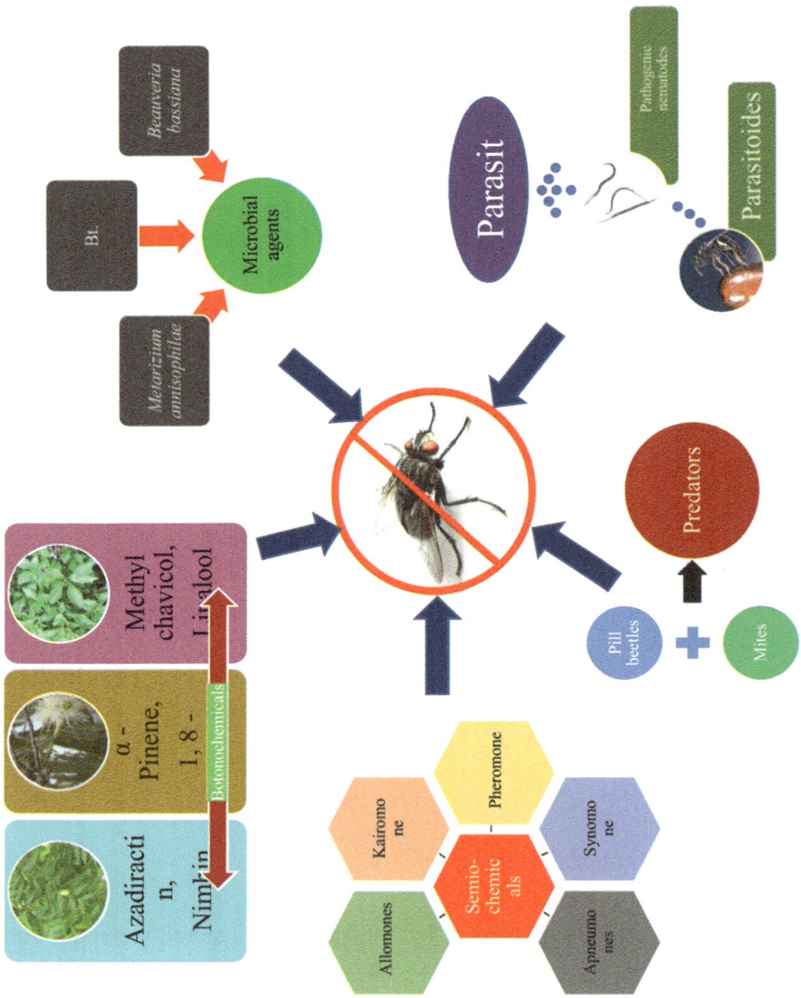

Fig. 12.1 Environmentally safe alternative tools in housefly *Musca domestica* L. management

However, several phytochemicals from various plant species have been reported to exhibit pernicious effects on houseflies and mosquitoes (Malik et al. 2007).

Recent studies have shown the probability of using plant oils and their wide spectrum action against bio-stages of *M. domestica* and other insect pests (Table 12.1). The effect of plant derived products against houseflies varies with the developmental stage, sex, fly strain, and the mode of application. The use of essential oils (EOs) as fumigants are due to their low toxicity to warm-blooded mammals, high volatility, easy availability, and relatively low cost. However, many researchers reported the chemical and therapeutic properties of plant oils responsible for pest control. The extracted plant oil from the flower of *C. odorata* by hydrodistillation perform as toxins, feeding deterrents, and oviposition deterrents against a wide variety of insect pests (Burdock and Carabin 2008). On the other hand, fewer studies have reported increased toxicity of plant oils due to the blends of its active ingredients (Subaharan et al. 2021).

Moreover, these aromatic medicinal plants proved to be toxic to different species of insects and ticks. Sinthusiri and Soonwera (2013) have reported that the insecticidal effects of 20 plant essential oils evaluated against housefly adult and found the most effective were shown by *C. citratus* oil, *M. piperita* oil and *L. angustifolia* oil, showing 100% mortality at 24 h. Soonwera (2015) has revealed that the essential oil of *S. aromaticum* showed promising larvicidal and oviposition repellence activity against female house flies than cypermethrin. Many reports are available as the bio-efficacy of essential oils from *Cymbopogon citratus, C. Pogostemon cablin, winterianus, Citrus sinensis, Ocimum basilicum,* etc., were evaluated for their insecticidal activity against house fly.

12.4.1.1 Compounds in Essential Oil

Insecticidal activities of plant essential oils are due to the presence of various bioactive components such as complex mixtures of monoterpenes, sesquiterpenes, hydrocarbons, and their oxygenated derivatives (alcohol, aldehyde, and ketones). The difference in the efficacy and mechanism of action of plant oil are due to the constituents of various compounds present in the oil. Recent studies have suggested that the synergising of EOs or its components has more efficacy towards targeted pest and not to the non-target organisms. Essential oils and their constituents induce neurotoxicity by various mechanisms and disrupt the endocrinologic balance, thereby disarray the normal process of morphogenesis in insects. Analysing the action mechanism of terpene on lepidopteran larvae showed that it blocks the stimulatory effects of glucose and inositol on chemosensory receptors present on their mouthparts (Gershenzon and Dudareva 2007). Further electrophysiological and molecular studies are needed to explore the mode of action of these oils against the housefly olfactory system to develop new generation pesticides and repellents (Senthoorraja et al. 2021). The use of botanical insecticides is another control option because they are safe to the environment and leave less residue due to

Table 12.1 Overview of studies have proven on housefly, *M. domestica* L. management by eco-friendly methods

Plant name	Part used	Active compound	Mode of action	*M. domestica* L. bio stage studied	Reference
Botanochemicals and its active components against *M. domestica* L. control					
Boesenbergia rotunda	Root	α-Pinene, camphene, β-myrcene, M-cymene, Citral	Topical application	Adult	Suwannayod et al. (2019)
Citrus hystrix	Fruit	δ-Cadinene, α-copaene, β-caryophyllene, β-citronellol, germacrene D			
Curcuma longa	Root	α-Curcumene, β-cymene, O-cymol, Limonene, norsabinene			
Ocimum gratissimum	Seed	δ-3-carene, O-Cymol, γ-terpinene, Pinene, (S)-phellandral			
Zanthoxylum limonella	Seed	Sabinene, O-cymol, β–ocimene, γ-terpinen, δ-carene			
E. globulus	Leaf	1,8-cineole, α-pinene, α-terpineol, D-limonene, linalool acetate, carvone	Repellent, Oral toxicity, Contact, Fumigant	Larvae, Pupae, Adult	Kumar et al. (2012), Chauhan et al. (2018), Soonwera and Sittichok (2020), Rossi and Palacios (2015)
Cymbopogon citratus	Leaf	Citral, 1,8-cineole, geranyl acetate, α-pinene, geraniol, linalool	Fumigant Contact Repellent	Larvae, pupae, adult	Chauhan et al. (2018), Soonwera and Sittichok (2020), Rani et al. (2019)
Camellia sinensis	Leaf	Limonene, myrcene, carveol	Repellent	Adult	Chauhan et al. (2018)
Microbial agents against housefly *M. domestica* L. control					
Entomofungal pathogens					

Fungal species	Strain	Mode of action	*M. domestica* L. bio stage studied	References
Beauveria bassiana	(Bb01), ARSEF 1564, ARSEF 8891, HQ917687, HF23, GHA	Contact toxicity; Oral toxicity;	Adult, larvae, pupae	Kanwar et al. (2017), Farooq et al. (2018), Mishra et al. (2016), Weeks et al. (2016)
Metarhizium anisopliae	(Ma4.1) M10 M 16 M54 M92 M93	Contact toxicity, Oviposition repellency	Adults, larvae, pupae	Kanwar et al. (2017), Baker et al. (2018), Baker et al. 2020)
Metarhizium robertsii	(ARSEF 1057)	Contact toxicity	Adult	Moloinyane et al. (2019)
Metarhizium brunneum	F52	Contact toxicity	Larvae, Pupae	Weeks et al. (2016)
Isaria fumosorosea	(If03)	Contact toxicity	Adult	Kanwar et al. (2017)
Bacterial agents				
Species	**Strain**	**Mode of action**	***M. domestica* L. bio stage studied**	**References**
Bacillus thuringiensis	Encompasses cry1, cry2, cry4, cry10, cry11, cry19, cry20, cry24, cry27, cry30, cry39, cry44, cry47, cry50, cry54, cry56, mpp60, tpp80, cyt1 and cyt2. Cry1Bc1, cpbA, cpbB, CHRD, pmpl,	Contact toxicity	Larvae, Adult	Johnson et al. (1998), Marche et al. (2017), Contreras et al. (2019), Valtierra-de-Luis et al. (2020)
Bacillus subtilis	1381, 933	Contact toxicity	Larvae, Adult	Valtierra-de-Luis et al. (2020)
Bacillus ser israelensis	(Cry4Aa, Cry4Ba, Cry10Aa, Cry11Aa, and Mpp60A/ Mpp60B (formerly Cry60A/Cry60B)	Contact toxicity	Larvae, Adult	Valtierra-de-Luis et al. (2020)

(continued)

Table 12.1 (continued)

Parasites against housefly *M. domestica* L. control			
Entomopathogenic nematodes			
Species	**Mode of action**	***M. domestica* L. bio stage studied**	**References**
Steinernema feltiae	Contact toxicity	Egg, larvae, pupae	Archana et al. (2017), Arriaga and Cortez-Madrigal (2018)
Steinernema carpocapsae	Contact toxicity	Egg, larvae, pupae	Arriaga and Cortez-Madrigal (2018), Archana et al. (2017)
Steinernema glaseri	Contact toxicity	Egg, larvae, pupae	Archana et al. (2017), Arriaga and Cortez-Madrigal (2018)
Steinernema abbasi	Contact toxicity	Egg, larvae, pupae	Archana et al. (2017), Arriaga and Cortez-Madrigal (2018)
Heterorhabditis indica	Contact toxicity	Egg, larvae, pupae	Archana et al. (2017), Arriaga and Cortez-Madrigal (2018)
Parasitoids			
Parasitoid species	**Mode of action**	***M. domestica* L. bio stage studied**	**References**
Spalangia endius	Parasitism	Pupae	Burgess et al. (2017, 2018)
Spalangia cameroni	Parasitism	Pupae	Geden and Hogsette (2006)
Muscidifura similadanacus	Parasitism	Pupae	Gao et al. (2020), Doganlar (2007)
Muscidifura sinesensilla	Parasitism	Pupae	Gao et al. (2020)
Muscidifurax raptor	Parasitism	Pupae	Fabritius (1981), Doganlar (2007)
Muscidifurax zaraptor	Parasitism	Pupae	Doganlar (2007)
Muscidifurax raptorellus	Parasitism	Pupae	Coats (1976), Petersen and Currey (1996), Floate et al. (2000), Geden and Hogsette (2006), Doganlar (2007)
Muscidifurax uniraptor	Parasitism	Pupae	Thomazini and Berti Filho (2000)

Predators (mites)			
Species	**Mode of action**	***M. domestica* L. bio stage studied**	**References**
Macrocheles embersoni	Feeding	Egg, larvae	Azevedo et al. (2018)
Macrocheles muscaedomesticae	Feeding	Egg, larvae	Abo-Taka et al. (2014), Azevedo et al. (2018), Farahi et al. (2018)
Macrocheles robustulus	Feeding	Egg, larvae	Azevedo et al. (2018)

Semiochemicals formulations in housefly control					
Semiochemical type	**Compound used**	**Formulation**	**Mode of action**	***M. domestica* L. bio stage studied**	**Reference**
Pheromone alone	Z-9 Tricosene	Plywood sticky trap	Attraction	Adult	Sundar et al. (2013)
Pheromone and synomone	Z-9 Tricosene + fish meal	Barrix trap	Attraction	Adult	Kannan et al. (2020)
Apneumones	Wheat bran diet materials	Bait	Oviposition attraction	Adult	Tang et al. (2016)
Chemical compound	(±)-1-octen-3-ol, 2-pentanone, R-(+)-limonene	Vapour	Attractant	Adult	Kelling et al. (2002)
Pheromone and hydrocarbon	Female housefly ovary extraction and wheat bran diet material	Bait	Oviposition attraction	Adult	Jiang et al. (2002)
Apneumones	Animal excreta	Formulation	Oviposition repellency	Adult	Lam et al. (2010)

Integrated approaches in housefly *M. domestica* L. control			
Type of control agents used	**Mode of action**	***M. domestica* L. bio stage studied**	**Reference**

(continued)

Table 12.1 (continued)

Granular baits + cultural method + biological agents	Attraction and kill	Egg, larvae, pupae, adult	Crespo et al. (1998)
Cultural method + biological agents	Semiochemicals mediated attraction	Egg, larvae, pupae, adult	Kaufman et al. (2002, 2005)
Biocontrol agents combination	Parasitism and predation	Egg, larvae, pupae	Axtell (1970)

their biodegradability. Botanical pesticides are cost-effective, species-specific, ease handling at end-users, and eco-friendly.

12.4.2 Microbial Agents in Housefly Control

12.4.2.1 Entomofungal Pathogen

Fungal infections on houseflies are very common, *Entomophthora muscae*, *Metarhizium anisopliae*, *Tolypocladium cylindrosporum*, and *Beauveria bassiana* etc., are some common fungal species that affects housefly (Malik et al. 2007). Fungi infect insects by breaching the host cuticle and also by ingestion or mutual contact of house flies. Some entomopathogenic fungi affect whole bio-stage of the host, hence it can be considered for viable and long-term control of the housefly density. *M. anisopliae* at a dose of 10^7–10^8 conidia/mL prevents housefly emergence. Certain vegetable and mineral oils and adjuvants are used as a carrier to formulate this fungal pathogen. For *M. anisopliae* conidia, soya bean oil and linseed were suggested as most effective carrier, especially linseed oil could achieve complete mortality in 3 days while water takes 6 days. To obtain desirable results in the lab and field considering the external factors such as, temperature, humidity etc., are essential.

B. bassiana (Balsamo) Vuillemin is an important pathogenic fungi species against more than 200 insects with zero reports on toxicity to humans and non-target organisms. Furthermore, various studies have been reported that the high potential of *B. bassiana* for housefly control (Table 12.1). Though *Entomophthora muscae* (Chon) Fresenius and *Beauveria bassiana* (Balsamo) Vuillemin are successfully controlled housefly populations, they are limited to specific environments. *Beauveria bassiana* has high activity against the adult and larval house flies (Malik et al. 2007). Various entomopathogenic fungi were shown effective against housefly control at lab and field conditions. Entomopathogenic fungi has low mammalian toxicity compare with synthetic chemical insecticides. In addition, it has great potential for controlling housefly populations (Khan et al. 2012). Many studies have been reported the rapid killing and high infection rates of *B. bassiana* (Bals.) Vuill., *M. anisopliae* (Metsch.) Sorok against housefly populations (Sharififard et al. 2011). However, research efforts are still required to explore the suitable fungal isolates for various geographic locations that can compete with synthetic pesticides (Dhanapal et al. 2020a). The carrier materials used in the formulation should be supported and increase its efficacy against housefly.

12.4.2.2 Bacterial Agents

The entomopathogenic bacteria were employed effectively against housefly control. After ingestion of the bacterial insecticides in target pests, the gut becomes paralyzed due to the toxins then it leads to death of the host. *Bacillus thuringiensis* is the best known entomopathogenic bacteria. Entomopathogenic bacteria has been widely known from 1901 to use against various major insect pests of agriculture, forestry, public and veterinary health (Sharma et al. 2019. Adult house flies were not susceptible against solubilized parasporal crystals of *Bt. israelensis* (H-14). But, the activated insecticidal crystal protein of *Bt.* species (H-14, HD-11, HD-1 and HD-3) are highly effective against adults but not for the larvae. But, *Bt. kurstaki* H3a3b3c was found highly effective (50–80% mortality) against *M. domestica* larvae (Malik et al. 2007).

Several *Bacillus* were isolated from the soil samples at different areas of Lahore and studied their toxicity against houseflies. Among these species, *B. megaterium and B. thuringiensis* are responsible for 71% and 82% mortality in *M. domestica*, respectively (Shakoori et al. 1999). Similarly, additive interaction of *Bt var. israelensis* was also highly effective against housefly. The housefly susceptibility against various species of Bt. led researchers to find alternate bacterial agent such as *Pseudomonas fluorescens.* Padmanabhan et al. (2005) examined the toxic effect of *P. fluorescens* (VCRC B426) against second instar housefly larvae at various concentrations and observed the mortality. *B. thuringiensis* were used at two formulations by wettable powder and liquid concentrations against housefly in synthetic fly breeding media and chicken feces. It has shown substantiate decline of all bio-stages of housefly especially liquid formulation showed more virulent than wettable powder (Labib and Rady 2001). *B. thuringiensis* and *Brevibacillus laterosporus* have been reported as useful biocontrol agents of houseflies due to the production of insecticidal endotoxins such as, delta-endotoxin from *Bt. israelis* (*Bti*).

12.4.3 Parasites

12.4.3.1 Entomopathogenic Nematode

For insect pest management, Nematode parasites are well known from seventeenth century; they carry a symbiotic bacterium that provides vital nourishment to the nematodes by infecting a host or by its multiplication inside the host organism. These nematodes feed on bacteria and decayed insect tissues. The infective juvenile that enters into the host through the mouth, anus, spiracles, or by direct penetration through the cuticle and reaches the hemocoel of a host, then it releases the bacteria, which multiply frequently in the haemolymph. The insect dies usually within 1–3 days after the invasion of nematodes. Though the bacterium is basically responsible for most insects' mortality, the nematode also releases some toxic materials which is lethal against the insect (Malik et al. 2007). Heterorhabditidae and Steinernematidae have received special attention among entomopathogenic nematodes (EPNs) family on pest control aspects., High pathogenicity occurs in the genera of Xenorhabdus and Photorhabdus due to the mutualistic relationship between EPNs and bacteria. Several studies reported that the pathogenicity of EPNs against housefly larvae and adults at in and outside of the laboratory (Geden 2012). The local nematode species have to be used in their climatic zone as they have a higher adaptation to those conditions to increase the success rate of *M. domestica* control (Arriaga and Cortez-Madrigal 2018).

Female houseflies are more susceptible to nematode infection than males, which offers higher probability in housefly control as a self-spread strategy. As to, the attractants bearing bait traps are highly effective against houseflies. So, the trap entices the fly enter into it and contaminates it with entomopathogens. Once the fly exit, they can propel it to other housefly populations (Vega et al. 2007). If their potential can be released or passed through special devices to strew the nematode by female flies into ovipositing and breeding sites, they would allow together with the adult death and eventual larval infection that leads to keep the *M. domestica* population under check at their breeding sites itself. The study carried out by Archana et al. (2017) examined and found that the *H. indica* species is the most promising one (100%) out of five nematode species studied. It has been recorded as the first time that Heterorhabditidae has a higher capability in *M. domestica* control especially on adult and larval stage.

There were difficulties in entering nematodes into larvae and adult *M. domestica* due to the morphological characteristics such as a spongy type of mouth parts and lesser spiracles which are reducing the entry. The species under Heterorhabditis genera have higher motility and probing potential (Pinnock and Mullens 2007), which aid to act as more possible to infect houseflies and cause mortality. In addition to that, the tiny size of *H. indica* favours the entry housefly adults and immatures. Mated female nematode *Paraiotonchium autumanalis* infect the host fly at larval stage by entering through the cuticle. The female parasite that oviposits in the hemocoel *of M. domestica* adult and start produce parthenogenetic females, which produce gametogenesis females that damages the fly ovaries, which leads them to produce less number of eggs. Though the infection of flies caused by pathogenicity of parasites, the environmental factors and the carrier substances also play a vital role as a promising housefly control. Renn and Wright (2000) examined and found that silver sand, potting compost, Fibertex SF250 and Poplin cotton were the most effective artificial substrates as a carrier for *S. feltiae* against housefly adults. The least efficient material was found as Perlite, which required much higher LD_{50}, and porosities materials were found as the most efficient at the range of 0.81–0.14.

The cost of Production, limited regional availability, and lack of knowledge till the end-users are the important factors which limits the broad level usages of entomopathogenic nematodes as biocontrol agents in pest management. Entomopathogenic nematodes can form a symbiotic association with the different bacterial members that showed toxic effects on insects. Entomopathogenic nematodes can be reared through both in vitro or in vivo conditions. Nematodes rearing by in-vitro method is cost effective and too complexity. It first require high start-up capital to establish, as the symbiotic bacteria and nematodes are need to be cultured separately then inoculate them with the medium along with bacteria and nematodes in the fermentation tanks. Where in In-vivo production is simple and it is being done oftenly by small scale industries as a quick production (Shapiro-Ilan et al. 2014).

Formulation of nematodes will uplift them in a market for large scale usage. There are different formulations available on the market, in which only a few are highly successful. For example, bait formulation of *Heterorhabditis megidis* and *S. feltiae* were evaluated against house fly along with control and methomyl, a commercial bait for comparison in a pig farm manure and found *S. feltiae* spray reduced fly population than methomyl bait. While apply with encapsulated *S. feltiae* no significant difference was observed. Entomopathogenic nematodes have fruitful advantages than the chemical pesticides as housefly control agents (Table 12.1). Although, some countries do not allow non-indigenous species release, nematodes are eco-friendly and safe and could be accepted after some essential evaluation. Parasitic nematodes are found to having the potential as classical biological agent against various stages of *M. domestica* especially at larval stage it cause deterioration of body and leads to death or parasitic castration on adult hosts (Malik et al. 2007). Further researches are needed to focus on effect of external factor for the successful implementation of nematodes in various locations.

12.4.4 Parasitoids

Adoption of natural enemies for biological control plays indispensable role in housefly control (Senthoorraja et al. 2020). For a latest decade many studies have proved the existence, enrichment and spreading of natural enemies that employed to reduce housefly population. In Pondicherry, during February 1984 to January 1985, Srinivasan and Balakrishnan (1989) were studied four parasitoid species Viz., *Pachycrepoideus vindemmiae, Spalangia cameroni*, *S. nigroaenea*, and *Dirhinus himalayanus* from different breeding sites such as the cattle shed, piggery farms, poultry farm and first time they recorded *D. himalayanus* in south India. In dairy farms, *S. cameroni* and *S. nigroaenea* were found as most abundant parasitoid species. *D. himalayanus* was observed only at poultry farms while *P. vindemmiae* was noticed in all the habitats. They, even suggested that release of parasitoids during post rainy season for effective housefly control. Srinivasan and Panicker (1988) examined the fecundity, efficiency, longevity, stinging behaviour, host feeding, and parasitism rate of *D. himalayanus* against housefly. 75% parasitism rate was found as the maximum level in 24–48 h aged pupae.

The behaviour and affluence of the parasitoid (Hymenoptera: Pteromalidae and Ichneumonidae) affects *M. domestica* pupae drastically. There are several pteromalid species, which attack housefly pupae for their reproduction. Especially, *Spalangia cameroni* and *Muscidifurax raptor* were found as the most effective and capable parasitoids in the evaluated regions via., Dairy farm, Poultry farm, and Pig farm. *Nasonia vitripennis* has been reported for the first time as an effective BCA (Biocontrol agents) against *M. domestica* in Iran. Parasitoids from the genus Brachymeria Westwood, Chalcididae family, Hymenoptera are important against insect pests such as muscoid flies (Marchiori et al. 2002). The family Chalcididae has been reported from Iran (*B. podagrica* from Kerman and East Azerbaijan) in 2012 as a parasitoid of various families against Muscoid, Calliphoridae, and Sarcophagidae. Chiel and Kuslitzky (2015) have also investigated 26 various habitats where housefly abundantly breed. Such as dairy, poultry farms, and goat farms throughout Israel and identified nine parasitoid species viz., *Spalangia cameroni, Spalangia endius, Spalangia drosophilae, Spalangia gemina, Spalangia nigroaenea, Pachycrepoideus vindemmiae, Muscidifurax raptor, Muscidifurax zaraptor,* and *Dirhinus giffardii* are effective against *M. domestica.* It has also shown that these collected species of parasitic wasp act synergistically and are species-specific. Therefore, exploiting these natural enemies of medically and important veterinary flies for their control is highly recommended (Akbarzadeh et al. 2017). Skovgard and Nachman (2004) reduced *M. domestica* and stable flies population using inundate releases of *S. cameroni,* a predominant parasitoid. The use of parasitoids are a promising solution for housefly control and it is less expensive than other control strategies. Such approaches need to be confirmed in the prior studies at lab and field before recommended to the end users.

12.4.5 Predators

Predation is one of the effective biological tools in housefly management. Many researchers evaluated the predation rate for immature houseflies, Predation rate of *Carcinops pumilio* per-day against adults (21–49) and in larvae (13–26); *Ophyra aenescens* (Muscid fly) on third instar (7–18), and macrochelid mite *Macrocheles muscaedomesticae* females (10–21) also caused remarkable predation against *M. domestica.* Meanwhile, the predation rate of histerid *Dendrophilus xavieri* adults (5–13), macrochelid *Glyptholapsis confusa* females (4–10), parasitic mite *Poecilochirus sp. Deutonymphs* (2–5), *Poecilochirus sp.* females (2–1) were shown less significant activity in predation. Larvae of the flies *Hydrotaea aenescens* (*Ophyra aenescens*) and *Ophyra capensis* (17 preys/predator) are effective predators against *M. domestica* larvae (Malik et al. 2007). *Careinops pumilio* adults and *M. museaedomesticae* females are potential to impairing 104 and 21 *M. domestica* immatures per day, respectively. In the laboratory condition, preys are not limiting, but in field condition the efficacy is reduced due to the predator crowding (*M. museaedomestieae*), prior feeding (*C. pumilio*). Under laboratory conditions, the predator/prey ratio ranging between 1:6 and 1:12 shown significant control over of 70% house fly population reduction and the same were assumed at field level.

H. aenescens were unable to establish their generations in the field level at wet poultry manure due to heavier wet condition in pullet-house manure as it generally have more than 80% moisture, which is significantly higher than the *H. aenescens* reared in the colony. However, *H. aenescens* culture were established at caged-layer poultry unit which was 0.3 km near from the release farm, it provided sources for larval development. As the biocontrol agents easily affect in the field condition by their sensitivity and susceptibility against the surrounding parameters. So, the whole areas need to be accentuated to execute the possible desired environmental conditions to achieve the projected field results (Geden et al. 1988). In non-predatory methods, Dipterans are also one of the effective biocontrol agents. Such as, black soldier fly, *Hermetia illucens* (L.) (Diptera: Stratiomyiidae) larvae outcompete with *M. domestica* larvae for food resources in the breeding habitat and reduce housefly's larval development.

12.4.6 Pheromone-Based Housefly Management

Insects confide naturally on their well-systemized and higher urbane olfactory system to evaluate odour signals produced from their mates, host and non-host substances, oviposition substrates, food and their breeding sources. (Z)-9-tricosene (muscalure) is a female housefly pheromone, which begins producing 2 days after the adult's emergence, but they are either absent or available at scanty level in the male flies. However, the housefly can produce high quantities of sex pheromone, which leads to investigate the pheromone and its biosynthesis process. (Z)-9-Tricosene is formed from oleic acid, and the methyl groups of the methyl alkanes arise from propionate and they are synthesized by its epidermal cells located in the abdomen of the female insect. Pheromones are highly active at even in very pinch amounts (nanogram level); that itself is highly expensive even if their chemical structure is less intricate to synthesize. Pest monitoring, mass trapping, and mating disruption are commonly used strategies in pest management. Mating disruption is one of the pheromones based successful tactics in pest management that can reduce the farmers usage of chemical pesticides and its negative impact. Developing a controlled releasing device for slow delivery of pheromones with long lasting capacity at low cost, will help end users to control the pest population at quick and effective manner, as there is a difficulty in release of pheromones at a constant rate over an extended period.

Lure and kill is one of the eco-friendly methods by the adoption of pheromone based traps such as yellow sticky trap, plywood sticky trap, rat glue trap, gel, and granule bait etc., to keep the *M. domestica* density under check at in and outside of lab condition and found more quantum of houseflies were attracted and trapped towards Z-9-Tricosene used traps than the control traps. Similar observations were also made by Butler et al. (2007) in animal husbandries and poultry units that the fly catches were significantly higher in traps made with (Z)-9-Tricosene and food bait than the traps without food baits; therefore, food baits were also found to increase the trap catches. Chin et al. (2008) reported that, the rat glue trap or plywood sticky traps were working well against housefly control. But it is ineffective, due to the saturation of glue and lack of space to catch hold the flies on the trap after 6 h. Therefore, (Z)-9-Tricosene based non-saturated traps are need to be made to attract and kill with long lasting potential. Paraffin emulsions and sprayable formulation are also viable technique as direct approach of pheromone based formulation for mating disruption and lure and kill trap against houseflies.

12.4.7 Integrated Pest Management Strategies (IPM)

In the latest years, integrated pest management tactics are used extensively against house flies. A combination of various control strategies such as semiochemical, biological method, physical method, and Nano-technological tools conferring an elegant option for controlling fly density while zero down the environmental deterioration (Dhanapal et al. 2020b). These agenda led a remarkable reduction in house fly population and solve the arising insect resistance issues in houseflies against chemical pesticides (Malik et al. 2007). Semiochemicals and biocontrol agents (BCAs) are a vital part of integrated pest management. The potential of insect pheromones and semiochemicals from plants and other sources have been used to invite and keep the natural enemies at in and outside of the lab conditions through planning, such as trap crops, attract-repel mechanisms and utilization of semiochemicals to attract and spreading of entomopathogenic microbials to keep the pest population under limit. The combination of Bt and *B. bassiana* has increased efficacies in pest control than they were used alone reported against various insects such as, *Malacosama neustria, Tuta absoluta, Leucinodes orbonalis, Earias vittella, T. ni,* etc. Devi et al. (2020) have found further and long lasting positive effects of the combined use of Bt with Bb and Nr. The effectiveness of natural parasitoid, *N. vitripennis,* in association with IGRs, has also been tested and proven for controlling house fly populations in the United Kingdom (Vazirianzadeh et al. 2008).

Pheromones and semiochemicals are currently used to reduce the insect pest of agriculture by observing, ensnaring in the mass, mating disruption, lure-and-kill, and push-pull strategies. Various type of kairomones, allomones, and synomones are available today, and their full information can be seen on Pherobase and Pherolist websites. Interspecific semiochemicals can be exploited to entice the natural enemies as a trap crop to protect from enemies as push-pull mechanisms (Dhanapal et al. 2021). Host-derived semiochemicals, non-host volatiles, Synthetic repellents, oviposition deterrent pheromones, alarm pheromones, antifeedants, anti-aggregation pheromones, and visual cues act as push factor and the pull components are host volatiles, visual stimulants, sex and aggregation pheromones, gustatory and oviposition stimulants (Sharma et al. 2019). Few studies have been reported that the tritrophic interaction of synthesized semiochemicals and plant or other natural volatiles such as livestock manure odours have been evaluated as attractants of microbes, parasitoids, and predators against housefly.

12.5 Conclusion

For a decade, attempts were made to identify clean and green technologies for housefly management. Plant secondary metabolites have the potential as cidal, repellence, and growth-regulating effect. Plant derived products were formulated and used to bring down the pest population under control. This can be harmoniously integrated with the bioagents that are used for the management of flies. In addition to plant-derived parts and biocontrol agents, the chemoecological approaches can also be used to exploit the olfactory cues to scale down the fly counts. It is difficult to control the housefly using a single method; its best contained using the integrated pest management method that blends in harmoniously with the environment. On this line, we need to make a tool to meet the demands for quick, simple, to solve the use complexities at various places like indoor and outdoor and cost savings in comparison to the base case without involvement of pocket burning expense by use of layman to prepare and apply for green rating and future study should focused on: association between known semiochemicals and biocontrol agents; exploration of more efficient microbial products (Eg. granular formulations of Entomopathogenic Fungus and environment-friendly trap materials); semiochemical induced predator/parasitoid visit and trap designing incorporation with semiochemicals. Utilization of nanotechnology to increase the efficacy and reduce the cost effects to the end-users. Finally, species-specific approaches should be taken up on where the pest association that needs to be controlled.

References

Abo-Taka M, Heikal H, Abd El-Raheem A (2014) Macrochelid mite, *Macrocheles muscaedomesticae* (Acarina: Macrochelidae) as a biological control agent against house fly, *Musca domestica* (Diptera: Muscidae) in Egypt. Int J Zool Res 10:30–36

Akbarzadeh K, Mirzakhanlou AA, Lotfalizadeh H et al (2017) Natural parasitism associated with species of Sarcophagidae family of Diptera in Iran. Ann Trop Med Public Health 10(1):134–137

Archana MD, Souza PE, Patil J (2017) Efficacy of entomopathogenic nematodes (Rhabditida: Steinernematidae and Heterorhabditidae) on developmental stages of house fly, *Musca domestica*. J Parasit Dis 41(3):782–794. https://doi.org/10.1007/s12639-017-0889-3

Arriaga AAM, Cortez-Madrigal H (2018) Susceptibility of *Musca domestica* larvae and adults to entomopathogenic nematodes (Rhabditida: Heterorhabditidae, Steinernematidae) native to Mexico. J Vec Ecol 43(2):312–320. https://doi.org/10.1111/jvec.12315

Axtell RC (1970) Integrated fly control program for caged-layer poultry houses. J Econ Entomol 63(2):400–455. https://doi.org/10.1093/jee/63.2.400

Azevedo LH, Ferreira MP, Castilho R et al (2018) Potential of Macrocheles species (Acari: Mesostigmata: Macrochelidae) as control agents of harmful flies (Diptera) and biology of *Macrocheles embersoni* Azevedo, Castilho and Berto on *Stomoxys calcitrans* (L.) and *Musca domestica* L. (Diptera: Muscidae). Biol Control 123:1–8. https://doi.org/10.1016/j.biocontrol.2018.04.013

Baker DK, Rice SJ, Leemon DM, James PJ (2018) Horizontal transmission of *Metarhizium anisopliae* (Hypocreales: Clavicipitacea) and the effects of infection on oviposition rate in laboratory populations of *Musca domestica* (Diptera: Muscidae). Pest Manag Sci 74(4):987–991. https://doi.org/10.1002/ps.4799

Baker D, Rice S, Leemon D, Godwin R, James P (2020) Development of a Mycoinsecticide bait formulation for the control of house flies, *Musca domestica* L. Insects 11(1):47. https://doi.org/10.3390/insects11010047

Burdock GA, Carabin IG (2008) Safety assessment of ylang-ylang (*Cananga* spp.) as a food ingredient. Food Chem Toxicol 46:433–445. https://doi.org/10.1016/j.fct.2007.09.105

Burgess ER, Kremer A, Elsawa SF et al (2017) Sublethal effects of imidacloprid exposure on *Spalangia endius*, a pupal parasitoid of filth flies. Biol Control 62(1):53–60. https://doi.org/10.1007/s10526-016-9776-6

Burgess ER, Watkins SM, King BH et al (2018) Dissemination of imidacloprid through dairy cattle manure and its effect on the biological control agent, *Spalangia endius* (Hymenoptera: Pteromalidae), and a filth fly host, *Musca domestica* (Diptera: Muscidae). J Econ Entomol. https://doi.org/10.1093/jee/toy386

Butler SM, Gerry AC, Mullens BA (2007) House fly (Diptera: Muscidae) activity near baits containing (Z)-9-Tricosene and efficacy of commercial toxic fly baits on a Southern California Dairy. J Econ Entomol 100(4):1489–1495. https://doi.org/10.1093/jee/100.4.1489

Chauhan N, Malik A, Sharma S (2018) Repellency potential of essential oils against housefly, *Musca domestica* L. Environ Sci Pollut Res 25(5):4707–4714. https://doi.org/10.1007/s11356-017-0363-x

Chiel E, Kuslitzky W (2015) Diversity and abundance of house fly pupal parasitoids in Israel, with first records of two *Spalangia* species. Environ Entomol 45(2):283–291. https://doi.org/10.1093/ee/nvv180

Chin HC, Sulaiman S, Othman HF (2008) Evaluation of Noepeace, Neopeace-F101 and Malaysia assurance rats glue for trapping *Musca domestica* (Diptera: Muscidae) in the field. J Trop Med Parasitol 31:1–5

Coats SA (1976) Life cycle and behavior of *Muscidifurax zaraptor* (Hymenoptera: Pteromalidae). Ann Entomol Soc Am 69(4):772–780. https://doi.org/10.1093/aesa/69.4.772

Contreras E, Masuyer G, Qureshi N, Chawla S et al (2019) A neurotoxin that specifically targets *Anopheles* mosquitoes. Nat Commun 10:1–10. https://doi.org/10.1038/s41467-019-10732-w

Crespo DC, Lecuona RE, Hogsette JA (1998) Biological control: an important component in integrated management of *Musca domestica* (Diptera: Muscidae) in caged-layer poultry houses in Buenos Aires, Argentina. Biol Control 13(1):16–24. https://doi.org/10.1006/bcon.1998.0632

Devi PSV, Duraimurugan P, Poorna Chandrika KSV et al (2020) Novel formulations of *Bacillus thuringiensis var.* kurstaki: an eco-friendly approach for management of lepidopteran pests. World J Microbiol Biotechnol 36(5):1–7. https://doi.org/10.1007/s11274-020-02849-8

Dhanapal R, Kumar DVSR, Lakshmipathy R, Sandhya Rani C, Manoj Kumar R (2020a) Exploration of indigenous strains of the green muscardine fungus from soils and their pathogenicity against the tobacco caterpillar, *Spodoptera litura* (Fabricius) (Lepidoptera: Noctuidae). Egypt J Biol Pest Control 30:34. https://doi.org/10.1186/s41938-020-00239-8

Dhanapal R, Kumar DVSR, Lakshmipathy R, Sandhya Rani C, Manoj Kumar R (2020b) Isolation of indigenous strains of the white halo fungus as a biological control agent against 3rd instar larvae of tobacco caterpillar, *Spodoptera litura* (Fabricius) (Lepidoptera: Noctuidae). Egypt J Biol Pest Control 30:89. https://doi.org/10.1186/s41938-020-00286-1

Dhanapal R, Singh RN, Raghuraman M, Mohan M, Subaharan K, Hemavathi M (2021) Evaluation of predatory potential and prey stage preference of mirid bug, *Nesidiocoris tenuis* on tomato pinworm, *Tuta absoluta*. Biologia 76:2965–2971. https://doi.org/10.1007/s11756-021-00786-x

Doganlar M (2007) A new species of Muscidifurax Girault & Sanders, 1910 (Hymenoptera: Pteromalidae) from Adana province, Turkey. Turk Entomol Derg 31:243–252

Fabritius K (1981) Combaterea biologica dirijata a mustei sinantrope *Musca domestica* L. (dipteraMuscidae) cu ajutorul parazitului *Muscidifurax raptor* Gir. Et Sand. (hymenopteraPteromalidae). Studii si Cercetari de Biologie 33:89–92

Farahi S, Shishehbor P, Nemati A (2018) Bisexual and oedipal reproduction of *Macrocheles muscaedomesticae* (Acari, Macrochelidae) feeding on *Musca domestica* (Diptera, Muscidae) eggs. Acarologia 58:430–441. https://doi.org/10.24349/acarologia/20184251

Farooq M, Steenberg T, Hojland DH, Freed S, Kristensen M (2018) Impact of sequential exposure of *Beauveria bassiana* and imidacloprid against susceptible and resistant strains of *Musca domestica*. BioControl 63(5):707–718. https://doi.org/10.1007/s10526-018-9892-6

Floate K, Coghlin P, Gibson G (2000) Dispersal of the filth fly parasitoid *Muscidifurax raptorellus* (Hymenoptera: Pteromalidae) following mass releases in cattle confinements. Biol Control 18(2):172–178. https://doi.org/10.1006/bcon.2000.0822

Gao M-Q, Cheng J-Q, Hu W, Pan D, Xiao H, Hu H-Y (2020) Life tables of two newly discovered parasitoid wasps, *Muscidifura similadanacus* and *M. sinesensilla* (Hymenoptera: Pteromalidae), reared on *Musca domestica* (Diptera: Muscidae). Biocontrol Sci Technol 30(8):779–794. https://doi.org/10.1080/09583157.2020.1766418

Geden CJ (2012) Status of biopesticides for control of house flies. J Biopest 5:1–11

Geden CJ, Hogsette JA (2006) Suppression of house flies (Diptera: Muscidae) in Florida poultry houses by sustained releases of *Muscidifurax raptorellus* and *Spalangia cameroni* (Hymenoptera: Pteromalidae). Environ Entomol 35(1):75–82. https://doi.org/10.1603/0046-225X-35.1.75

Geden CJ, Stinner RE, Axtell RC (1988) Predation by predators of the house fly in poultry manure: effects of predator density, feeding history, interspecific interference, and field conditions. Environ Entomol 17:320–329. https://doi.org/10.1093/ee/17.2.320

Gershenzon J, Dudareva N (2007) The function of terpene natural products in the natural world. Nat Chem Biol 3(7):408–414. https://doi.org/10.1038/nchembio.2007.5

Jiang Y et al (2002) Semiochemicals from ovaries of gravid females attract ovipositing female houseflies, *Musca domestica* L. J Insect Physiol 48(10):945–950. https://doi.org/10.1016/S0022-1910(02)00162-2

Johnson C, Bishop AH, Turner CL (1998) Isolation and activity of strains of *Bacillus thuringiensis* toxic to larvae of the housefly (Diptera: Muscidae) and tropical blowflies (Diptera: Calliphoridae). J Invertebr Pathol 71:138–144

Kannan S, Lokesh M, Kolarath R (2020) Evaluation of (Z)-9-tricosene pheromone and food bait for house flies, *Musca domestica* L. (Diptera: Muscidae) attraction using Domo trap. J Entomol Zoo Stud 8(5):1071–1074

Kanwar WA, Shoaib F, Rana FS (2017) Efficacy of entomopathogenic fungi and botanicals on development of *Musca domestica*. J Entomol Zoo Stud 5(2):593–599

Kaufman PE, Burgess M, Rutz DA, Glenister C (2002) Population dynamics of manure inhabiting arthropods under an Integrated Pest Management (IPM) program in New York poultry facilities-3 case studies. J Appl Poult Res 11(1):90–103. https://doi.org/10.1093/japr/11.1.90

Kaufman PE, Reasor C, Rutz DA, Ketzis JK, Arends JJ (2005) Evaluation of *Beauveria bassiana* applications against adult house fly, *Musca domestica*, in commercial caged-layer poultry facilities in New York state. Biol Control 33(3):360–367. https://doi.org/10.1016/j.biocontrol.2005.03.01

Keiding J (1986) The housefly-biology and control. Training and information guide (advanced level). Geneva, World Health Organization, 1986 (unpublished document WHO/VBC/ 86.937; available on request from Division of Control of Tropical Diseases, World Health Organization, 1211 Geneva 27, Switzerland)

Kelling FJ, Ialbnti F, DenOtter CJ (2002) Background odour induces adaptation and sensitization of olfactory receptors in the antennae of houseflies. Med Vet Entomol 16(2):161–169. https://doi.org/10.1046/j.1365-2915.2002.00359.x

Khan S, Guo LH, Maimaiti Y et al (2012) Entomopathogenic fungi as microbial biocontrol agent. Mol Plant Breed 3:63–79. https://doi.org/10.5376/mpb.2012.03.0007

Kumar P, Mishra S, Malik A, Satya S (2012) Compositional analysis and insecticidal activity of *Eucalyptus globulus* (family: Myrtaceae) essential oil against housefly (*Musca domestica*). Acta Trop 122(2):212–218. https://doi.org/10.1016/j.actatropica.2012.01.015

Labib IM, Rady M (2001) Application of *Bacillus thuringiensis* in poultry houses as a biological control agent against the housefly, *Musca domestica* sorbens. J Egypt Soc Parasitol 31:531–544

Lam K, Tsang M, Labrie A, Gries R, Gries G (2010) Semiochemical-mediated oviposition avoidance by female house flies, *Musca domestica*, on animal feces colonized with harmful fungi. J Chem Ecol 36(2):141–147. https://doi.org/10.1007/s10886-010-9741-2

Malik A, Singh N, Satya S (2007) House fly (*Musca domestica*): a review of control strategies for a challenging pest. J Environ Sci Health B 42:453–469. https://doi.org/10.1080/03601230701316481

Marche MG, Mura ME, Falchi G, Ruiu L (2017) Spore surface proteins of *Brevibacillus laterosporus* are involved in insect pathogenesis. Sci Rep 7:1–10

Marchiori CH, Pereira LA, Silva Filho OM et al (2002) *Brachymeria Podagrica* Fabricius, 1787 (Hymenoptera: Chalcididae) as parasitoids of *Sarcodexia lambens* (Wiedemann) (Diptera: arcophagidae) in Brazil. Arq Inst Biol Sao Paulo 69:121–122

Mehta R, Nambiar RG, Singh SK et al (2002) Livestock industrialization, trade and social-health-environment impacts in developing countries. Washington (D.C): IFPRI-FAO project. 2002 Livestock, Environment and Development (LEAD) initiative at FAO; Report No: IFPRI2033

Mishra S, Kumar P, Malik A (2016) Microscopic investigation to determine the effect of *Beauveria bassiana* (Bals.) Vuill. and *Cymbopogon citratus* (DC.) Stapf. treatment on different life stages of *Musca domestica* (L.). J Parasit Dis 41(2):543–550. https://doi.org/10.1007/s12639-016-0848-4

Moloinyane S, Addison P, Achiano AK, Nchu F (2019) Association between chemical properties of vineyard soils and occurrence of entomopathogenic fungi causing different levels of mortality in *Planococcus ficus*. Biol Control. https://doi.org/10.1007/s10526-019-09989-3(0123456789

Padmanabhan V, Prabakaran G, Paily KP et al (2005) Toxicity of a mosquitocidal metabolite of *Pseudomonas fluorescens* on larvae & pupae of the house fly, *Musca domestica*. Indian J Med Res 121:116–119

Petersen J, Currey D (1996) Timing of releases of gregarious *Muscidifurax raptorellus* (Hymenoptera: Pteromalidae) to control flies associated with confined beef cattle. J Agri Entomol 13:55–63

Pinnock DE, Mullens BA (2007) Application and evaluation of entomopathogens for control of livestock and poultry pests. In: Lacey LA, Kaya HK (eds) Field manual of techniques in invertebrate pathology. Springer, Dordrecht, pp 713–734

Rani N, Ponnudurai G, Harikrishnan TJ (2019) In vitro insecticidal activities of essential oil of Lemon grass against house fly: *Musca domestica* L. J Entomol Zoo Stud 7(1):206–209

Renn N, Wright E (2000) The effect of artificial substrates on the pathogenicity of *Steinernema feltiae* to adult *Musca domestica*. Nematol 2:217–222. https://doi.org/10.1163/156854100508953

Rossi YE, Palacios SM (2015) Insecticidal toxicity of *Eucalyptus cinerea* essential oil and 1,8-cineole against *Musca domestica* and possible uses according to the metabolic response of flies. Indus Crop Prod 63:133–137. https://doi.org/10.1016/j.indcrop.2014.10.019

Senthamarai Selvan P, Senthoorraja R, Ramesh V, Jebanesan A (2021) Field evaluation of toxicity of plant extracts against vector of filariasis Culex quinquefasciatus Say, 1823 (Diptera: Culicidae). S Afr J Bot 139:58–66. https://doi.org/10.1016/j.sajb.2021.01.027

Senthoorraja R, Subaharan K, Manjunath S, Pragadheesh VS, Bakthavatsalam N, Mohan MG, Senthil-Nathan S, Basavarajappa S (2021) Electrophysiological, behavioural and biochemical effect of Ocimum basilicum oil and its constituents methyl chavicol and linalool on Musca domestica L. Environ Sci Pollut Res Int 8:1–14. https://doi.org/10.1007/s11356-021-14282-x

Senthoorraja R, Subaharan K, Gupta A, Basavarajappa S, Lalitha Y, Bakthavatsalam N, Chandran KP (2020) Host factors influencing the parasitism by *Nesolynx thymus* (Girault) (Hymenoptera: Eulophidae) on housefly, *Musca domestica* L. J Biol Control 34(3):200–207. https://doi.org/10.18311/jbc/2020/26205

Shah RM, Azhar F, Shad SA et al (2016) Effects of different animal manures on attraction and reproductive behaviors of common house fly, *Musca domestica* L. Parasitol Res 115(9): 3585–3598. https://doi.org/10.1007/s00436-016-5124-0

Shakoori AR, Iftikhar N, Khurshed N (1999) Evaluation of different species of *Bacillus* isolated from soil samples as bio-insecticide against housefly, *Musca domestica*. Pakistan J Zool 31: 379–383

Shapiro-Ilan DI, Han R, Qiu X (2014) Production of entomopathogenic nematodes. In: Morales-Ramos JA, Rojas MG, Shapiro-Ilan DI (eds) Mass production of beneficial organisms. Academic, San Diego, pp 321–355

Sharififard M, Mossadegh MS, Vazirianzadeh B et al (2011) Laboratory evaluation of pathogenicity ofentomopathogenic fungi, *Beauveria bassiana* (Bals.) Vuill. And *Metarhizium anisopliae* (Metch.) Sorok. to larvae and adults of the house fly, *Musca domestica* L. (Diptera: Muscidae). Asian J Biol Sci 4:128–137

Sharma A, Shandi RK, Reddy GVP (2019) A review of interactions between insect biological control agents and semiochemicals. Insects 10(12):439. https://doi.org/10.3390/insects10120439

Sinthusiri J, Soonwera M (2013) Efficacy of herbal essential oils as insecticides against the house fly, *Musca domestica* L. Southeast Asian J Trop Med Pub Health 44:188–196

Skovgard H, Nachman G (2004) Biological control of house flies *Musca domestica* and stable flies *Stomoxys calcitrans* (Diptera: Muscidae) by means of inundative releases of *Spalangia cameroni* (Hymenoptera: Pteromalidae). Bull Entomol Res 94:555–567. https://doi.org/10.1079/BER2004322

Soonwera M (2015) Larvicidal and oviposition deterrent activities of essential oils against house fly (*Musca domestica* L.; Diptera: Muscidae). J Agri Technol 11(3):657–667

Soonwera M, Sittichok S (2020) Adulticidal activities of *Cymbopogon citratus* (Stapf.) and *Eucalyptus globulus* (Labill.) essential oils and of their synergistic combinations against *Aedes aegypti* (L.), *Aedes albopictus* (Skuse), and *Musca domestica* (L.). Environ Sci Pollut Res 27:20201–20214. https://doi.org/10.1007/s11356-020-08529-2

Srinivasan R, Balakrishnan N (1989) Preliminary note on the parasitoids of *Musca domestica* (Diptera: Muscidae) in Pondicherry. Entomon 14:349–352

Srinivasan R, Panicker KN (1988) Studies on *Dirhinus himalayanus* (Hymenoptera: Chalcididae) a pupal parasitoid of *Musca domestica*. Entomon 13:279–281

Subaharan K, Senthoorraja R, Manjunath S, Thimmegowda GG, Pragadheesh VS, Bakthavatsalam N, Mohan MG, Senthil-Nathan S, David KJ, Basavarajappa S, Ballal C (2021) Toxicity, behavioural and biochemical effect of Piper betle L. essential oil and its constituents against housefly, Musca domestica L. Pestic Biochem Physiol 174:104804. https://doi.org/10.1016/j.pestbp.2021.104804

Sundar STB, Latha BR, Vijayashanthi R, Pandian SS (2013) (Z)-9-Tricosene based *Musca domestica* lure study on a garbage dump yard using plywood sticky trap baited with fish meal. J Parasit Dis 40(1):32–35. https://doi.org/10.1007/s12639-013-0400-8

Suwannayod S, Sukontason KL, Pitasawat B, Junkum A, Limsopatham K, Jones MK et al (2019) Synergistic toxicity of plant essential oils combined with pyrethroid insecticides against blow flies and the house fly. Insects 10(6):178. https://doi.org/10.3390/insects10060178

Tang R, Zhang F, Kone N, Chen JH, Zhu F, Han RC et al (2016) Identification and testing of oviposition attractant chemical compounds for *Musca domestica*. Sci Rep 6(1):1–9. https://doi.org/10.1038/srep33017

Thomazini MJ, Berti Filho E (2000) Fertility life table of *Muscidifurax uniraptor* Kogan & Legner (Hymenoptera: Pteromalidae) on pupae of *Musca domestica* L. (Diptera: Muscidae). Anais da Sociedade Entomológica do Brasil 29(4):715–721. https://doi.org/10.1590/S0301-80592000000400011

Valtierra-de-Luis D, Villanueva M, Berry C, Caballero P (2020) Potential for *Bacillus thuringiensis* and other bacterial toxins as biological control agents to combat dipteran pests of medical and agronomic importance. Toxins 12(12):1–28. https://doi.org/10.3390/toxins12120773

Vazirianzadeh B, Kidd NAC, Moravej SA (2008) Side effects of IGR cyromazine on *Nasonia vitripennis* (Hymenoptera: Pteromalidae), a parasitic wasp of house fly pupae. J Arthropod Born Dis 2(2):1–6

Vega FE, Dowd PF, Lacey LA et al (2007) Dissemination of beneficial microbial agents by insects. In: Lacey LA, Kaya HK (eds) Field manual of techniques in invertebrate pathology. Springer, Dordrecht, pp 127–146

Weeks ENI, Machtinger ET, Gezan SA, Kaufman PE, Geden CJ (2016) Effects of four commercial fungal formulations on mortality and sporulation in house flies (*Musca domestica*) and stable flies (*Stomoxys calcitrans*). Med Vet Entomol 31(1):15–22. https://doi.org/10.1111/mve.12201

The manufacturer's authorised representative in the EU is Springer Nature Customer Serv ce Centre GmbH, Europaplatz 3, 69115 Heidelberg, Germany. If you have any concerns regarding our products, please contact ProductSafety@springernature.com

Printed and bound by CPI Group (UK) Ltd, Croydon, CR0 4YY

30/09/2026

02225329-0011